# MECHANISMS OF LYMPHOCYTE ACTIVATION AND IMMUNE REGULATION V

## Molecular Basis of Signal Transduction

# ADVANCES IN EXPERIMENTAL MEDICINE AND BIOLOGY



---

## Recent Volumes in this Series

Volume 357
LACTOFERRIN: Structure and Function
Edited by T. William Hutchens, Sylvia V. Rumball, and Bo Lönnerdal

Volume 358
ACTIN: Biophysics, Biochemistry, and Cell Biology
Edited by James E. Estes and Paul J. Higgins

Volume 359
TAURINE IN HEALTH AND DISEASE
Edited by Ryan J. Huxtable and Dietrich Michalk

Volume 360
ARTERIAL CHEMORECEPTORS: Cell to System
Edited by Ronan G. O'Regan, Philip Nolan, Daniel S. McQueen, and David J. Paterson

Volume 361
OXYGEN TRANSPORT TO TISSUE XVI
Edited by Michael C. Hogan, Odile Mathieu-Costello, David C. Poole, and Peter D. Wagner

Volume 362
ASPARTIC PROTEINASES: Structure, Function, Biology, and Biomedical Implications
Edited by Kanji Takahashi

Volume 363
NEUROCHEMISTRY IN CLINICAL APPLICAITON
Edited by Lily C. Tang and Steven J. Tang

Volume 364
DIET AND BREAST CANCER
Edited under the auspices of the American Institute for Cancer Research;
Scientific Editor: Elizabeth K. Weisburger

Volume 365
MECHANISMS OF LYMPHOCYTE ACTIVATION AND IMMUNE REGULATION V:
Molecular Basis of Signal Transduction
Edited by Sudhir Gupta, William E. Paul, Anthony DeFranco, and Roger Perlmutter

---

A Continuation Order Plan is available for this series. A continuation order will bring delivery of each new volume immediately upon publication. Volumes are billed only upon actual shipment. For further information please contact the publisher.

# MECHANISMS OF LYMPHOCYTE ACTIVATION AND IMMUNE REGULATION V

## Molecular Basis of Signal Transduction


Edited by

Sudhir Gupta
University of California
Irvine, California

William E. Paul
National Institute of Allergy and Infectious Diseases
National Institutes of Health
Bethesda, Maryland

Anthony DeFranco
University of California
San Francisco, California

and

Roger Perlmutter
University of Washington
Seattle, Washington



Springer Science+Business Media, LLC

Library of Congress Cataloging-in-Publication Data

Mechanisms of lymphocyte activation and immune regulation V : molecular basis of signal transduction / edited by Sudhir Gupta ...[et al.].
p. cm. -- (Advances in experimental medicine and biology ; v. 365)
"Proceedings of the fifth International Conference on Mechanisms of Lymphocyte Activation and Immune Regulation, held February 4-6, 1994, in Newport Beach, California"--T.p. verso.
Includes bibliographical references and index.

1. Lymphocyte transformation--Congresses. 2. Cellular signal transduction--Congresses. 3. Immune response--Regulation--Congresses. I. Gupta, Sudhir. II. International Conference on Mechanisms of Lymphocyte Activation and Immune Regulation (5th : 1994 : Newport Beach, Calif.) III. Series.
QR185.8.L9M44 1994
616.07'9--dc20 94-31128
CIP

Proceedings of the Fifth International Conference on Mechanisms of Lymphocyte Activation and Immune Regulation, held February 4–6, 1994, in Newport Beach, California

DOI 10.1007/978-1-4899-0987-9

## PREFACE

Signaling through antigen receptor initiates a complex series of events resulting in the activation of genes that regulate the development, proliferation and differentiation of lymphocytes. During the past few years, rapid progress has been made in understanding the molecular basis of signaling pathways mediated by antigen and cytokine receptors. These pathways involve protein tyrosine kinases which are coupled to downstream regulatory molecules, including small guanine nucleotide binding proteins (e.g. $p21^{ras}$), serine threonine kinases (e.g., members of the ERK family), and a large group of transcription factors. More recently, there have been breakthroughs in elucidating the genetic defects underlying three X-linked primary immunodeficiency diseases in humans. This volume surveys aspects of these rapidly developing areas of research.

The book is divided into 5 different sections. Section I deals with signaling pathways in B lymphocytes. It includes a contemporary assessment of B cell antigen receptor structures, and discussion of the role of Ig-α/Ig-ß polypeptides in linking the antigen receptor to intracellular signal transduction pathways. The role of accessory molecules in the regulation of signaling by the B cell antigen receptor is also considered. Section II adopts a similar approach to the analysis of the antigen receptor on T lymphocytes. The importance of specialized signaling motifs in the CD3 polypeptides, mechanisms whereby these motifs may interact with the lymphocyte-specific protein tyrosine kinases, and the downstream consequences of these interactions are reviewed. In addition, the role of antigen-induced apoptosis in the generation of immunological tolerance is discussed. Section III considers in more detail the various signaling components of lymphocytes. Included in this discussion are chapters on the calcium currents activated by the T cell antigen receptor, the role of non-receptor protein tyrosine kinases in lymphopoiesis and lymphocyte activation, the importance of the CD45 phosphotyrosine phosphatase in T cell activation, and the structure of a key regulator of lymphokine gene activation, the transcription factor NF-AT. Section IV provides summaries of signal transduction events mediated by Fc receptors and cytokine receptors. Contributions from noted experts consider the functional relationships among Fc receptors, and the mechanism whereby receptor aggregation triggers cellular reponses. In addition, the signaling responses initiated by IL-2 and IL-4 are reviewed. A detailed dissection of sequences in the common β subunit shared by the IL-3, IL-5, and GM-CSF receptors is also included. Finally, in Section V, the genetic defects that underlie primary immunodeficiency diseases are examined. These include the loss of IL-2Rγ chain expression in X-linked severe combined immunodeficiency, mutations in the Bruton tyrosine kinase (*btk*) gene in X-linked agammaglobulinemia, and the defective expression of CD40 ligand in the X-linked hyper-IgM syndrome. There is also a brief discussion of a phosphotyrosine phosphatase deficiency which has recently been shown to underlie the phenotype observed in the "motheaten" mouse.

This book should be of interest to basic immunologists and molecular biologists.

The Editors wish to thank Miss Nancy Doman for her excellent secretarial assistance.

Sudhir Gupta
William Paul
Anthony DeFranco
Roger Perlmutter

CONTENTS

## SECTION I: SIGNALING BY ANTIGEN RECEPTORS OF B CELLS



## SECTION II: SIGNALING BY ANTIGEN RECEPTORS OF T CELLS

## SECTION III: SIGNALING COMPONENTS OF LYMPHOCYTES



## SECTION IV: SIGNALING VIA Fc RECEPTORS AND CYTOKINE RECEPTORS

# SIGNALING AND INTERNALISATION FUNCTION OF THE B CELL ANTIGEN RECEPTOR COMPLEX

Heinrich Flaswinkel, Peter Weiser, Kwang-Myong Kim, and Michael Reth

Max-Planck-Institut für Immunbiologie
Stubeweg 51
79108 Freiburg, Germany

## INTRODUCTION

The mouse B-cell antigen receptor (BCR) is a multi-component transmembrane protein complex. This complex is comprised of the membrane-bound immunoglobulin (mIg) and the disulfide-linked Ig-$\alpha$ and Ig-ß heterodimer which is non-covalently associated with all classes of mIg molecules,[1] (for review see references 2-4). Ig-$\alpha$ and Ig-ß are glycoproteins of Mr 34000 and Mr 39000,[5,6,1] which are encoded by the B-cell specific genes mb-1[7] and B29,[8] respectively. Both proteins carry extracellularly a glycosylated Ig-like domain, a single transmembrane region of 22 amino acids, and a cytoplasmic portion of either 61 or 48 amino acids. The transmembrane and the cytoplasmic part of Ig-$\alpha$ and Ig-ß is strongly conserved between the mouse and the human proteins.[9-13]

The cytoplasmic sequences of Ig-$\alpha$ and Ig-ß contain the conserved amino acid sequence motif D/ExxxxxxD/ExxYxx LxxxxxxxYxxL/I.[14] This motif, which is known as tyrosine based activation motif TAM[15] or antigen receptor homology 1 (ARH1),[16] is also present in the cytoplasmic tail of the components of the T cell receptor (TCR) and of Fc receptors.[17-20] This motif consists of two negatively charged amino acids and two tyrosines followed with precise spacing by either a leucine or isoleucine residue. This motif is also found in the cytoplasmic tail of several viral transmembrane proteins.[21,22] The presence of TAM in components of the TCR is required for their function as signal-transducing molecules and it is thought that the antigen receptors on T and B cells are coupled to protein tyrosine kinases (PTK) via this motif.

Two types of PTKs are activated via the BCR, namely the src-related PTKs lyn, fyn, blk, or lck[23-26] and the cytoplasmic enzyme PTK72[27] which might be the mouse homologue of syk.[28] Law *et al.* have shown that PTK72, lyn, and fyn bind to the antigen receptor.[29] These PTKs are responsible for the diverse substrate phosphorylation seen after the cross-linking of the BCR as well as the phosphorylation of the BCR components Ig-$\alpha$ and Ig-ß. How and in what order these PTKs become activated after cross-linking of the BCR is not known at present. We have tested point mutations of the TAM/ARH1 motif of Ig-$\alpha$ in the context of chimeric CD8/Ig-$\alpha$ molecules as well as in the context of the

*Mechanisms of Lymphocyte Activation and Immune Regulation V*
Edited by S. Gupta *et al.*, Plenum Press, New York, 1994

complete BCR. We have also analysed two different forms of the IgG2a-BCR. Together the results of these analyses show that the Ig-α/Ig-ß heterodimer is required for the signaling function, but not for the internalisation function of the BCR.

## MUTATIONAL ANALYSIS OF THE TAM/ARH1 MOTIF OF IG-α

The cytoplasmic sequence of Ig-α is comprised of 61 amino acids containing four tyrosine residues (see Figure 1). With a PCR approach, we mutated the first three of these tyrosine residues into phenylalanine (see Table 1). In mutation M5, only the first tyrosine (Y17) is mutated, whereas M3 and M4 carry a point mutation of the TAM/ARH1 tyrosines Y34 and Y23, respectively. Two of these mutations are combined in M1 and M2. The mutation M6 is not a point mutation but an internal deletion of 52 cytoplasmic amino acids which includes all four tyrosines of the Ig-α tail. The mutations of the cytoplasmic sequence of Ig-α were tested for their signaling function either in the context of chimeric CD8/Ig-α molecules or in the context of the complete BCR.

**Table 1.** Calcium response and substrate phosphorylation after crosslinking of chimeric CD8/Ig-α or the complete BCR

| | | CD8/Ig-α | | BCR | |
|---|---|---|---|---|---|
| Mutation | Type | $Ca^{++}$ | Substrate[1] | Ig-α[2] | Substrate |
| M1 | Y23F; Y34F | - | - | - | +/- |
| M2 | Y17F; Y34F | - | - | + | ++ |
| M3 | Y34F | - | - | + | ++ |
| M4 | Y23F | - | - | - | ++ |
| M5 | Y17 | ++ | ++ | + | ++++ |
| M6 | Del 6-57[3] | - | - | - | +/- |
| WT[4] | None | + | + | + | +++ |

[1]Tyrosine phosphorylation of dominant substrate proteins (p55, p65, p80).
[2]Tyrosine phosphorylation of TAM/ARH1 of Ig-α.
[3]Deletion of Ig-α cytoplasmic sequence. Numbering starts with the first amino acid in the cytoplasm.
[4]The analysis of the wild-type (WT) Ig-α is given for comparison.

## TEST OF MUTATED CD8/Ig-α CHIMERIC RECEPTORS

In a previous study, we have shown that after their cross-linking chimeric CD8/Ig-α or CD8/Ig-ß receptors can signal on the surface of the B-lymphoma cell line K46.[30] K46 cells expressing CD8/Ig-α chimeric molecules were used to study the signaling capacity of Ig-α independently of Ig-ß. The CD8/Ig-α chimeric molecules carried either the cytoplasmic sequence of the wild-type Ig-α (WT) or of the mutations M1-M6. All transfectants (WT, M1-M6) expressed similar levels of chimeric CD8/Ig-α homodimer.

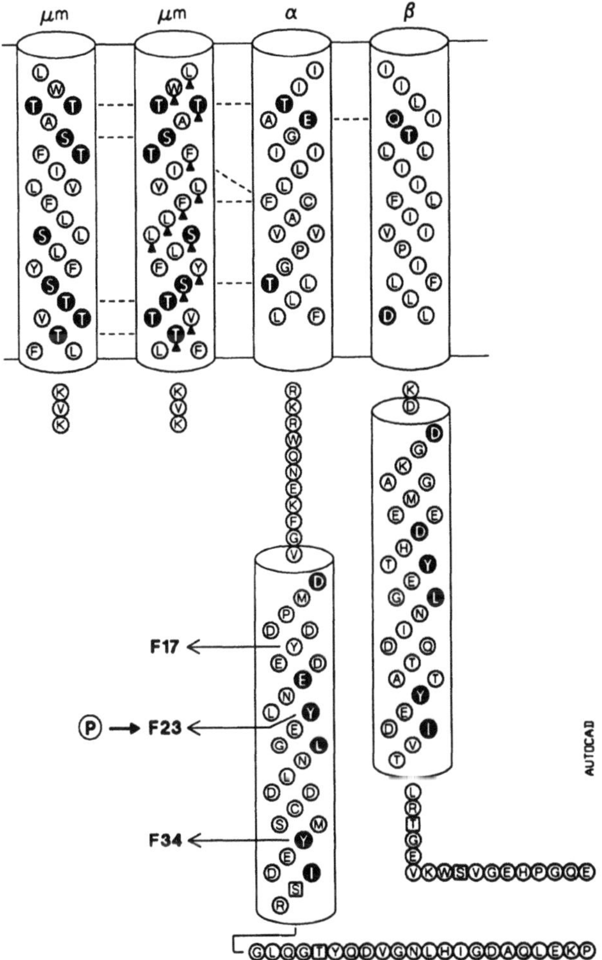


**Figure 1.** Mutations and phosphorylation site on the cytoplasmic tail of Ig-α

The outcome of this study was clear cut. Cross-linking of the WT and M5 CD8/Ig-α chimeric molecules results in increased substrate phosphorylation and rapid release of intracellular calcium ions. None of these responses were seen after the cross-linking of CD8/Ig-α mutations M1-M4 and M6. In all the defective mutations one or both tyrosines of TAM/ARH1 were mutated to phenylalanine. This analysis suggests that PTK activation and release of calcium via the Ig-α molecule require the presence of both tyrosines of the TAM. Studies of chimeric CD8/Ig-α molecules on K46 transfectants thus yield similar conclusions to those of previous studies of chimeric components of the TCR[31] which also are defective in signaling as soon as one of the tyrosines of the TAM/ARH1 is mutated. The data are also compatible with those derived from IgM/Ig-α or IgM/Ig-ß chimeric receptors.[32,39]

## TEST OF MUTATED Ig-α IN THE COMPLETE BCR

The myeloma cell line J558Lμm expresses the membrane form of IgM intracellularly but, due to the lack of Ig-α, fails to transport it onto the cell surface.[33] Previous experiments have shown that upon transfection of an mb-1 expression vector coding for Ig-α, IgM is expressed at high levels on the cell surface of J558Lmm/mb-1.[34,1] In addition, the BCR expressed on the surface of J558Lμm/mb-1 upon transfection of mb-1 can transduce signals into the cytoplasm by activating PTK.[35] Cross-linking of the BCR on J558Lμm/mb-1 cells with either the antigen NIP-BSA or anti-IgM antibodies results in rapid phosphorylation of several dominant PTK substrate proteins (p55, p65 and p80) as well as in tyrosine phosphorylation of Ig-α and Ig-ß.[35] The phosphorylation of Ig-α and Ig-ß is not detected after the cross-linking of BCRs carrying the M1, M4 or M6 mutation of Ig-α. This shows that in the activated BCR only the first tyrosine of the TAM-ARH1 motif (Y23) becomes phosphorylated. The second tyrosine of TAM/ARH1 (Y34) can be mutated without interfering with the phosphorylation of Y23. In comparison to the WT BCR the phosphorylation of other substrate proteins is only mildly reduced in the BCRs carrying the point mutation of either Y23 or Y34. The phosphorylation of the TAM/ARH1 of Ig-α is therefore not a prerequisite for substrate phosphorylation in activated B cells. Substrate phosphorylation is much more drastically reduced in cells expressing the M1 and M6 BCR carrying the Y23F/Y34F double mutation and the deletion of most of the cytoplasmic residues of Ig-α, respectively. This shows that the substitution of the two conserved tyrosines of TAM/ARH1 of Ig-α can disturb PTK activation via the BCR even in the presence of a wild-type Ig-ß.

## *IN VITRO* KINASE ASSAY WITH MUTATED Ig-α

From the four src-related PTK (lyn, blk, fyn, lck) implicated in BCR signaling, only fyn and lyn are expressed in J558L cells. We have employed an *in vitro* kinase assay to test the phosphorylation of mutated Ig-α. For this purpose, fusion proteins consisting of the glutathione-S-transferase (GST) and cytoplasmic portions of the WT or M1-M6 Ig-α proteins produced in bacteria, were incubated for 30 minutes in a kinase buffer with human fyn kinase expressed from a recombinant baculovirus. While the fusion proteins carrying the cytoplasmic portion of WT, M2, M3 or M5 Ig-α show a strong tyrosine phosphorylation, there is no detectable phosphorylation in fusion proteins carrying the cytoplasmic portion of M1, M4 or M6 Ig-α. The same result was obtained when src instead of fyn was used in the *in vitro* kinase assay. Together the *in vitro* and *in vivo* analyses gave the same result and showed that Y23 of Ig-α is the dominant target for phosphorylation by presumably a src-related PTK.

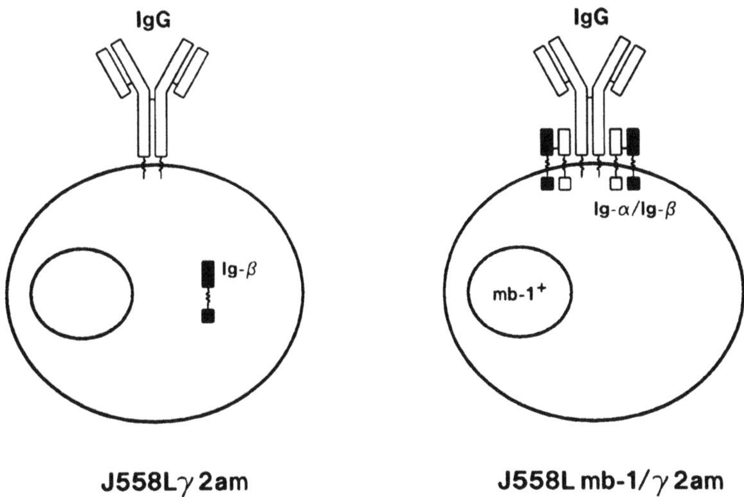


**Figure 2.** Two forms of the IgG2a-BCR expressed on the cell surface of the myeloma line J558L.

## TWO FORMS OF THE IgG2a ANTIGEN RECEPTOR

We have transfected the expression vector pSV2gptG2am into the λ1 light chain-producing myeloma lines J558L and J558Lmb-1, only the latter of which expresses the mb-1 gene. From both transfections we obtained cell lines which express the mIgG2a on their surface. The J558Lmb-1/γ 2am carries the complete IgG2a BCR complex on the cell surface while on J558Lγ 2am cells we could detect only the mIgG2a molecule and no Ig-α/Ig-ß heterodimer. This result shows that the mIgG2a molecule can be transported to the surface of cells which do not produce the Ig-α/Ig-ß heterodimer.

In the absence of an Ig-α/Ig-ß heterodimer, the IgD molecule can be expressed on the cell surface either as a transmembrane protein[1] or as a glycosyl phosphatidylinositol (GPI)-linked molecule.[36] We tested for a GPI linkage of IgG2a and found that on J558Lγ2am cells the IgG2a molecule was a transmembrane protein. These data indicate that the IgG2a-BCR can be expressed on the B cell surface in two different forms, only one of which contains the Ig-α/Ig-ß heterodimer. We then compared the signaling and internalisation function of the two different forms of the IgG2a BCR. After cross-linking of the IgG2a BCR with anti-IgG antibodies, no enhanced substrate phosphorylation is detected in J558Lγ2am cells, whereas activated J558Lmb-1/γ2am transfectants shows a strong increase of substrate phosphorylation after 1 min of stimulation. This shows that PTKs require the Ig-α/Ig-ß for their activation via the IgG2a BCR.

Another important function of the BCR is endocytosis of bound antigen which is processed intracellularly and finally presented as MHC-bound peptides to T cells.[37-39] It has been proposed that the cytoplasmic tails of Ig-α and Ig-ß are also involved in the internalisation of the BCR.[40] To test for internalisation we incubated J558Lγ2am and J558Lmb-1/γ2am cells for different times with biotinylated antigen or anti-IgG2a antibodies and measured the amount of biotin on the cell surface and inside the cell after different times. This experiment shows that an IgG2a-BCR which lacks the Ig-α/Ig-ß heterodimer is still efficiently internalised. In contrast to mIgM and mIgD, the mIgG2a molecule has a cytoplasmic tail of 28 amino acids. A mutant carrying a deletion of this part of the mIgG2a molecule still can become internalised in an Ig-α/Ig-ß independent fashion. Furthermore, we also found that mIgD molecules expressed on the cell surface without Ig-α/Ig-ß heterodimer are efficiently internalised. Molecules controlling the internalisation process must therefore interact directly with the mIg-molecule either extracellularly or in the membrane. We have started a search for such mIg interacting molecules and have detected some promising protein bands.

In summary, our studies show that the Ig-α/Ig-ß heterodimer is a true signaling subunit of the BCR which via its TAM/ARH1 motif couples the receptor complex to PTKs. The second major function, namely the internalisation of the BCR, seems, however, to be mostly controlled by other yet unknown molecules. The data presented here are partially published.[41,42]

## REFERENCES


1. A.R. Venkitaraman, G.T. Williams, P. Dariavach, and M.S. Neuberger, The B-cell antigen receptor of the five immunoglobulin classes, *Nature* 352:777 (1991).
2. M. Reth, Antigen receptors on B lymphocytes, *Annual Rev Immunol* 10:97 (1992)
3. N. Sakaguchi, T. Matsuo, J. Nomura, K. Kuwahare, H. Igarashi, and S. Inui, Immunoglobulin receptor-associated molecules, *Adv Immunol* 54:337 (1993).
4. A.L. DeFranco, Structure and function of the B cell antigen receptor, *Annual Rev Cell Biol* 9:377-410 (1993).
5. D. Hombach, F. Lottspeich, and M. Reth, Identification of the genes encoding the IgM-α and Ig-ß components of the IgM antigen receptor complex by aminoterminal sequencing, *Eur J Immunol* 20:2795 (1990).
6. K.S. Campbell and J.C. Cambier, B lymphocyte antigen receptors (mIg) are non-covalently associated with a disulfide linked inducible phosphorylated glycoprotein complex, *EMBO J* 9:441 (1990).
7. N. Sakaguchi, S. Kashiwamura, M. Kimoto, P. Thalmann, and F. Melchers, B lymphocyte lineage-restricted expression of mb-1, a gene with CD3-like structural properties, *EMBO J* 7:3457 (1988).
8. G.G. Hermanson, D. Eisenberg, P.W. Kincade, and R. Wall, B29: a member of the immunoglobulin gene superfamily exclusively expressed on B-lineage cells, *Proc Natl Acad Sci USA* 85:6890 (1988).
9. L.M. Yu and T.W. Chang, Human mb-1 gene: complete cDNA sequence and its expression in B cells bearing membrane Ig of various isotypes, *J Immunol* 148:633 (1992).
10. H.J. Ha, H. Kubagawa, and P.D. Burrows, Molecular cloning and expression pattern of a human gene homologous to the murine mb-1 gene, *J Immunol* 148:1526 (1992).
11. H. Flaswinkel and M. Reth, Molecular cloning of the Ig-α subunit of the human B-cell antigen receptor complex, *Immunogenetics* 36:266 (1992).
12. B. Müller, L. Cooper, and C. Terhorst, Cloning and sequencing of the cDNA encoding the human homologue of the murine immunoglobulin-associated protein B29, *Eur J Immunol* 222:1621 (1992).
13. S. Hashimoto, P.K. Gregersen, and N. Chiorazzi, The human Ig-ß cDNA sequence, a homologue of murine B29, is identical in B cell and plasma cell lines producing all the human Ig isotypes, *J Immunol* 150:491 (1993).
14. M. Reth, Antigen receptor tail clue, *Nature* 338:383 (1989).
15. L.E. Samelson and R.D. Klausner, Tyrosine kinases and tyrosine-based activation motifs, *J Biol Chem* 267:24913 (1992).
16. J.C. Cambier, Signal transduction by T- and B-cell antigen receptors: converging structures and concepts, *Curr Opin Immunol* 4:257 (1992).

17. F. Letourneur and R.D. Klausner, T-cell and basophil activation through the cytoplasmic tail of T-cell-receptor zeta family proteins, *Proc Natl Acad Sci USA* 88:8905 (1991).
18. A.-M.K. Wegener, F. Letourneur, A. Hoeveler, T. Brocker, F. Luton, and B. Malissen, The T cell receptor/CD3 complex is composed of at least two autonomous transduction molecules, *Cell* 68:83 (1992).
19. C. Romeo, M. Amiot, and B. Seed, Sequence requirements for the induction of cytolysis by the T cell antigen/Fc receptor $\zeta$ chain, *Cell* 68:889 (1992).
20. B.A. Irving, A.C. Chan, and A. Weiss, Functional characterization of a signaling transducing motif present in the T cell antigen receptor $\zeta$ chain, *J Exp Med* 177:1093 (1993).
21. G. Alber, K.M. Kim, P. Weiser, C. Riesterer, R. Carsetti, and M. Reth, Molecular mimicry of the antigen receptor signaling motif by transmembrane proteins of the Epstein-Barr virus and the bovine leukaemia virus, *Current Biology* 3:333 (1993).
22. P. Beaufils, D. Choquat, R.Z. Mamoun and B. Malissen, The (YXXL/I)2 signalling motif found in the cytoplasmic segments of the bovine leukaemia virus envelope protein and the Epstein-Barr virus latent membrane protein 2A can elicit early and late lymphocyte activation events, *EMBO J* 12:5105 (1993).
23. Y. Yamanashi, T. Kakiuchi, J. Mizuguchi, T. Yamamoto, and K. Toyoshima, Association of B cell antigen receptor with protein tyrosine kinase Lyn, *Science* 251:192 (1991).
24. A.L. Burkhardt, M. Brunswick, J.B. Bolen, and J.J. Mond, Anti-immunoglobulin stimulation of B lymphocytes activates src-related protein-tyrosine kinases, *Proc Natl Acad Sci USA* 88:7410 (1991).
25. J. Lin and L.B. Justement, The MB-1/B29 heterodimer couples the B cell antigen receptor to multiple src family protein tyrosine kinases, *J Immunol* 149:1548 (1992).
26. M.A. Campbell and B.M. Sefton, Association between B-lymphocyte membrane immunoglobulin and multiple members of the Src family of protein tyrosine kinases, *Mol Cell Biol* 12:2315 (1992).
27. J.E. Hutchcroft, M.L. Harrison, and R.L. Geahlen, Association of the 72-kDa protein-tyrosine kinase PTK72 with the B cell antigen receptor, *J Biol Chem* 267:8613 (1992).
28. T. Taniguchi, T. Kobayashi, T. Kondo, J. Takahashi, K. Nakamura, H. Suzuki, J. Nagai, T. Yamada, S. Nakamura, and H. Yamamura, Molecular cloning of a porcine gene syk that encodes a 72-kDa protein-tyrosine kinase showing high susceptibility to proteolysis, *J Biol Chem* 266:15790 (1991).
29. D.A. Law, V.W.F. Chan, S.K. Datta, and A.L. DeFranco, B-cell antigen receptor motifs have redundant signalling capabilities and bind the tyrosine kinases PTK72, Lyn and Fyn, *Current Biology* 3:645 (1993).
30. K.M. Kim, G. Alber, P. Weiser, and M. Reth, Differential signaling through the Ig-$\alpha$ and Ig-ß components of the B cell antigen receptor, *Eur J Immunol* 23:911 (1993).
31. F. Letourneur and R.D. Klausner, Activation of T cells by a tyrosine kinase activation domain in the cytoplasmic tail of CD3$\epsilon$, *Science* 255:79 (1992).
32. M. Sanchez, Z. Misulovin, A.L. Burkhardt, S. Mahajan, T. Costa, R. Franke, J.B. Bolen, and M. Nussenzweig, Signal transduction by immunoglobulin is mediated through Ig-$\alpha$ and Ig-ß, *J Exp Med* 178:1049 (1993).
33. J. Hombach, F. Sablitzky, K. Rajewsky, and M. Reth, Transfected plasmacytoma cells do not transport the membrane form of IgM to the cell surface, *J Exp Med* 167:652 (1988).

34. J. Hombach, T. Tsubata, L. Leclercq, H. Stappert, and M. Reth, Molecular components of the B cell antigen receptor complex of the IgM class, *Nature* 343:760 (1990).
35. K.M. Kim, G. Alber, P. Weiser, and M. Reth, The B-cell antigen receptor complex, *Immunol Reviews* 132:125 (1993).
36. J. Wienands and M. Reth, Glycosyl-phosphatidylinositol linkage as a mechanism for cell-surface expression of immunoglobulin D, *Nature* 356:246 (1992).
37. J.C. Antoine, S. Avrameas, N.K. Gonatas, A. Stieber, and J.O. Gonatas, Plasma membrane and internalized immunoglobulins of lymph node cells studied with conjugates of antibody or its Fab fragments with horseradish peroxidase, *J Cell Biol* 63:12 (1974).
38. E.R. Unanue, Cellular events following binding of antigen to lymphocytes, *Amer J Pathol* 77:2 (1974).
39. A. Lanzavecchia, Receptor-mediated antigen uptake and its effect on antigen presentation to class II-restricted T lymphocytes, *Annual Rev Immunol* 8:773 (1990).
40. S. Amigorena, J. Salamero, J. Davoust, W.H. Fridman, and C. Bonnerot, Tyrosine-containing motif that transduces cell activation signals also determines internalization and antigen presentation via type III receptors for IgG, *Nature* 358:337 (1992).
41. H. Flaswinkel and M. Reth, Dual role of the tyrosine activation motif of the Ig-$\alpha$ protein during signal transduction via the B cell antigen receptor, *EMBO J* 13:83 (1994).
42. P. Weiser, C. Riesterer, and M. Reth, The internalization of the IgG2a antigen receptor does not require the association with Ig-$\alpha$ and Ig-ß but the activation of protein kinases does, *Eur J Immunol* 24:665 (1994).

# MECHANISM OF B CELL ANTIGEN RECEPTOR FUNCTION: TRANSMEMBRANE SIGNALING AND TRIGGERING OF APOPTOSIS

Anthony L. DeFranco[1,2,3], Paul R. Mittelstadt[1,3], Jonathan H. Blum[2,3], Tracy L. Stevens[1,3], Debbie A. Law[1,3], Vivien W.-F. Chan[2,3], Shaun P. Foy[4], Sandip K. Datta[1,3] and Linda Matsuuchi[4]

[1]Departments of Microbiology & Immunology and [2]Biochemistry & Biophysics and the [3]G. W. Hooper Foundation, University of California, San Francisco, and the [4]Department of Zoology, University of British Columbia, Vancouver

## INTRODUCTION

The B cell antigen receptor (BCR) plays a key role in regulating B cell development, activation, and inactivation[1]. The transmembrane form of μ is required for proper B cell development[2], as is the surrogate light chain encoded by the λ5 gene[3]. Once the immunoglobulin (Ig) genes have been rearranged properly, the B cell precursor expresses the conventional form of the BCR. If this immature B cell contacts antigen before leaving the bone marrow, it continues to rearrange the Ig light chain genes[4, 5]. This response, referred to as receptor editing, is presumably an effort to change antigen specificity away from what would usually be self-reactivity. If unsuccessful in altering its antigen specificity, this auto-reactive immature B cell is inactivated either by cell death[6-8] or clonal anergy[9].

In contrast, stimulation of mature B cells through their BCR promotes proliferation and differentiation to antibody production, provided adequate T cell help is provided[10]. In addition, the BCR also plays an important role in the survival of germinal center B cells following somatic mutation of Ig genes[11]. Thus, the BCR plays a critical role in the physiology of B cells at virtually every stage of their life.

The BCR is a complex of membrane immunoglobulin (mIg) and a heterodimer of transmembrane polypeptides called Ig-α and Ig-β. Ig-α and Ig-β each have one extracellular Ig-like domain, a single transmembrane domain, and a moderate size cytoplasmic domain (61 and 48 amino acids, respectively)[12]. The sites of contact between mIg and the Ig-α/Ig-β heterodimer are not completely known, although sequences within the transmembrane domain

*Mechanisms of Lymphocyte Activation and Immune Regulation V*
Edited by S. Gupta *et al.*, Plenum Press, New York, 1994

of μ are important for assembly of this complex. In addition, the membrane proximal Ig-like domain of the mIg heavy chain ($C_H4$ in the case of μ chain) probably interacts with the Ig-like domains of Ig-α, Ig-β or both. Experiments related to these issues are described below.

## BCR ASSEMBLY AND TRAFFICKING

Expression of IgM heavy and light chains in non-lymphoid cells leads to expression of mIgM in the endoplasmic reticulum (ER), but not on the cell surface[13]. Co-expression of mIgM with Ig-α and Ig-β, but neither one of these by itself, results in formation of complete BCR complexes and cell surface expression[14, 15]. Chimeric heavy chains in which the 41 C-terminal amino acids of the μ chain (representing the extracellular spacer region, the transmembrane region and the cytoplasmic tail) have been replaced with the corresponding region from MHC class I or CD8 molecules go to the cell surface in the absence of Ig-α and Ig-β[13]. Thus, the C-terminal region of the μ heavy chain is necessary for intracellular retention in the absence of complex formation with the Ig-α/Ig-β heterodimer.

Each of the heavy chain isotypes has a transmembrane region with a high fraction of hydrophilic amino acid residues. For example, 10 out of 26 residues of the μ transmembrane domain have side chains with hydroxyl groups. Many of these hydrophilic residues are conserved in the different heavy chain transmembrane domains (Fig. 1)[12]. The μ transmembrane domain has two prominent polar patches with four out of five and five out of six adjacent residues having hydroxylated side chains. Clustered conservative mutation of either the N-terminal[13] or C-terminal[16] of these two regions abrogates the ER retention function of the transmembrane domain. In addition, we have found that mutation of two adjacent residues (YS) in the more C-terminal of these regions to hydrophobic residues (VV) allowed cell surface expression in the non-lymphoid AtT20 cells[17]. These observations suggest that the two polar patches are important for the interaction between mIgM and one or more proteins that are localized to the ER. This component could be a molecular chaperone that participates in the assembly of mIg with Ig-α/Ig-β. According to this hypothesis, the ER resident component would hold mIgM in the ER until one or more Ig-α/Ig-β heterodimers bind and displace it. The identity of this presumed chaperone is not known, although Bip and calnexin are plausible candidates[16, 18, 19].

The transmembrane regions of Ig heavy chains also likely interact with the Ig-α/Ig-β heterodimer. This conclusion is suggested by the fact that the BCR complex is only stable in very mild detergents such as digitonin and CHAPS[12]. Slightly harsher non-ionic detergents such as Triton X-100 and NP-40 disrupt this complex, as assessed by lack of co-immunoprecipitation under these conditions. Moreover, the replacement of the C-terminal region of μ chain with the equivalent region of other proteins prevents assembly with Ig-α/Ig-β [20, 21]. For example, a μ/CD8 chimeric heavy chain assembled properly with κ light chains into a $H_2L_2$ mIgM-type structure that was expressed well on the cell surface. These cell surface mIgM molecules failed to coprecipitate with Ig-α/Ig-β heterodimers, however, indicating that mIgM/CD8 chimeras were present on the cell surface without these accessory proteins bound to them[21]. Thus, the extracellular Ig domains of mIgM are not sufficient for stable assembly with Ig-α and Ig-β. Moreover, the YS/VV mutant form of murine mIgM assembles poorly with Ig-

| | | | |
|---|---|---|---|
| murine μ | EGFE | NLWTTASTFIVLFLLSLFYSTTVTLF | KVK |
| murine δ | LEEE | NGLWPTMCTFVALFLLTLLYSGFVTFI | KVK |
| murine γ1 | GELD | GLWTTITIFISLFLLSVCYSAAVTLF | KVKW |
| murine γ2α | GELD | GLWTTITIFISLFLLSVCYSASVTLF | KVKW |
| murine γ2β | GELD | GLWTTITIFISLFLLSVCYSASVTLF | KVKW |
| murine γ3 | GELD | GLWTTITIFISLFLLSVCYSASVTLF | KVKW |
| murine ε | ELEE | LWTSICVFITLFLLSVSYGATVTVL | KVKW |
| murine α | EEAPGASLWPTTVTFLTLFLLSLFYSTALTVTTV | | RGPF |
| CONSENSUS | | LWTTxxxFIxLFLLS(L/V)xYSxxVTxx | KVK |

Figure 1. Homology of immunoglobulin heavy chain transmembrane domains. The sequences of the murine Ig heavy chain transmembrane domains are shown in the single letter amino acid code. The regions believed to be just outside the membrane are also shown (the outside of the cell is on the left and the inside of the cell on the right). Some Ig heavy chains have residues at the boundary that are not charged or strongly hydrophobic (e.g., the N in μ), and thus could either be in the bilayer or outside of it.

α and Ig-β[17]. Five- to ten-fold less Ig-α co-precipitates with this mutant mIgM compared to wild type mIgM. This mutant mIgM molecule is also no longer retained intracellularly in the absence of Ig-α/Ig-β[17]. It could be that the associations between mIgM and Ig-α/Ig-β are not changed by this mutation, but that decreased assembly with Ig-α/Ig-β is due to premature exit from the ER. Alternatively, the YS/VV mutation could affect mIgM interactions with both the retention factor and with Ig-α/Ig-β.

The importance of the transmembrane region of the μ heavy chain in BCR assembly and function is also apparent from the effects of two other mutations of μ. In the first, removal of the three amino acid long cytoplasmic domain has been found to abrogate signaling ability[21, 22]. Polypeptides containing a hydrophobic transmembrane region without a charged anchor for the cytoplasmic domain following it are often converted to a glycosyl-phosphatidylinositol-linked protein, and that is what happens with this tailless mutant[21, 23]. In the process, the transmembrane domain is removed. The resulting glycosyl-phosphatidylinositol-linked form of mIgM is not associated with Ig-α/Ig-β[21, 24]. The cytoplasmic tail of μ may not be important for BCR assembly or signaling, as several different point mutations that retain the charged character of this region but change its sequence extensively do not affect signaling function[21, 22]. Thus, it appears to be the transmembrane domain of μ that is required for the assembly of functional BCR complexes. In agreement with this conclusion, alteration of the transmembrane domain by addition or deletion of a single amino acid residue results in intracellular retention in most cases (Blum, Stevens, and DeFranco, manuscript in preparation). These mutations alter the alignment of amino acid side chains along the transmembrane α-helix. Any interactions between another protein and the μ transmembrane domain that involves regions on both sides of the insertion or deletion would be severely affected by these mutations. The most likely possibility is that these mutations affect binding of Ig-α/Ig-β to μ. In any case, these mutations also indicate the transmembrane domain of μ plays an important role in BCR assembly.

The other structural regions of mIgM may also play important roles in the association with Ig-α/Ig-β. For example, a chimeric molecule with the transmembrane and cytosolic domains of μ chain fused to the extracellular domain of the class I MHC molecule $K^k$ can come to the cell surface in transfected WEHI-231 cells and in transgenic B cells, but it fails to signal[25]. Association of this chimeric mIgM with Ig-α/Ig-β was not examined directly, but the lack of signaling function suggests that this chimera may fail to assemble into a BCR complex. The most attractive hypothesis is that the membrane proximal Ig-like domain of the heavy chain is also required to bind to Ig-α/Ig-β. The chimeric $K^k$-μ molecule is expressed relatively weakly on the cell surface of the transgenic B cells, perhaps reflecting some intracellular retention.

## SIGNAL TRANSDUCTION FUNCTION OF THE BCR

The BCR creates intracellular signals by activating one or more protein tyrosine kinases[12, 26]. Targets of BCR-induced tyrosine phosphorylation include a number of known signaling components. Among these are phospholipases Cγ1 and γ2, which catalyze the hydrolysis of phosphatidylinositol 4,5-bisphosphate ($PIP_2$)[27-31]. This releases the second messengers diacylglycerol, which activates protein kinase C, and inositol trisphosphate ($IP_3$), which causes elevation of intracellular free calcium.

The signaling function of the BCR is contained in the cytoplasmic domains of Ig-α and Ig-β. Chimeric proteins in which the C-terminal region of the μ heavy chain is replaced by equivalent regions of other proteins fail to assemble with Ig-α and Ig-β and also fail to initiate intracellular protein tyrosine phosphorylation[21, 32, 33]. Similarly, the YS/VV double point mutant described above that exhibits decreased assembly with Ig-α and Ig-β exhibits similarly decreased signaling ability[17]. Interestingly, when the YS/VV mutation was put into the human μ gene and then introduced into a murine B cell line, assembly of this mutant mIgM with Ig-α/Ig-β was not detected and signal transduction was almost completely deficient[22, 24, 34]. Presumably, human mIgM has a somewhat weaker interaction with murine Ig-α/Ig-β than does murine mIgM, and the YS/VV mutation further weakens this interaction, with the result that mIgM-Ig-α/Ig-β complexes do not form with this mutated human form or are much less stable. The mutant mIgM molecules that have a deletion of the cytoplasmic three amino acids of μ also exhibit a correlation between ability to assemble with Ig-α/Ig-β and ability to induce intracellular signaling events. As described above, mutations of this type result in proteolytic removal of the μ transmembrane domain, addition of a glycosyl-phosphatidylinositol anchor to the extracellular portion of the μ heavy chain, and transit to the cell surface in the absence of association with Ig-α/Ig-β. This mutation also abolishes signal transduction function of mIgM[21, 22, 24]. Thus, association of mIgM with the accessory proteins Ig-α/Ig-β is closely correlated with signaling function.

## SIGNAL TRANSDUCTION CAPABILITY OF THE Ig-α AND Ig-β CYTOPLASMIC DOMAINS

The cytoplasmic domains of Ig-α and Ig-β each contain one copy of a sequence also found in the cytoplasmic domains of components of the T cell antigen receptor, of components of certain Fc receptors and of two viral glycoproteins. These sequences have been referred to as the antigen receptor homology 1 (ARH1) motif, the tyrosine-based activation motif (TAM), or the antigen receptor activation motif (ARAM)[35-37]. This motif contains two YxxL/I sequences (Y: tyrosine, L: leucine, I: isoleucine, x: any amino acid) with a characteristic spacing and with leucine or isoleucine residues found in the third positions following the tyrosines (Fig. 2). Chimeric proteins containing ARAMs from T cell antigen receptor ζ or CD3 ε chains have signaling capability similar to that exhibited by the intact T cell antigen receptor[38-43]. Similarly, chimeric proteins containing the entire cytoplasmic domain of Ig-α are functional for inducing tyrosine phosphorylation and calcium elevation in B cells[24, 44]. In our experiments, chimeric proteins containing the ARAMs and local surrounding sequences of Ig-α or Ig-β are fully capable of triggering all of the BCR signaling events examined to date[45]. In the experiments of some other groups, chimeras containing the entire cytoplasmic domain of Ig-β were not fully active for signaling[24, 44]. This may reflect a difference in the nature of the cell lines where these chimeras were tested for function, although the reason for this difference is not yet apparent.

There are several ways in which the ARAM could contribute to signaling. One possibility is suggested by recent insights into the signaling mechanism of the platelet-derived growth factor (PDGF) receptor. This receptor has intrinsic protein tyrosine kinase activity. PDGF binding induces receptor dimerization, after which each receptor molecule transphosphorylates sites on the other molecule. These phosphorylated sites on the receptor then act as binding sites for signaling components activated by the receptor. In the case of the PDGF receptor, these include phospholipase Cγ1, phosphatidylinositol 3-kinase (PI 3-kinase), the GTPase activating protein of Ras (RasGAP), and Grb-2, a linker protein that is also involved in regulating Ras[46-50]. These signaling components have SH2 (Src-homology 2) domains that bind to the autophosphorylated receptor in a phosphorylation-dependent manner. Next, the signaling components are activated, either as a direct consequence of binding or secondarily due to tyrosine phosphorylation by the PDGF receptor. Thus, for receptors like the PDGF receptor, the receptor autophosphorylation sites are responsible for determining the specificity of the receptor for downstream signaling components.

Crosslinking chimeric proteins containing the ARAMs of Ig-α or Ig-β induces tyrosine phosphorylation of the chimeric proteins themselves [45]. The only cytoplasmic tyrosines in the Ig-β-containing chimera are those of the motif, so clearly the motif is being phosphorylated in this case. The Ig-α chimera contains an additional tyrosine in its cytoplasmic domain, but mutation of this tyrosine to phenylalanine does not interfere with signaling or crosslinking-induced tyrosine phosphorylation of the chimera (Law and DeFranco, unpublished observations). Thus, for the Ig-α chimera as well as for the Ig-β chimera, it is the motif tyrosines that are phosphorylated. If the ARAMs of Ig-α and of Ig-β serve the same purpose as the PDGF receptor autophosphorylation sites, then the function of different motifs in the same receptor would likely be complementary in that the tyrosine phosphorylation sites would be expected to serve as binding sites for distinct signaling targets. On the contrary, chimeras containing the Ig-α and Ig-β motifs are equivalent in signaling capability, rather than being incomplete and complementary. For example, chimeric proteins containing the Ig-α or Ig-β motifs are each capable of activating phosphoinositide breakdown and of inducing phosphorylation of PI 3-kinase and of Vav[45].

These observations suggest that the ARAMs play roles in antigen receptor signaling that involves activation of tyrosine phosphorylation *per se*, and are not involved in the choice of targets for the activated tyrosine kinases. Interestingly in this regard, crosslinking of chimeric

| | |
|---|---|
| <u>B cell antigen receptor</u> | |
| murine Ig-α | ENL YEGL NLDDCSM YEDI |
| human Ig-α | ENL YEGL NLDDCSM YEDI |
| murine Ig-β | DHT YEGL NIDQTAT YEDI |
| human Ig-β | DHT YEGL DIDQTAT YEDI |
| <u>T cell antigen receptor</u> | |
| murine CD3-γ | EQL YQPL KDREYDQ YSHL |
| murine CD3-δ | EQL YQPL RDREDTQ YSRL |
| murine CD3-ε | NPD YEPI RKGQRDL YSGL |
| murineTCR-ζ1 | NQL YNEL NLGRREE YDVL |
| murine TCR-ζ2 | EGV YNAL QKDKMAEAYSEI |
| murine TCR-ζ3 | DGL YQGL STATKDT YDAL |
| <u>Fc receptors</u> | |
| murine FcεRI-β | DRL YEEL NVYSPI YSEL |
| murine FcεRI-γ | DAV YTGL NTRSQET YETL |
| Consensus | $^{D}_{E}$xx Yxx$^{L}_{I}$ xxxxxxx Yxx$^{L}_{I}$ |

Figure 2. The antigen receptor activation motif (ARAM). The ARAM of human and murine Ig-α, and Ig-β and other antigen receptor polypeptides is shown. It should be noted that besides the motif per se, there is additional conservation between Ig-α and Ig-β not seen in the other antigen receptor polypeptides. In addition, in the entire cytoplasmic tails of Ig-α or Ig-β, there is only one difference between the mouse and human homologs. These sequence conservations imply an important role for amino acids other than those of the consensus motif.

proteins containing the parts of Ig-α or Ig-β cytoplasmic domains leads to association with the protein tyrosine kinases PTK72/Syk, p53/56*lyn* and p59*fyn*. Although some p59*fyn* could be detected associated with the chimeric proteins prior to stimulation, the amount of p59*fyn* bound to the chimeras increased upon crosslinking[45]. In contrast, no phosphorylated Syk or p53/56*lyn* was detected associated with the chimeras prior to stimulation. Each of these tyrosine kinases has one or more SH2 domains, so it is likely that phosphorylation of the motif tyrosines promotes binding of the tyrosine kinases via their SH2 domains. In this regard it is interesting that the SH2 domains of the *src*-family members prefer to bind to sequences containing a leucine or isoleucine in the third position after the phosphorylated tyrosine[51], and this is a conserved feature of the ARAM. It is unclear at this time how BCR crosslinking activates these tyrosine kinases, but this could result from the binding of the tyrosine kinase to the motif or from a subsequent phosphorylation of the kinase [52, 53].

The following model for signal initiation by the BCR is based on the observations described above (Fig. 3). Crosslinking of BCR molecules clusters the cytoplasmic domains of Ig-α and Ig-β. Most BCR molecules do not have any tyrosine kinases bound to them prior to stimulation, but a few may have a tyrosine kinase bound. For example, some p59*fyn* is seen associated with the chimeras prior to stimulation[45]. Thus, crosslinking of BCR molecules causes to clustering of many ARAMs near a few tyrosine kinase molecules. These tyrosine kinase molecules then phosphorylate Ig-α and Ig-β on one or more likely both motif tyrosines. The phosphorylated ARAMs are now good binding sites for the tyrosine kinases PTK72/Syk, p53/p56*lyn*, and p59*fyn*. The tyrosine kinases thus recruited to the BCR complex can phosphorylate more ARAM tyrosines, providing a positive feedback loop leading to extensive phosphorylation and clustering of active tyrosine kinase molecules. The clustering of tyrosine kinase molecules could promote their activation, perhaps by tyrosine phosphorylation. Indeed, anti-Ig treatment of B cells does lead to tyrosine phosphorylation of PTK72/Syk[45, 52]. Moreover, clustering of chimeric transmembrane proteins that have Syk attached to their cytoplasmic domains triggers phosphoinositide breakdown and cytolytic effector function in T cells[54]. These observations suggest that PTK72/Syk is the effector kinase responsible for phosphorylating downstream signaling targets. This hypothesis suggests that the *src*-family tyrosine kinases (p53/p56*lyn* and p59*fyn*) are regulatory kinases that phosphorylate the clustered cytoplasmic domains of Ig-α and Ig-β, thereby initiating the antigen receptor signaling cascade. Interestingly, crosslinking BCR molecules expressed in non-lymphoid AtT20 cells results in strong tyrosine phosphorylation of Ig-α and Ig-β cytoplasmic domains, but deficient signaling ability with regard to downstream targets[15]. These cells express p59*fyn*, but do not express PTK72/Syk (Matsuuchi, J. Richards and DeFranco, unpublished observations). We are currently introducing PTK72/Syk into these cells to see if it will be sufficient to confer signaling ability on the BCR in non-lymphoid cells.

If, as this model states, the ARAMs play key roles in initiation of BCR signaling, that leaves open the question of how downstream signaling targets are selected. One interesting possibility is that other transmembrane proteins cooperate and confer specificity for downstream targets on BCR signaling. Both CD19 and CD22 can be found associated with the BCR, and they both become tyrosine phosphorylated upon stimulation of B cells with anti-Ig[55-58]. Moreover, BCR stimulation leads to association between CD19 and PI 3-kinase, one of the signaling targets of the BCR[57]. Thus, CD19 and/or CD22 may act like the autophosphorylation sites on the PDGF receptor to select signaling targets for the BCR-activated tyrosine kinase(s).

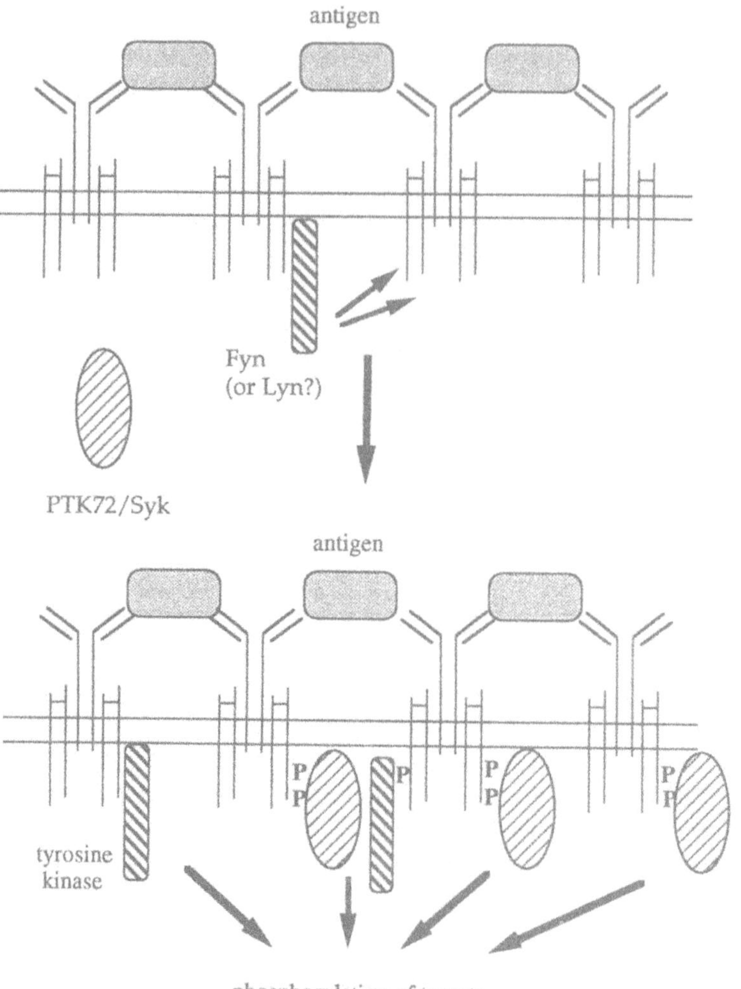


Figure 3. Model for B cell antigen receptor signal initiation. Crosslinking by antigen clusters a few pre-bound tyrosine kinases (probably of the Src-family) with a large number of cytoplasmic tails of Ig-α and Ig-β. The ARAM tyrosines of Ig-α and Ig-β then serve as substrates for the pre-bound tyrosine kinases. Phosphorylation of motif tyrosines permits binding by both Src-family tyrosine kinases and PTK72/Syk, probably via the SH2 domains of these kinases. This in turn leads to the activation of these kinases, either by binding or by subsequent phosphorylation. These kinases further phosphorylate the clustered ARAMs and phosphorylate downstream signaling targets.

## TRIGGERING OF APOPTOSIS BY THE BCR

One approach to understanding the roles of individual signaling reactions is to try to determine which intracellular events they regulate. One type of event that is regulated by receptor signaling is gene induction. Genes that are induced by receptor stimulation in the absence of new protein synthesis are referred to as "early response genes", and must be regulated by the alteration of the activity of pre-existing components rather than by prior changes in gene expression. Early response genes previously identified in B cells include *c-fos, c-myc, jun-B* and *egr-1*, all of which are transcription factors[1]. We examined a panel of seven early response genes from serum-stimulated fibroblasts and found that four of them (*nur77,nup475, pip92* and *3CH134*) are induced by anti-IgM stimulation of resting splenic B cells[59]. We found that each of these genes could be induced by phorbol esters, indicating a role for protein kinase C, with the exception that under some circumstances *3CH134* required calcium elevation for its induction. Thus, the induction of these genes appeared to be downstream of $PIP_2$ breakdown. This conclusion was supported by the observation that activating $PIP_2$ breakdown by a separate mechanism, stimulation of an introduced M1-muscarinic acetylcholine receptor, also induced expression of these genes in B cells. Similarly, phosphoinositide-derived second messengers seem to be responsible for the induction of the other early response genes, mentioned above[1]. As BCR signaling activates other signaling reactions in addition to $PIP_2$ breakdown, it is surprising that this latter event seems to mediate activation of all of the early response genes known to be induced by this receptor. It may be that these other signaling events contribute to the gene activations seen, or alternatively, that they are involved in mediating other types of events, such as regulation of the cytoskeleton. It should be emphasized that it is likely that other early response genes exist that have not yet been identified. We have tried to isolate such genes by screening of cDNA libraries with subtracted cDNA probes. However, all of the isolates that were identified in this way turned out to be one of the early response genes listed above (Mittelstadt and DeFranco, unpublished results). Thus, any currently unidentified early response genes are probably expressed at lower levels than the early response genes identified to date.

The early response genes are likely to be important mediators or regulators of BCR signaling. The earlier identified set of four early response genes all encode transcription factors and therefore are likely to be involved in inducing a second round of gene expression. Also in this category are *nur77*, which is a member of the steroid receptor family and *nup475*, the product of which has been localized to nuclei. The *pip92* gene product is a highly labile cytosolic protein of unknown function and the *3CH134* product is a phosphatase that apparently can act upon MAP kinase to inactivate it[60]. Two of these genes have recently been implicated in regulation of apoptosis: *c-myc* and *nur77* in T cells and fibroblasts[61-64]. We have also obtained evidence indicating that *c-myc* plays a role in the apoptosis and growth arrest induced by anti-IgM in WEHI-231 cells (P. R. Mittelstadt, G. Ramsay, J. M. Bishop, and A. L. DeFranco, submitted for publication). Antisense oligonucleotides directed against the first five codons of *c-myc* protect WEHI-231 cells from apoptosis. These concentrations of antisense reduce c-Myc protein levels by about 3-fold in unstimulated cells. Anti-IgM treatment of WEHI-231 cells causes a transient rise in c-Myc levels, followed by a decline which is quite dramatic by 18 hr.[65], a time which corresponds to growth arrest but precedes apoptosis substantially. The *c-myc* antisense oligonucleotides are effective even if added after the initial rise and fall in c-Myc levels, indicating that there is an essential role of *c-myc* in the apoptotic response after the early response gene stage. The late fall in c-Myc levels appears to be secondary to growth arrest and/or apoptosis, as it does not occur in the presence of *c-myc* antisense oligonucleotides or other conditions that prevent apoptosis (P. R. Mittelstadt, G. Ramsay, J. M. Bishop, and A. L. DeFranco, submitted for publication)[66, 67]. Thus, at late times after anti-Ig treatment, the *c-myc* antisense oligonucleotides appears to have stabilized c-Myc, but this could be indirect and the result of an earlier inhibition of c-Myc expression. Similar results have recently been presented by Fischer et al.[66], although they suggest that the *c-myc* antisense oligonucleotide is blocking apoptosis and growth arrest due to elevating c-Myc levels. This interpretation predicts that exogenously introduced c-Myc would protect cells from apoptosis induced by anti-IgM. On the contrary, we have introduced a regulatable form of c-Myc (a chimera between c-Myc and the estrogen-binding domain of the estrogen receptor, c-MycER) into WEHI-231 cells, and we find that activation of c-Myc induces apoptosis on its own, and that together with anti-IgM there is a still faster apoptosis (P. R. Mittelstadt, G. Ramsay, J. M. Bishop, and A. L. DeFranco, submitted for publication). These results are most compatible with the view that c-Myc plays an active role the triggering of apoptosis in WEHI-231 B lymphoma cells, as previously concluded for serum-starved fibroblasts and antigen-receptor stimulated T hybridoma cells. It is unclear from these experiments whether BCR signaling triggers apoptosis by increasing the activity of c-Myc or whether it does so by inducing other events that cooperate with a basal level of c-Myc activity.

## SUMMARY

The antigen receptor of B lymphocytes (BCR) plays important roles in virtually every stage in the development, inactivation, or activation of B cells. The BCR is a complex of membrane immunoglobulin (mIg) and a heterodimer of two transmembrane polypeptides called Ig-α and Ig-β. Site directed mutation of the μ immunoglobulin heavy chain has demonstrated that the μ transmembrane domain plays a key role in the assembly of mIgM with Ig-α/Ig-β. In addition, there is a strong correlation between the ability of various mutant mIgM molecules to associate with Ig-α/Ig-β and their ability to induce signal transduction reactions such as protein tyrosine phosphorylation and phosphoinositide breakdown. The cytoplasmic domains of Ig-α and Ig-β share a region of limited homology with each other and with components of the T cell antigen receptor and of the Fc receptor. The presence of regions of the cytoplasmic domains of Ig-α or Ig-β including this conserved amino acid sequence motif is sufficient to confer signaling function on chimeric transmembrane proteins. Both Ig-α and Ig-β chimeras are capable of inducing all of the BCR signaling events tested. Based on these and related observations, we propose that the motifs act to initiate the BCR signaling reactions by binding and activating tyrosine kinases. Among the important events mediated by BCR signaling is induced expression of a series of genes referred to as early response genes. In B cells these include transcription factors and at least one component that regulates signaling events. One of these genes, *c-myc*, appears to play an important role in mediating apoptosis in B cells stimulated via the BCR complex.

## ACKNOWLEDGEMENTS

This work was supported by U. S. Public Health Service Grant AI20038 and Grant MT-11528 from the Medical Research Council of Canada and funds from the University of British Columbia. T.L.S. and D.A.L. are recipients of Arthritis Foundation Postdoctoral awards. J.H.B. was supported by training grant GM07618, as well as by the Sussman Fund and a grant from Achievement Rewards for College Scientists.

## REFERENCES

1. A.L. DeFranco, Structure and function of the B cell antigen receptor, *Ann. Rev. Cell Biol.* 9: 377 (1993).
2. D. Kitamura, J. Roes, R. Kuhn, and K. Rajewsky, A B cell-deficient mouse by targeted disruption of the membrane exon of the immunoglobulin μ chain, *Nature* 350: 423 (1991).
3. D. Kitamura, A. Kudo, S. Schaal, W. Muller, F. Melchers, and K. Rajewsky, A critical role of λ5 protein in B cell development., *Cell* 69: 823 (1992).
4. D. Gay, T. Saunders, S. Camper, and M. Weigert, Receptor editing: an approach by autoreactive B cells to escape tolerance, *J. Exp. Med.* 177: 999 (1993).
5. S.L. Tiegs, D.M. Russell, and D. Nemazee, Receptor editing in self-reactive bone marrow B cells, *J. Exp. Med.* 177: 1009 (1993).
6. D.M. Russell, Z. Dembic, G. Morahan, J.F. Miller, K. Burki, and D.A. Nemazee, Peripheral deletion of self-reactive B cells, *Nature* 354: 308 (1991).
7. S.B. Hartley, J. Crosbie, R. Brink, A.B. Kantor, A. Basten, and C.C. Goodnow, Elimination from peripheral lymphoid tissues of self-reactive B lymphocytes recognizing membrane-bound antigens, *Nature* 353: 765 (1991).
8. M. Murakami, T. Tsubata, M. Okamoto, A. Shimizu, S. Kumagai, H. Imura, and T. Honjo, Antigen-induced apoptotic death of Ly-1 B cells responsible for autoimmune disease in transgenic mice, *Nature* 357: 77 (1992).
9. C.C. Goodnow, J. Crosbie, S. Adelstein, T.B. Lavoie, S.J. Smith-Gill, R. Brink, H. Pritchard-Briscoe, J.S. Wotherspoon, R.H. Loblay, K. Raphael, R.J. Trent, and A. Basten, Altered immunoglobulin expression and functional silencing of self-reactive B lymphocytes in transgenic mice, *Nature* 334: 676 (1988).
10. A.L. DeFranco, Molecular aspects of B-lymphocyte activation, *Ann. Rev. Cell Biol.* 3: 143 (1987).
11. Y.-J. Liu, D.E. Joshua, G.T. Williams, C.A. Smith, J. Gordon, and I.C.M. MacLennan, Mechanism of antigen-driven selection in germinal centres, *Nature* 342: 929 (1989).
12. M. Reth, Antigen receptors on B lymphocytes, *Ann. Rev. Immunol.* 10: 97 (1992).
13. G.T. Williams, A.R. Venkitaraman, D.J. Gilmore, and M.S. Neuberger, The sequence of the mu transmembrane segment determines the tissue specificity of the transport of immunoglobulin M to the cell surface, *J. Exp. Med.* 171: 947 (1990).
14. A.R. Venkitaraman, G.T. Williams, P. Dariavach, and M.S. Neuberger, The B cell antigen receptor of the five immunoglobulin classes, *Nature* 352: 777 (1991).
15. L. Matsuuchi, M.R. Gold, A. Travis, R. Grosschedl, A.L. DeFranco, and R.B. Kelly, The membrane IgM-associated proteins MB-1 and Ig-β are sufficient to promote surface expression of a partially functional B-cell antigen receptor in a nonlymphoid cell line, *Proc. Natl. Acad. Sci. USA* 89: 3404 (1992).
16. B.J. Cherayil, K. MacDonald, G.L. Waneck, and S. Pillai, Surface transport and internalization of the membrane IgM H chain in the absence of the Mb-1 and B29 proteins, *J. Immunol.* 151: 11 (1993).

17. T.L. Stevens, J.B. Blum, S.P. Foy, L. Matsuuchi, and A.L. DeFranco, A mutation of the μ transmembrane that disrupts ER retention: effects on association with accessory proteins and signal transduction, *J. Immunol.* 152: in press (1994).
18. E. Degen and D.B. Williams, Participation of a novel 88-kD protein in the biogenesis of murine class I histocompatability molecules, *J. Cell Biol.* 112: 1099 (1991).
19. F. Hochstenbach, V. David, S. Watkins, and M.B. Brenner, Endoplasmic reticulum resident protein of 90 kilodaltons associates with the T- and B-cell antigen receptors and major histocompatability antigens during their assembly, *Proc. Natl. Acad. Sci. USA* 89: 4734 (1992).
20. J. Hombach, T. Tsubata, L. Leclercq, H. Stappert, and M. Reth, Molecular components of the B-cell antigen receptor complex of the IgM class, *Nature* 343: 760 (1990).
21. J.H. Blum, T.L. Stevens, and A.L. DeFranco, Role of the μ immunoglobulin heavy chain transmembrane and cytoplasmic domains in B cell antigen receptor expression and signal transduction, *J. Biol. Chem.* 27238-27247: (1993).
22. A.C. Shaw, R.N. Mitchell, Y.K. Weaver, J. Campos-Torres, A.K. Abbas, and P. Leder, Mutations of immunoglobulin transmembrane and cytoplasmic domains: Effects on intracellular signaling and antigen presentation, *Cell* 63: 381 (1990).
23. R.N. Mitchell, A.C. Shaw, Y.K. Weaver, P. Leder, and A.K. Abbas, Cytoplasmic tail deletion converts membrane immunoglobulin to a phosphatidylinositol-linked form lacking signaling and efficient antigen internalization functions, *J. Biol. Chem.* 266: 8856 (1991).
24. M. Sanchez, Z. Misulovin, A.L. Burkhardt, S. Mahajan, T. Costa, R. Franke, J.B. Bolen, and M. Nussenzweig, Signal transduction by immunoglobulin is mediated through Ig-α and Ig-β, *J. Exp. Med.* 178: 1049 (1993).
25. W.K. Tsang, J. Mizuguchi, Y. Ishida, C. Watson, T. Chused, J. Inman, D.H. Margulies, and W.E. Paul, Failure of signaling through a chimeric class I-immunoglobulin molecule expressed on the surface of transfected B lymphoma cells and cells of transgenic mice, *Cell. Immunol.* 143: 80 (1992).
26. A.L. DeFranco, Tyrosine phosphorylation and the mechanism of signal transduction by the B-lymphocyte antigen receptor, *Eur. J. Biochem.* 210: 381 (1992).
27. R.H. Carter, D.J. Park, S.G. Rhee, and D.T. Fearon, Tyrosine phosphorylation of phospholipase C induced by membrane immunoglobulin in B lymphocytes, *Proc. Natl. Acad. Sci. USA* 88: 2745 (1991).
28. W.M. Hempel, R.C. Schatzman, and A.L. DeFranco, Tyrosine phosphorylation of phospholipase C γ2 upon crosslinking of membrane Ig on murine B lymphocytes, *J. Immunol.* 148: 3021 (1992).
29. K.M. Coggeshall, J.C. McHugh, and A. Altman, Predominant expression and activation-induced tyrosine phosphorylation of phospholipase C-γ2 in B lymphocytes, *Proc. Natl. Acad. Sci. USA* 90: 5660 (1992).
30. S.B. Kanner, J.P. Deans, and J.A. Ledbetter, Regulation of CD3-induced phospholipase C-gamma1 (PLCγ1) tyrosine phosphorylation by CD4 and CD45 receptors, *Immunology* 75: 441 (1992).
31. C.M. Roifman and G. Wang, Phospholipase C-γ1 and phospholipase C-γ2 are substrates of the B cell antigen receptor associated protein tyrosine kinase, *Biochem. Biophys. Res. Commun.* 183: 411 (1992).
32. C.F. Webb, C. Nakai, and P.W. Tucker, Immunoglobulin receptor signalling depends on the carboxyl terminus but not the heavy-chain class, *Proc. Natl. Acad. Sci. USA* 86: 1977 (1989).
33. P.M. Dubois, J. Stepinski, J. Urbain, and C.H. Sibley, Role of the transmembrane and cytoplasmic domains of surface IgM in endocytosis and signal transduction, *Eur. J. Immunol.* 22: 851 (1992).
34. S.A. Grupp, K. Campbell, R.N. Mitchell, J.C. Cambier, and A.K. Abbas, Signaling-defective mutants of the B lymphocyte antigen receptor fail to associate with Ig-α and Ig-β/γ, *J. Biol. Chem.* 268: 25776 (1993).

35. J.C. Cambier, Signal transduction by T- and B-cell antigen receptors: converging structures and concepts, *Curr. Opin. Immunol.* 4: 257 (1992).

36. L.E. Samelson and R.D. Klausner, Tyrosine kinases and tyrosine-based activation motifs, *J. Biol. Chem.* 267: 24913 (1992).

37. A. Weiss, T cell antigen receptor signal transduction: A tale of tails and cytoplasmic protein-tyrosine kinases, *Cell* 73: 209 (1993).

38. B. Irving and A. Weiss, The cytoplasmic domain of the T cell receptor ζ chain is sufficient to couple to receptor-associated signal transduction pathways, *Cell* 64: 891 (1991).

39. B.A. Irving, A.C. Chan, and A. Weiss, Functional characterization of a signal transducing motif present in the T cell antigen receptor ζ chain, *J. Exp. Med.* 177: 1093 (1993).

40. C. Romeo and B. Seed, Cellular immunity to HIV activated by CD4 fused to T cell or Fc receptor polypeptides, *Cell* 64: 1037 (1991).

41. C. Romeo, M. Amiot, and B. Seed, Sequence requirements for induction of cytolysis by the T cell antigen/Fc receptor ζ chain, *Cell* 68: 889 (1992).

42. A.-M.K. Wegener, F. Letourneur, A. Hoeveler, T. Brocker, F. Luton, and B. Malissen, The T cell receptor/CD3 complex is composed of at least two autonomous transduction molecules, *Cell* 68: 83 (1992).

43. F. Letourneur and R.D. Klausner, Activation of T cells by a tyrosine kinase activation domain in the cytoplasmic tail of CD3 ε, *Science* 255: 79 (1992).

44. K.-M. Kim, G. Alber, P. Weiser, and M. Reth, Differential signaling through the Ig-α and Ig-β components of the B cell antigen receptor, *Eur. J. Immunol.* 23: 911 (1993).

45. D.A. Law, V.W.F. Chan, S.K. Datta, and A.L. DeFranco, B-cell antigen receptor motifs have redundant signalling capabilities and bind the tyrosine kinases PTK72, Lyn and Fyn, *Curr. Biol.* 3: 645 (1993).

46. W.J. Fantl, J.A. Escobedo, G.A. Martin, C.W. Turck, M. del Rosario, F. McCormick, and L.T. Williams, Distinct phosphotyrosines on a growth factor receptor bind to specific molecules that mediate different signaling pathways, *Cell* 69: 413 (1992).

47. A. Kazlauskas, A. Kashishian, J.A. Cooper, and M. Valius, GTPase-activating protein and phosphatidylinositol 3-kinase bind to a distinct region of the platelet-derived growth factor receptor β subunit, *Mol. Cell. Biol.* 12: 2534 (1992).

48. M. Rozakis-Adcock, J. McGlade, G. Mbamalu, G. Pelicci, R. Daly, W. Li, A. Batzer, S. Thomas, J. Brugge, P.G. Pelicci, J. Schlessinger, and T. Pawson, Association of the Shc and Grb2/Sem5 SH2-containing proteins is implicated in activation of the Ras pathway by tyrosine kinases, *Nature* 360: 689 (1992).

49. E.J. Lowenstein, R.J. Daly, A.G. Batzer, W. Li, B. Margolis, R. Lammers, A. Ullrich, E. Skolnik, D. Bar-Sagi, and J. Schlessinger, The SH2 and SH3 domain-containing protein GRB2 links receptor tyrosine kinases to ras signaling, *Cell* 70: 431 (1992).

50. J.P. Olivier, T. Raabe, M. Henkemeyer, B. Dickson, G. Mbamalu, B. Margolis, J. Schlessinger, E. Hafen, and T. Pawson, A Drosophila SH2-SH3 adapter protein implicated in coupling the sevenless tyrosine kinase to an activator of Ras guanine nucleotide exchange, Sos, *Cell* 73: 179 (1993).

51. Z. Songyang, S.E. Shoelson, M. Chaudhuri, G. Gish, T. Pawson, W.G. Haser, F. King, T. Roberts, S. Ratnofsky, R.J. Lechleider, B.G. Neel, R.B. Birge, J.E. Fajardo, M.M. Chou, H. Hanafusa, B. Schaffhausen, and L.C. Cantley, SH2 domains recognize specific phosphoprotein sequences, *Cell* 72: 767 (1993).

52. J.E. Hutchcroft, M.L. Harrison, and R.L. Geahlen, B lymphocyte activation is accompanied by phosphorylation of a 72-kDa protein-tyrosine kinase, *J. Biol. Chem.* 266: 14846 (1991).

53. J.E. Hutchcroft, M.L. Harrison, and R.L. Geahlen, Association of the 72-kDa protein-tyrosine kinase PTK72 with the B cell antigen receptor, *J. Biol. Chem.* 267: 8613 (1992).
54. W. Kolanus, C. Romeo, and B. Seed, T cell activation by clustered tyrosine kinases, *Cell* 74: 171 (1993).
55. J.M. Pesando, L.S. Bouchard, and B. McMaster, CD19 is functionally and physically associated with surface immunoglobulin, *J. Exp. Med.* 170: 2159 (1989).
56. R.J. Schulte, M.-A. Campbell, W.H. Fischer, and B.M. Sefton, Tyrosine phosphorylation of CD22 during B cell activation, *Science* 258: 1001 (1992).
57. D.A. Tuveson, R.H. Carter, S.P. Soltoff, and D.T. Fearon, CD19 of B cells as a surrogate kinase insert region to bind phosphatidylinositol 3-kinase, *Science* 260: 986 (1993).
58. C. Leprince, K.E. Draves, R.L. Geahlen, J.A. Ledbetter, and E.A. Clark, CD22 associates with the human surface IgM-B cell antigen receptor complex, *Proc. Natl. Acad. Sci. USA* 90: 3236 (1993).
59. P. Mittelstadt and A.L. DeFranco, Induction of early-response genes by cross-linking membrane immunoglobulin on B lymphocytes, *J. Immunol.* 150: 4822 (1993).
60. H. Sun, C.H. Charles, L.F. Lau, and N.K. Tonks, MKP-1 (3CH134), an immediate early gene product, is a dual specificity phosphatase that dephosphorylates MAP kinase in vivo, *Cell* 75: 487 (1993).
61. Y. Shi, J.M. Glynn, L.J. Guilbert, T.G. Cotter, R.P. Bissonette, and D.R. Green, Role for c-myc in activation-induced apoptotic cell death in T cell hybridomas, *Science* 257: 212 (1992).
62. G.I. Evans, A.H. Wyllie, C.S. Gilbert, T.D. Littlewood, H. Land, M. Brooks, C.M. Waters, L.Z. Penn, and D.C. Hancock, Induction of apoptosis in fibroblasts by c-myc protein, *Cell* 69: 119 (1992).
63. J.D. Woronicz, B. Calnan, V. Ngo, and A. Winoto, Requirement for the orphan steroid receptor Nur77 in apoptosis of T-cell hybridomas, *Nature* 367: 277 (1994).
64. Z.-G. Liu, S.W. Smith, K.A. McLaughlin, L.M. Schwartz, and B.A. Osborne, Apoptotic signals delivered through the T-cell receptor of a T-cell hybrid require the immediate-early gene *nur77*, *Nature* 367: 281 (1994).
65. J.E. McCormack, V.H. Pepe, R.B. Kent, M. Dean, A. Marshak-Rothstein, and G.E. Sonenshein, Specific regulation of *c-myc* oncogene expression in a murine B-cell lymphoma, *Proc. Natl. Acad. Sci. USA* 81: 5546 (1984).
66. G. Fischer, S.C. Kent, L. Joseph, D.R. Green, and D.W. Scott, Lymphoma models for B cell activation and tolerance. X. Anti-μ mediated growth arrest and apoptosis of murine B cell lymphomas is prevented by the stabilization of myc, *J. Exp. Med.* 179: 221 (1994).
67. U. Hibner, L.E. Benhamou, P.-A. Cazenave, and P. Sarthou, Signaling of programmed cell death induction in WEHI-231 B lymphoma cells, *Eur. J. Immunol.* 23: 2821 (1993).

# B-CELL ACTIVATION BY WILD TYPE AND MUTANT IG-β CYTOPLASMIC DOMAINS

John A. Taddie, Tamara R. Hurley, and Bartholomew M. Sefton

Molecular Biology and Virology Laboratory
The Salk Institute
P.O. Box 85800
San Diego, CA 92186

## ABSTRACT

In B lymphocytes, the cytoplasmic domains of the membrane immunoglobulin-associated heterodimeric Ig-α and Ig-β proteins link membrane immunoglobulin to intracellular signalling molecules. We constructed chimeric genes encoding the extracellular and transmembrane domain of human CD8α and the cytoplasmic domain of Ig-α or Ig-β and examined the ability of the chimeric proteins to induce signalling in the murine B-cell lymphoma A20. Crosslinking of CD8/Ig-α or CD8/Ig-β induced both calcium mobilization and protein tyrosine phosphorylation, although induction by CD8/Ig-α was somewhat stronger. We also carried out mutagenesis of residues within the "Reth" motif of the CD8/Ig-β cytoplasmic domain and determined the effects of these mutations on signalling in the murine B-cell hybridoma LK 35.2. Mutants in which alanine was substituted for glutamine 202, threonine 205, and isoleucine 209 retained the ability to induce protein tyrosine phosphorylation and calcium mobilization. In contrast, substitution of alanine for leucine 198 abrogated these responses, suggesting a critical role for this residue in interaction with cytoplasmic signalling proteins.

*Mechanisms of Lymphocyte Activation and Immune Regulation V*
Edited by S. Gupta *et al.*, Plenum Press, New York, 1994

## INTRODUCTION

The antigen receptor of B lymphocytes contains antigen-specific membrane immunoglobulin (mIg) in a noncovalent association with a disulfide-linked transmembrane heterodimer of the Ig-α and Ig-β polypeptides.[1-5] Engagement of the antigen receptor with either antigen or antibodies to mIg triggers a complex signal transduction cascade that begins with an increase in protein tyrosine phosphorylation.[6,7] Two lines of evidence have indicated that Ig-α and Ig-β couple mIg to intracellular signalling molecules. First, mutations in the transmembrane domain of mIg that prevent the association of mIg with Ig-α/Ig-β also prevent signalling through mIg.[8,9] Second, crosslinking of chimeric proteins consisting of irrelevant extracellular and transmembrane domains and the cytoplasmic domain of Ig-α or Ig-β induces protein tyrosine phosphorylation and calcium ($Ca^{++}$) mobilization.[8,10-13] The ability of Ig-α and Ig-β to signal is due to the presence of a so-called "Reth motif," a 26-amino acid sequence also found in the cytoplasmic domains of the T-cell antigen receptor (TCR) chains CD3-δ, -ε, -γ, -η and -ζ, several Fc receptor chains, the Epstein-Barr virus (EBV) LMP2A protein and the bovine leukemia virus (BLV) gp30 protein.[14-17] All or part of this motif has been shown to be necessary and sufficient to induce signalling when expressed as the cytoplasmic domain of heterologous proteins.[11,18-20]

In all B-cell lines analyzed, chimeric proteins containing the cytoplasmic domain of Ig-α induce both $Ca^{++}$ mobilization and dramatic protein tyrosine phosphorylation.[8,10,11,13] In contrast, while Ig-β chimeras induce $Ca^{++}$ mobilization consistently, their ability to induce protein tyrosine phosphorylation has varied. For example, crosslinking of Ig-β chimeras was found to induce substantial protein tyrosine phosphorylation in the B-cell lymphoma 2PK3 and the B-cell hybridoma LK 35.2[11,13] but not in the B-cell lymphoma K46.[10] Interestingly, in the B-cell lymphoma A20, the Ig-β cytoplasmic domain was observed by one group of investigators to induce dramatic protein tyrosine phosphorylation,[12] while another laboratory observed no induction.[8] In this report, we also compared the ability of the cytoplasmic domains of Ig-α and Ig-β to induce signalling in A20 cells, utilizing chimeric genes encoding the extracellular and transmembrane domains of the human CD8α glycoprotein and the cytoplasmic domain of Ig-α or Ig-β. In addition, we have used site-directed mutagenesis to examine the importance of four residues within the Reth motif of Ig-β for signalling in LK 35.2 cells.

## MATERIALS AND METHODS

### DNA Constructs and Mutagenesis

Chimeric genes encoding the extracellular and transmembrane domains of human CD8α and the cytoplasmic domain of Ig-α, Ig-β, or vesicular stomatitis virus glycoprotein G (VSV G) were constructed using the polymerase chain reaction (PCR) and introduced into the retroviral expression plasmid LXSN as described.[13] PCR was also used to generate CD8/Ig-β chimeras with point mutations conferring the substitution of alanine for the cytoplasmic amino acids corresponding to the following residues of wild type (wt) Ig-β:

leucine 198 (L198A), glutamine 202 (Q202A), threonine 205 (T205A), or isoleucine 209 (I209A). The procedure, described in detail elsewhere,[13] utilized the following pairs of complementary primers to introduce the indicated mutations: L198A, 5'-CTATGAGGGC-GCGAACATTGAC-3' and 5'-GTCAATGTTCGCGCCCTCATAG-3'; Q202A, 5'-GAAC-ATTGACGCGACAGCCACC-3' and 5'-GGTGGCTGTCGCGTCAATGTTC-3'; T205A, 5'-CCAGACAGCCGCCTATGAAGA-3' and 5'-TCTTCATAGGCGGCTGTCTGG-3'; I209A, 5'-CTATGAAGACGCAGTGACTCTT-3' and 5'-AAGAGTCACTGCGTCTTCA-TAG-3'. The complete mutant chimeras were also introduced into LXSN.

### Cell Lines

The murine B-cell lymphoma A20[21] was grown in RPMI 1640 supplemented with 10% fetal bovine serum (FBS, Intergen), $5 \times 10^{-5}$ M β-mercaptoethanol, 1x non-essential amino acids, and 1x sodium pyruvate (RPMI B-cell medium). The murine B-cell hybridoma LK 35.2[22] was grown in Dulbecco-Vogt modified Eagle's medium (DMEM, Cellgro) containing the same supplements (DMEM B-cell medium). NK, an IL-2- and IL-4-dependent cell line,[23] was grown in DMEM B-cell medium supplemented with 5% conditioned medium from EL-4 cells. COSm-6 cells were grown in DMEM supplemented with 10% iron-supplemented calf serum (Gemini).

### Generation of B Cells Expressing Chimeric Proteins

LXSN constructs encoding CD8/Ig-α, CD8/Ig-β (wt and mutants), or CD8/VSV G were transfected individually with a viral helper plasmid, SV-$\Psi^-$-EMLV,[24] into COSm-6 cells on 5 cm plates as previously described.[25] At 48 h post-transfection, $5 \times 10^5$ A20 cells or $10^6$ LK 35.2 cells were co-cultivated for 24 h with the transfected cells in 2.5 ml B-cell medium containing 3 μg/ml or 6 μg/ml polybrene, respectively, to allow infection of these cells by the retroviruses produced by the transfected COSm-6 cells. The infected A20 cells were diluted to approximately $5 \times 10^4$/ml and seeded into 12 wells. The infected LK 35.2 cells were diluted to approximately $8 \times 10^4$/ml and seeded into 24 wells. After 24 h, the drug G418 (GIBCO/BRL) was added to a final concentration of 0.8 mg/ml for A20 cells and 1.2 mg/ml for LK 35.2 cells. G418-resistant A20 and LK 35.2 cell pools were analyzed for surface expression of CD8 chimeras by flow cytometry using biotinylated OKT8 antibodies followed by FITC-conjugated streptavidin (Pierce).

### Analysis of Tyrosine Phosphorylation by Crosslinking of CD8 Chimeras

$2 \times 10^6$ A20 or LK 35.2 cells expressing CD8/VSV G, CD8/Ig-α, or CD8/Ig-β (wt or mutant forms) were incubated for 15 min on ice in the appropriate B-cell medium (1% FBS) containing approximately 25 μg/ml biotinylated OKT8 antibodies or approximately 25 μg/ml biotinylated F(ab') fragment of goat anti-mouse IgG (Cappel). The cells were then washed and resuspended in cold RPMI 1640 (A20) or DMEM (LK 35.2). The cells were warmed to 37°C for 30 sec, and crosslinking was induced with avidin D (Vector Laboratories) at a final concentration of 40 μg/ml for 2.5 min at 37°C. Cells were lysed immediately at $10^7$ per ml by the addition of an equal volume of 2x sodium dodecyl sulfate-gel (SDS) sample buffer, and samples were boiled for 10 min. Lysate derived from $2 \times 10^5$ cells was loaded per lane, and the phosphotyrosine-containing proteins were analyzed by Western blotting with anti-phosphotyrosine antibodies as described previously.[26]

### Measurement of Intracellular $Ca^{++}$

Cells were loaded with the $Ca^{++}$-binding dye Indo-1 AM (Molecular Probes), and fluorescence emission following UV excitation was measured at 405 nm and 485 nm on a FACSTAR Plus (Becton Dickinson) as described.[13] After 20 sec of analysis, cell samples were injected with buffer alone, OKT8 antibodies (40 µg/ml final) or goat anti-mouse IgG/IgA/IgM (40 µg/ml final, Cappel), and analysis was continued for a total of 102 sec. The data are obtained as a a ratio of the fluorescence emission of Indo-1 AM at 405 nm (with $Ca^{++}$ bound) to the emission at 485 nm (without $Ca^{++}$ bound) versus time.

## RESULTS

### Signalling by CD8/Ig-α and CD8/Ig-β in A20 cells

In A20 cells, Sanchez *et al.*[8] observed that crosslinking of a chimeric protein encoding the extracellular and transmembrane domains of human IgM heavy chain and the cytoplasmic domain of Ig-α induced protein tyrosine phosphorylation, whereas crosslinking of the analogous Ig-β chimera had little effect. In contrast, Williams *et al.*[12] observed dramatic induction of protein tyrosine phosphorylation upon crosslinking of a similar Ig-β chimera in A20 cells. We have also compared the ability of the cytoplasmic domains of Ig-α and Ig-β to induce protein tyrosine phosphorylation in A20 cells. Chimeric genes encoding the extracellular and transmembrane domains of human CD8α and the cytoplasmic domain of murine Ig-α or Ig-β or the vesicular stomatitis virus (VSV) glycoprotein G (as a negative control) were placed into the retroviral expression vector LXSN[13] and introduced into A20 cells by retroviral infection. G418-resistant cell pools expressed significant and comparable levels of the CD8 chimeric proteins at the cell surface (Fig. 1A). The chimeras are presumably expressed as homodimers on the cell surface, since human CD8α dimerizes via a disulfide-linkage between extracellular domains.[27]

To examine the ability of the CD8/Ig-α and CD8/Ig-β chimeras to induce protein tyrosine phosphorylation, biotin-conjugated OKT8 (a mouse monoclonal antibody specific for human CD8), or a biotin-conjugated goat F(ab') antibody fragment specific for mouse IgG, was prebound at 4°C to uninfected A20 cells or cells expressing the CD8 chimeras and then crosslinked with avidin for 2.5 minutes at 37°C. Tyrosine phosphorylation of cellular proteins was detected by Western blotting of cell lysates with anti-phosphotyrosine antibodies. Stimulation with OKT8 did not induce protein tyrosine phosphorylation above background in uninfected cells or cells expressing CD8/VSV G (Fig. 1B). In contrast, crosslinking of the CD8/Ig-α chimera induced protein tyrosine phosphorylation as efficiently

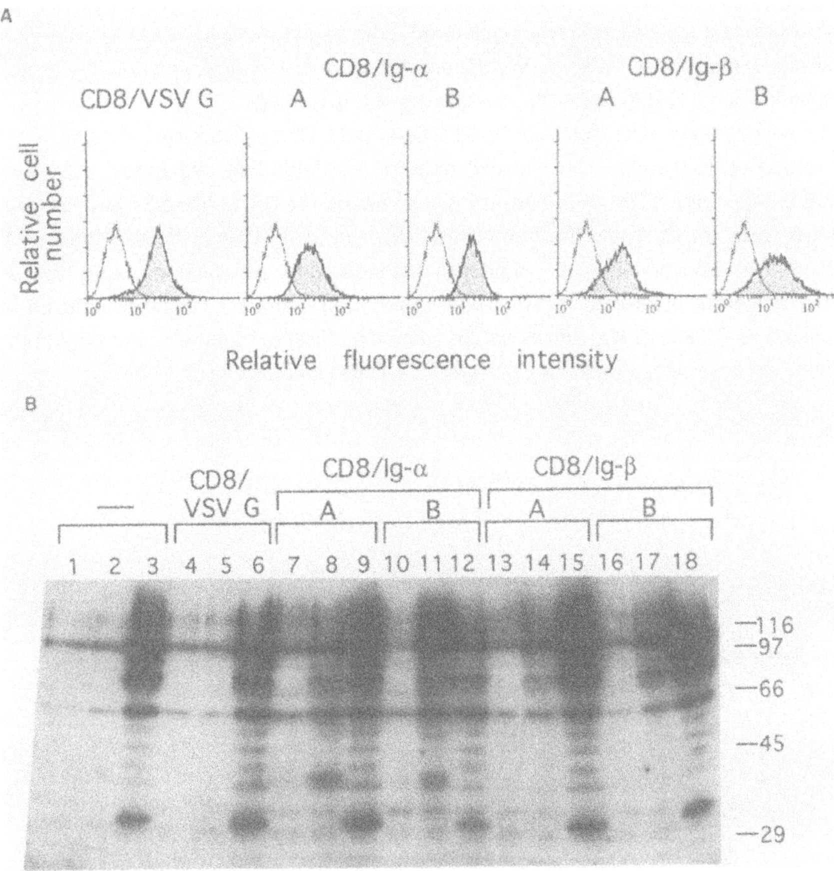


**Figure 1.** (A) Flow cytometric analysis of surface expression of CD8 chimeric proteins on A20 cells. Cells were prebound with biotinylated anti-CD8 antibodies and stained with FITC-conjugated streptavidin. Fluorescence of individual pools (A and B) expressing the indicated chimera is shown (shaded curves) relative to the fluorescence of uninfected cells (open curves). (B) Induction of protein tyrosine phosphorylation following crosslinking of CD8/Ig-α or CD8/Ig-β in A20 cells. Uninfected cells (-) or pools expressing the indicated chimera were preincubated with medium (lanes 1, 4, 7, 10, 13, and 16), biotinylated anti-CD8 antibodies (lanes 2, 5, 8, 11, 14, and 17), or biotinylated F(ab') fragment of goat anti-mouse IgG antibodies (lanes 3, 6, 9, 12, 15 and 18), and crosslinking was induced with avidin D for 2.5 minutes at 37°C as described in Materials and Methods. Cell lysates were fractionated by electrophoresis on an SDS-polyacrylamide gel and subjected to Western blotting with anti-phosphotyrosine antibodies as described.[26] Positions of molecular weight standards (in kDa) are indicated on the right.

as crosslinking of mIg. One notable difference was that only crosslinking mIg induced tyrosine phosphorylation of a 34 kDa substrate that is probably endogenous Ig-α, while only crosslinking CD8/Ig-α induced tyrosine phosphorylation of a 40 kDa substrate that comigrates with the chimera itself (data not shown). Crosslinking of the CD8/Ig-β chimera also induced significant protein tyrosine phosphorylation in A20 cells. While the pattern was also similar to that induced by crosslinking of mIg, the magnitude of the response was not as great as that induced by CD8/Ig-α. We did not detect tyrosine phosphorylation of a protein corresponding to CD8/Ig-β following crosslinking of this chimera.

We next compared the ability of the CD8/Ig-α and CD8/Ig-β chimeras to induce $Ca^{++}$ mobilization in A20 cells. Cells loaded with the $Ca^{++}$-binding dye Indo-1 AM were incubated with anti-CD8 antibodies or goat anti-mouse Ig antibodies and assayed fluorometrically for $Ca^{++}$ mobilization. Crosslinking of the CD8/Ig-α chimera induced an increase in free intracellular [$Ca^{++}$] similar to that induced by crosslinking of mIg (Fig. 2). Like the induction of protein tyrosine phosphorylation, the $Ca^{++}$ response induced by crosslinking of CD8/Ig-β was somewhat less than that observed following crosslinking of CD8/Ig-α. As expected, crosslinking of CD8/VSV G had no effect on $Ca^{++}$ levels.

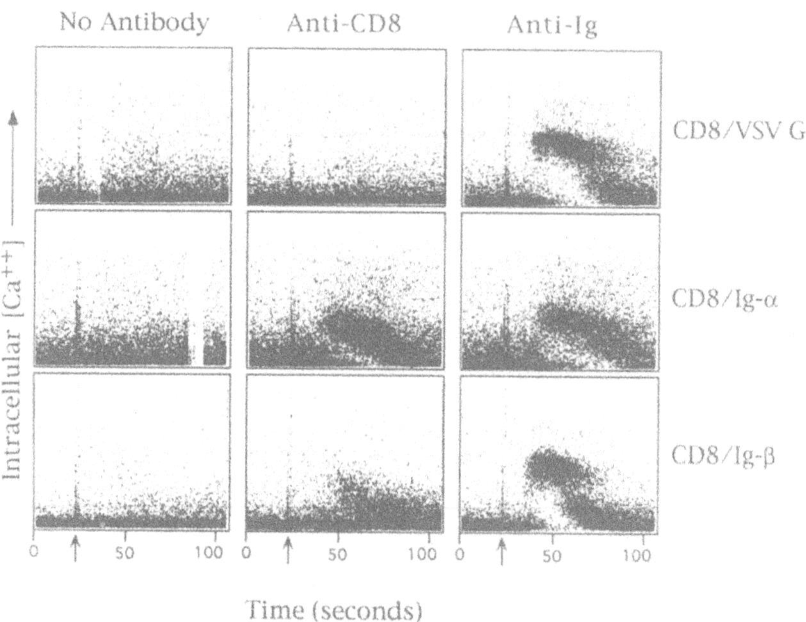


**Figure 2.** Crosslinking of CD8/Ig-α or CD8/Ig-β induces calcium mobilization in A20 cells. Cells expressing the indicated chimera were loaded with the $Ca^{++}$-binding dye Indo-1 AM and stimulated with PBS (No antibody), OKT8 (Anti-CD8), or goat anti-mouse Ig (Anti-Ig) at the time denoted by the arrows. The ratio of the fluorescence of $Ca^{++}$-bound dye to the fluorescence of the free dye in individual cells was measured by flow cytometry as described[13] and is displayed on the y-axis as relative intracellular [$Ca^{++}$]. Each dot represents a single cell.

Previously, the cytoplasmic domains of both Ig-α and Ig-β chimeras have been shown to induce production of the lymphokine interleukin-2 (IL-2) in T-cell lines.[13,28] Interestingly, crosslinking of a CD8 chimera containing the cytoplasmic tail of either the bovine leukemia virus (BLV) gp30 protein or the Epstein-Barr virus (EBV) LMP2A protein, both of which harbor a Reth motif, induces IL-2 production by IIA1.6 cells, an FcγRII-negative variant of A20.[17] Therefore, we also examined whether crosslinking of CD8/Ig-α or CD8/Ig-β could induce IL-2 production by A20 cells (assayed as described[13]). While crosslinking of CD8/Ig-α in the T-cell hybridoma DO-11.10 induced secretion of 0.8 units of IL-2/ml, crosslinking of CD8/Ig-α, CD8/Ig-β or CD8/VSV G did not induce detectable production (<0.024 units/ml) of IL-2.

## Signalling by Mutant Ig-β Cytoplasmic Domains in LK 35.2 Cells

Crosslinking of either CD8/Ig-α or CD8/Ig-β chimeras in murine B-cell hybridoma LK 35.2 cells stimulates protein tyrosine phosphorylation and induces $Ca^{++}$ mobilization to an extent essentially indistinguishable from that following crosslinking of mIg.[13] The signal-transducing capability of the cytoplasmic domains of Ig-α and Ig-β is attributed to their 26-amino acid Reth motif D-$X_7$-D/E-X-X-Y-X-X-L-$X_7$-Y-X-X-I (Fig. 3A). To begin to address the structural requirements of this motif for signalling, we used site-directed PCR mutagenesis to direct the substitution of alanine for several amino acids within the Ig-β cytoplasmic domain. Because of the documented requirement of the conserved tyrosines of the Reth motif for signalling[8,12,17-19,29,30] and the demonstrated signalling ability of the last 17 or 18 residues of a Reth motif of the T-cell receptor ζ chain,[19,20] we focussed our initial analysis on residues within the submotif Y-X-X-L-$X_7$-Y-X-X-I. Specifically, the alanine-substituted residues we studied were leucine 198 (L198A), glutamine 202 (Q202A), threonine 205 (T205A), and isoleucine 209 (I209A)(Fig. 3A). The resulting mutant CD8/Ig-β chimeras, as well as wt CD8/Ig-β and CD8/VSV G, were introduced into LK 35.2 cells as described. Equivalent surface expression of the chimeric proteins on G418-resistant cell pools was confirmed by flow cytometry (Fig. 3B).

Crosslinking of wt CD8/Ig-β and mutants Q202A, T205A, and I209A induced protein tyrosine phosphorylation to similar extents (Fig. 3C). However, the L198A mutant was severely defective in its ability to induce protein tyrosine phosphorylation above background levels, although phosphorylation of proteins of approximately 70 and 105 kDa did increase slightly following crosslinking. The mutant chimeras were also tested for their ability to induce $Ca^{++}$ mobilization in LK 35.2 cells. Again, crosslinking of wt CD8/Ig-β and mutants Q202A, T205A and I209A stimulated $Ca^{++}$ mobilization (Fig. 4), although the response of I209A was somewhat weaker. Similar to what was observed with protein tyrosine phosphorylation in cells expressing wt, Q202A, T205A and I209A CD8/Ig-β chimeras, the $Ca^{++}$ response induced by chimera crosslinking was also generally not as strong as that induced by mIg crosslinking. In contrast to the Q202A, T205A, and I209A mutants, crosslinking of the L198A chimera induced no detectable change in intracellular $Ca^{++}$ levels.

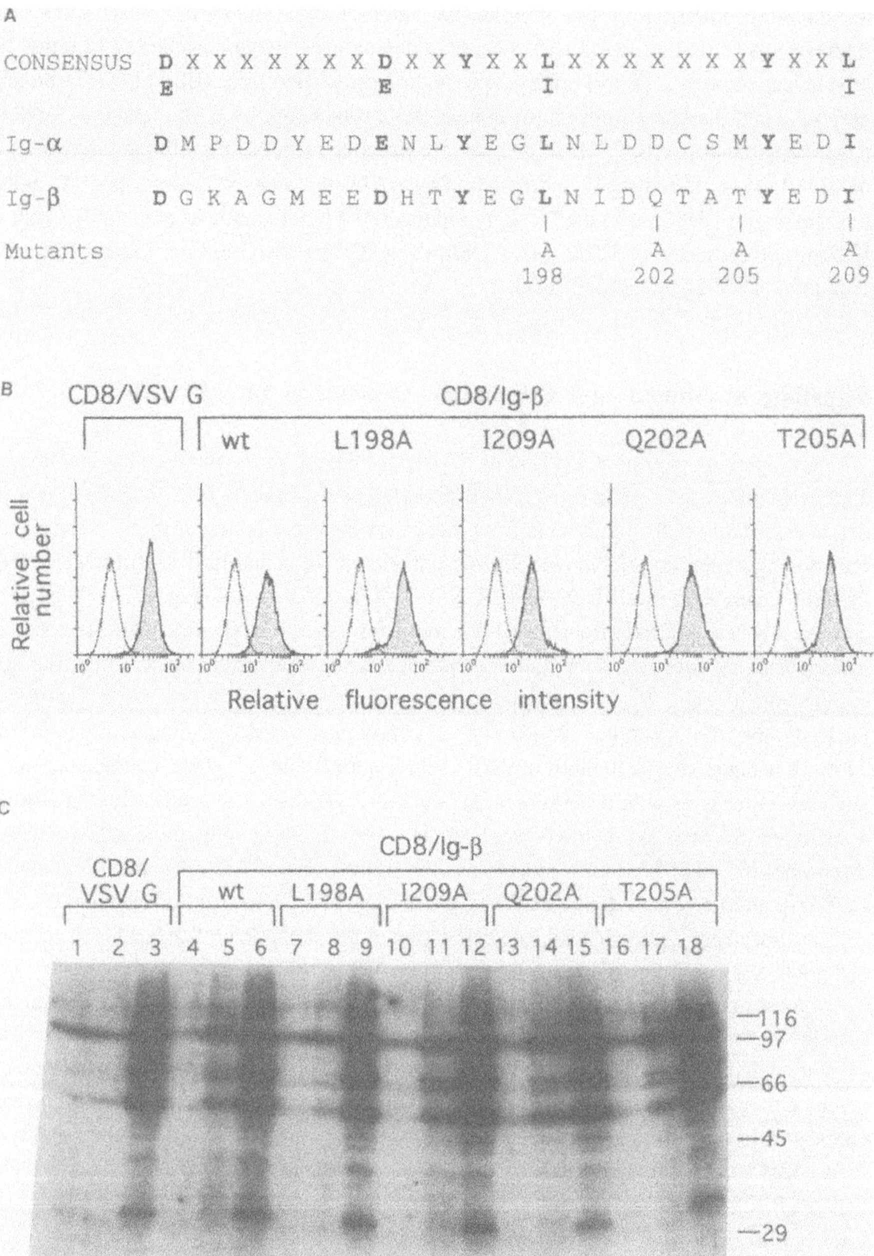


**Figure 3.** (A) The Reth motifs within the Ig-α and Ig-β cytoplasmic domains are shown below the consensus motif in the single letter amino acid code, with conserved residues indicated in bold. Shown beneath the Ig-β sequence are the numbered positions of alanine substitutions compared in this study. (B) Flow cytometric analysis of surface expression of CD8/VSV G and various CD8/Ig-β chimeric proteins on LK 35.2 cells. Immunolabeling was performed as in Fig. 1. Fluorescence of pools expressing the indicated chimera is shown (shaded curves) relative to the fluorescence of uninfected cells (open curves). (C) Induction of protein tyrosine phosphorylation following crosslinking of mutant CD8/Ig-β chimeras in LK 35.2 cells. Stimulations of cells expressing the indicated chimeras and lane designations are as described for Fig. 1.

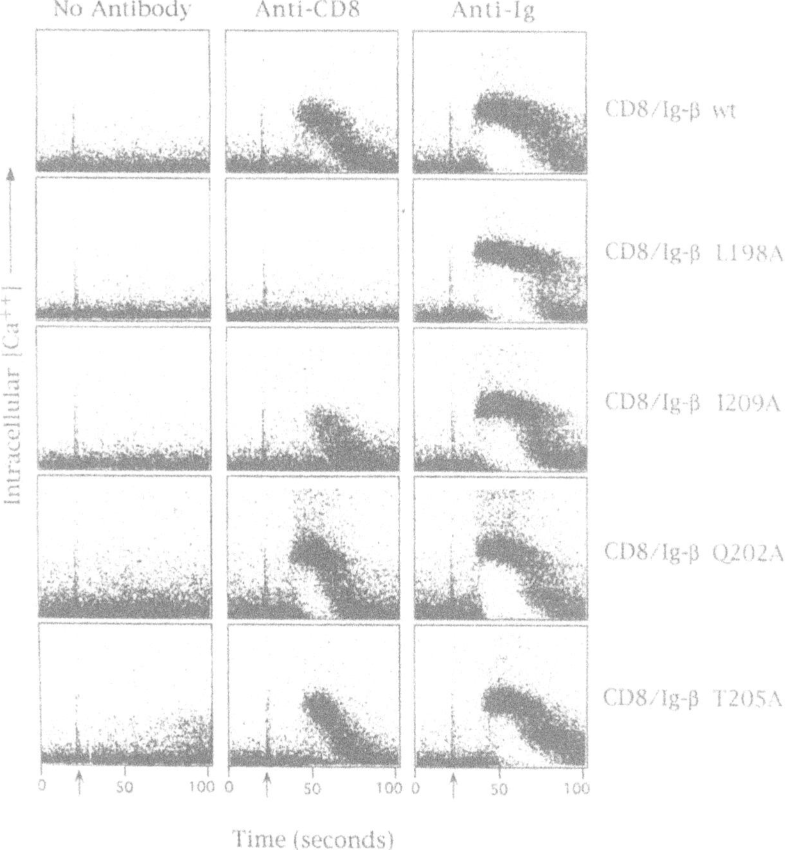


**Figure 4.** Induction of $Ca^{++}$ mobilization following crosslinking of mutant CD8/Ig-β proteins on LK 35.2 cells. Cells expressing the indicated chimera were assayed for changes in intracellular $Ca^{++}$ levels following crosslinking of the CD8 chimeras or mIg as described for Fig 2. The data shown for the L198A and I209A mutants were generated using different cell pools than those analyzed in Fig. 3B and 3C; both pools expressing a given mutant chimera yielded equivalent results.

## DISCUSSION

Using chimeric proteins, we have shown that the cytoplasmic tails of both Ig-α and Ig-β are capable of inducing significant protein tyrosine phosphorylation and $Ca^{++}$ mobilization in the B-cell lymphoma A20. The cytoplasmic tail of Ig-α did, however, appear a somewhat more potent inducer of both responses than that of Ig-β. Unlike Williams *et al.*,[12] we did not observe an equivalent induction of protein tyrosine phosphorylation by the cytoplasmic domain of Ig-β and mIg. However, as these investigators performed crosslinking for 10 minutes versus 2.5 minutes in our experiments, this difference may reflect the kinetics of signal transduction by the Ig-β cytoplasmic tail. Nevertheless, these findings contrast with those of Sanchez *et al.*,[8] who observed that only the cytoplasmic domain of Ig-α was capable of inducing protein tyrosine phosphorylation in A20 cells, even after 10 minutes of crosslinking. While these apparently conflicting observations might be attributed to differences in chimeric proteins utilized in the different studies, we do not favor this explanation as our CD8/Ig-α and CD8/Ig-β chimeras were equally capable of inducing protein tyrosine phosphorylation in both the murine B cell hybridoma LK 35.2 and the murine T-cell hybridoma DO-11.10.[13] It seems more likely that the observed differences in Ig-β signalling in A20 cells reflects differences in the A20 lines themselves. That independently-carried A20 lines may differ in biological responsiveness is further suggested by comparison of our observations with reports that a CD8/Ig-α chimera can induce IL-2 production by A20 cells[31] and that CD8 chimeras containing the cytoplasmic domains of the EBV LMP2A protein or the BLV gp30 protein can induce IL-2 production by the A20 derivative IIA1.6.[17] In contrast to these findings, neither CD8/Ig-α nor CD8/Ig-β could induce IL-2 production in our A20 cells.

One possible explanation for these disparate observations is that various A20 derivatives differ in their expression of proteins that mediate responses initiated through the cytoplasmic domain of Ig-α or Ig-β. Expression of some protein tyrosine kinases, for example, varies considerably between different B-cell lines.[32] Notably, the cytoplasmic domains of Ig-α and Ig-β exhibited differential binding to cytoplasmic effector molecules from extracts of the B-cell lymphoma K46,[31] in which a CD8/Ig-α but not a CD8/Ig-β chimera induced significant protein tyrosine phosphorylation.[10] While it is clear under many circumstances that the cytoplasmic domains of both Ig-α and Ig-β can induce lymphocyte activation equally, the fact that differential signalling is sometimes observed suggests that these molecules may play different signalling roles in different cellular contexts.

As we observed substantial induction of protein tyrosine phosphorylation upon CD8/Ig-β crosslinking in LK 35.2 B cells, we used these cells to examine the effects of alanine substitution for four amino acids within the Reth motif of Ig-β (Fig. 3A). One or both of the Reth submotifs Y-E-G-L and Y-E-D-I within Ig-β undergo tyrosine phosphorylation upon B-cell activation.[11] This is thought to induce binding of the SH2 domains of *src*- and/or *syk*-family protein tyrosine kinases, as these kinases bind preferentially to phosphotyrosine residues within the peptide sequence Y-E-E-I.[33] In view of the absolute conservation of the leucine and isoleucine residues and the requirement for both conserved tyrosines of the Ig-β Reth motif for signalling,[8,12] it was surprising that our I209A mutant retained signalling ability while our L198A mutant did not. This result suggests that the binding specificity of SH2 domains that interact with the Y-E-D-I motif may be less stringent than that of SH2 domains that interact with the Y-E-G-L motif. That the integrity of the upstream Y-X-X-L/I motif is more critical than that of the downstream motif is further supported by the observation that substitution by alanine of the upstream, but not the downstream, leucine residue within the BLV gp30 Reth motif destroys signalling ability.[17] Our Q202A and T205A mutants were not detrimental to Ig-β signalling, a finding perhaps consistent with the inability of substitutions at these positions within the TCR proteins CD3-ε and CD3-ζ to destroy signalling.[18,19] Thus, despite their proximity to the Y-X-X-L/I motifs, these residues do not appear critical for function of the Reth motif. Additional mutagenic analyses should aid greatly our understanding of the structural requirements for a functional signalling motif in B cells.

## ACKNOWLEDGEMENTS

We thank Robert Hyman for antibodies and Joseph Trotter for his help with flow cytometry. This work was supported by NIH grants CA14195 and CA17289. J.A.T. is supported by NIH postdoctoral fellowship AI08685.

## REFERENCES


1. J. Hombach, T. Tsubata, L. Leclercq, H. Stappert and M. Reth, Molecular components of the B-cell antigen receptor complex of the IgM class, *Nature* 343:760 (1990).
2. J. Hombach, F. Lottspeich and M. Reth, Identification of the genes encoding the IgMα and Igβ components of the IgM antigen receptor complex by amino terminal sequencing, *Eur. J. Immunol.* 20:2795 (1990).
3. K.S. Campbell, E.J. Hager, R.J. Friedrich and J.C. Cambier, IgM antigen receptor complex contains phosphoprotein products of B29 and *mb-1* genes, *Proc. Natl. Acad. Sci. USA* 88:3982 (1991).
4. A.R. Venkitaraman, G.T. Williams, P. Dariavach and M.S. Neuberger, The B-cell antigen receptor of the five immunoglobulin classes, *Nature* 352:777 (1991).
5. M. Reth, Antigen receptors on B lymphocytes, *Annu. Rev. Immunol.* 10:97 (1992).
6. M.-A. Campbell and B.M. Sefton, Protein tyrosine phosphorylation is induced in murine B lymphocytes in response to stimulation of anti-immunoglobulin, *EMBO J.* 9:2125 (1990).
7. M.R. Gold, D. Law and A.L. DeFranco, Stimulation of protein tyrosine phosphorylation by the B lymphocyte antigen receptor, *Nature* 345:810 (1990).
8. M. Sanchez, Z. Misulovin, A.L. Burkhardt, S. Mahajan, T. Costa, R. Franke, J.B. Bolen and M. Nussenzweig, Signal transduction by immunoglobulin is mediated through Igα and Igβ, *J. Exp. Med.* 178:1049 (1993).
9. S.A. Grupp, K. Campbell, R.N. Mitchell, J.C. Cambier and A.K. Abbas, Signaling-defective mutants of the B lymphocyte antigen receptor fail to associate with Ig-α and Ig-β/γ, *J. Biol. Chem.* 268:25776 (1993).
10. K.-M. Kim, G. Alber, P. Weiser and M. Reth, Differential signalling through the Ig-α and Ig-β components of the B cell antigen receptor, *Eur. J. Immunol.* 23:911 (1993).
11. D.A. Law, V.W.F. Chan, S.K. Datta and A.L. DeFranco, B-cell antigen receptor motifs have redundant signalling capabilities and bind the tyrosine kinases PTK72, Lyn, and Fyn, *Current Biology* 3:645 (1993).
12. G.T. Williams, C.J.G. Peaker, K.J. Patel and M.S. Neuberger, The α/β sheath and its cytoplasmic tyrosine are required for signaling by the B-cell antigen receptor but not for capping or for serine/threonine recruitment, *Proc. Natl. Acad. Sci. USA* 91:474 (1994).
13. J.A. Taddie, T.R. Hurley, B.S. Hardwick and B.M. Sefton, Activation of B and T cells by the cytoplasmic domains of the B-cell antigen receptor proteins Ig-α and Ig-β, *J. Biol. Chem.* In Press (1994).
14. M. Reth, Antigen receptor tail clue, *Nature* 338:383 (1989).
15. J.C. Cambier, Signal transduction by T- and B-cell antigen receptors: converging structures and concepts, *Curr. Opin. Immunol.* 4:257 (1992).
16. G. Alber, K.-M. Kim, P. Weiser, C. Riesterer, R. Carsetti and M. Reth, Molecular mimicry of the antigen receptor signalling motif by transmembrane proteins of the Epstein-Barr virus and the bovine leukemia virus, *Curr. Biol.* 3:333 (1993).
17. P. Beaufils, D. Choquet, R.Z. Mamoun and B. Malissen, The $(YXXL/I)_2$ signalling motif found in the cytoplasmic segments of the bovine leukaemia virus envelope protein and Epstein-Barr virus latent membrane protein 2A can elicit early and late lymphocyte activation events, *EMBO J.* 12:5105 (1993).
18. F. Letourneur and R.D. Klausner, Activation of T cells by a tyrosine kinase activation domain in the cytoplasmic tail of CD3 ε, *Science* 255:79 (1992).
19. C. Romeo, M. Amiot and B. Seed, Sequence requirements for induction of cytolysis by the T cell antigen/Fc receptor ζ chain, *Cell* 68:889 (1992).

20. B.A. Irving, A.C. Chan and A. Weiss, Functional characterization of a signal transducing motif present in the T cell antigen receptor ζ chain, *J. Exp. Med.* 177:1093 (1993).
21. K.J. Kim, C. Kanellopoulos-Langevin, R.M. Merwin, D.H. Sachs and R. Asofsky, Establishment and characterization of BALB/c lymphoma lines with B cell properties, *J. Immunol.* 122:549 (1979).
22. J. Kappler, J. White, D. Wegmann, E. Mustain and P. Marrack, Antigen presentation by $Ia^+$ B cell hybridomas to *H-2*-restricted T cell hybridomas, *Proc. Natl. Acad. Sci. USA* 79:3604 (1982).
23. G. Dennert, Cloned lines of natural killer cells, *Nature* 287:47 (1980).
24. N.R. Landau and D.R. Littman, Packaging system for rapid production of murine leukemia virus vectors with variable tropism, *J. Virol.* 66:5110 (1992).
25. T.R. Hurley, R. Hyman and B.M. Sefton, Differential effects of the CD45 tyrosine protein phosphatase on the tyrosine phosphorylation of the *lck*, *fyn*, and c-*src* tyrosine protein kinases, *Mol. Cell. Biol.* 13:1651 (1993).
26. T.R. Hurley and B.M. Sefton, Analysis of the activity and phosphorylation of the *lck* protein in lymphoid cells, *Oncogene* 4:265 (1989).
27. D.J. Leahy, R. Axel and W.A. Hendrickson, Crystal structure of a soluble form of the human T cell coreceptor CD8 at 2.6 Å resolution, *Cell* 68:1145 (1992).
28. A.L. Burkhardt, T. Costa, Z. Misulovin, B. Stealy, J.B. Bolen and M.C. Nussenzweig, Igα and Igβ are functionally homologous to the signaling proteins of the T-cell receptor, *Mol. Cell. Biol.* 14:1095 (1994).
29. W. Kolanus, C. Romeo and B. Seed, Lineage-independent activation of immune system effector function by myeloid Fc receptors, *EMBO J.* 11:4861 (1992).
30. H. Flaswinkel and M. Reth, Dual role of the tyrosine activation motif of the Ig-α protein during signal transduction via the B cell antigen receptor, *EMBO J.* 13:83 (1994).
31. M.R. Clark, K.S. Campbell, A. Kazlauskas, S.A. Johnson, M. Hertz, T.A. Potter, C. Pleiman and J.C. Cambier, The B cell antigen receptor complex: association of Ig-α and Ig-β with distinct cytoplasmic effectors, *Science* 258:123 (1992).
32. D.L. Law, M.R. Gold and A.L. DeFranco, Examination of B lymphoid cell lines for membrane immunoglobulin-stimulated tyrosine phosphorylation and src-family tyrosine kinase mRNA expression, *Mol. Immunol.* 29:917 (1992).
33. Z. Songyang, S.E. Shoelson, M. Chaudhuri, G. Gish, T. Pawson, W.G. Haser, F. King, T. Roberts, S. Ratnofsky, R.J. Lechleider, B.G. Neel, R.B. Birge, J.E. Fajardo, M.M. Chou, H. Hanafusa, B. Schaffhausen and L.C. Cantley, SH2 domains recognize specific phosphopeptide sequences, *Cell* 72:767 (1993).

# ACCESSORY MOLECULES THAT INFLUENCE SIGNALING THROUGH B LYMPHOCYTE ANTIGEN RECEPTORS

Edward A. Clark, Ingolf Berberich, Stephen J. Klaus, Che-Leung Law, and Svetlana P. Sidorenko

Department of Microbiology
University of Washington Medical Center SC-42
Seattle, WA 98195

## INTRODUCTION

The B lymphocyte antigen receptor complex, commonly called the B cell receptor (BCR) complex, consists of surface immunoglobulin (sIg) and heterodimers of the Igα (CD79a, mb1) and Igβ (CD79b, B29) phosphoglycoproteins. Recent reviews detail pertinent findings on the BCR complex.[1] The structure of the BCR complex and some cell-surface molecules that influence signaling via the BCR are shown in Figure 1. The clonotypic Ig receptor has only a very short cytoplasmic tail and therefore must rely on the invariant members of the BCR complex to transmit signals to the cytosol after receptor crosslinking. Igα/Igβ associate with sIgM and sIgD and are both necessary and sufficient for the expression of sIg. This heterodimer is analogous to the TCR CD3ε/δ or CD3ε/γ heterodimers,[2] which, like Igα/Igβ, (1) contain subunits with a single extracellular Ig-like domain; (2) are phosphorylated on tyrosine (CD3ε and CD3ζ) after crosslinking of antigen receptors; and (3) contain within their cytoplasmic tail a single antigen receptor homology 1 (ARH1) motif, D/E-X7-D/E-X2-Y-X2-L-X7-Y-X2-L/I (X = any amino acid). Regions within the transmembrane domain of sIgM are required for the release of $[Ca^{2+}]i$ or internalization of bound antigen.[3] Mutations within the transmembrane domain of sIgM that inhibit the activation of new protein tyrosine phosphorylation (PTP) and release of $[Ca^{2+}]i$ also uncouple sIgM from Igα/Igβ.[4] However, even though such a mutant sIgM does not associate with Igα/Igβ, when crosslinked it still induces some new PTP.[4] Both Sanchez et al.[4] and Kim et al.[5] reported that surface chimeric fusion proteins expressing the cytoplasmic tails of Igα vs. Igβ differ in their ability to transmit signals: the Igα but not the Igβ tail can induce new PTP, results consistent with studies suggesting that Igα and not Igβ strongly associates with the protein tyrosine kinase (PTK) p53/56Lyn (Lyn).[6] Matsuuchi et al.[7] found that sIgM expression could be reconstituted in a pituitary cell line with Igα/Igβ coexpression, but that Igα/Igβ were not sufficient to reconstitute a complete signal through sIgM. Thus, IgM interaction with Igα/Igβ is critical for signaling but other factors may also be required.

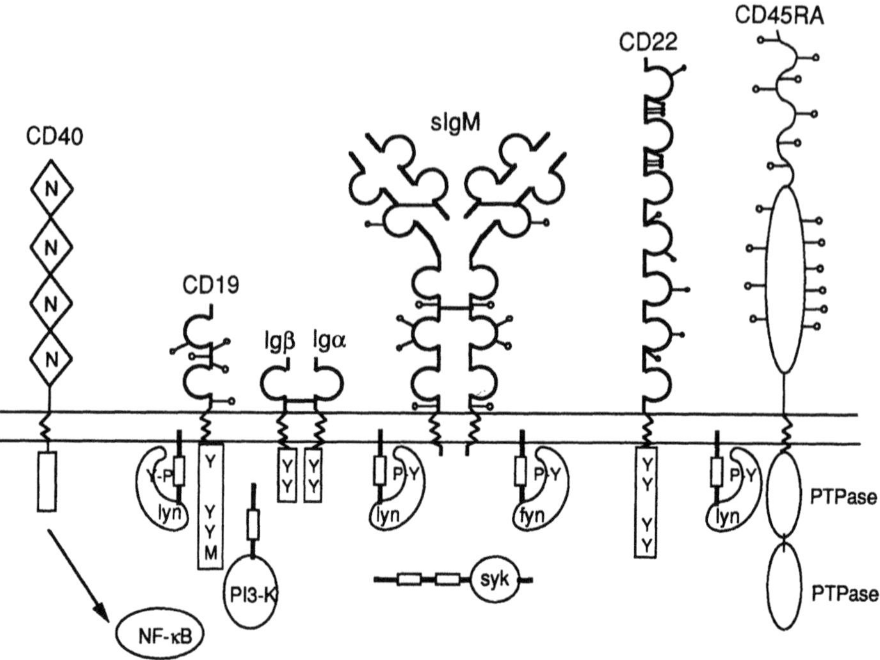


**Figure 1.** Cell-surface molecules that can interact with and/or influence signaling via the surface IgM BCR complex. The IgM BCR complex consists of sIgM and the Igα/Igβ heterodimer. Both Igα and Igβ have antigen-receptor homology motifs within their cytoplasmic tails; these motifs contain two spaced tyrosine residues capable of being phosphorylated (PY) by protein tyrosine kinases (PTK) such as Lyn. After BCR crosslinking, Igα and Igβ are phosphorylated on tyrosine residues (PY), which may then interact with PTK such as Lyn and Syk required for downstream signaling pathways. The CD45RA protein tyrosine phosphatase (PTPase) appears to be required for this initial BCR signal, perhaps to release inactive PTK to become active. The B-cell-specific surface molecules CD22 and CD19 both are rapidly phosphorylated on tyrosine after crosslinking of the BCR; the PY in CD19 have been shown to interact with PI-3 kinase, and the PY in CD22 may interact with Syk. Both CD19 and CD22 have been reported to physically interact with sIgM, and crosslinking either CD19 or CD22 has been shown to modulate proliferative signals via sIgM. Crosslinking the CD40 receptor prevents immature B cells from dying after the BCR is crosslinked. This may be related to the ability of CD40 signaling to activate the NF-kB transcription factor within minutes.

## THE BCR COMPLEX

### PTK Associated with the BCR Complex

A number of PTK and B-cell-associated surface receptors have been reported to interact with and contribute to antigen-specific signaling. Several members of the src-family PTK including Lyn, p59fyn (Fyn), p55blk (Blk), and p56lck (Lck), which are myristylated and therefore localized at the plasma membrane, have been reported to associate with the BCR complex.[2,8] The facts that (1) both Igα and Igβ are rapidly phosphorylated on tyrosine after Ig-crosslinking and (2) these tyrosines are critical for signaling function and contain consensus YXXL/I repeats predicted, when phosphorylated on tyrosine (PY), to interact with Src homology 2 (SH2) domains of Src-family kinases[9] suggest that Src-family kinases such as Lyn may bind to Igα/Igβ via their SH2 domains. However, non-phosphorylated cytoplasmic tails of Igα bind to Lyn in vitro,[6] apparently via an interaction with the SH3 domain of Lyn.[10] Lyn is associated with the BCR in B-cell lines and normal B cells, and after crosslinking sIg, the activity of Lyn is increased about three- to fivefold.[8] After sIg crosslinking, Lyn also associates with 70- and 75-kDa phosphoproteins in the Daudi human

B-cell line and with a 66-kDa protein in the WEHI-231 mouse B-cell line.[8] A 66-kDa form of the B-cell-associated 72-kDa PTK called spleen tyrosine kinase, or Syk, is activated in WEHI-231,[5] results consistent with our findings that Lyn interacts directly with Syk (see below).

The Syk PTK was originally described in porcine splenocytes.[11,12] A 72-kDa PTK, PTK72, similar biochemically to porcine Syk, was also identified in mice.[13-15] PTK72 is associated with the BCR complex in humans[16] and mice[15] and was recently confirmed to be the human and murine homologs of Syk[17] (R. Geahlen, personal communication). We[17] and others[18] have isolated full-length cDNAs encoding human Syk, and we have also isolated a cDNA encompassing the coding region of mouse Syk (unpublished data). Human, porcine and murine Syk are highly conserved PTK; they share a >90% identity in the overall amino acid sequence. The most conserved regions among them are the kinase domains, which have a 98% identity. The SH2(N) and the SH2(C) domains have 88–91% and 93–95% identity, respectively. The areas linking the SH2(C) domains and the kinase domains are the least conserved, at around 80% identity. Moreover, the human *SYK* locus maps to chromosome 9 at band q22, a region where no other genes influencing hematopoiesis have yet been localized.[17]

All 14 of the human Syk+ human B-cell lines that we have examined contain some detectable Syk associated with IgM receptor complexes (Law et al., in preparation). A functional role for Syk in B-cell-surface signaling has been suggested by the findings that (1) a chimeric protein containing Syk is sufficient for inducing release of $[Ca^{2+}]i$ in transfected cells[18] and (2) total cellular Syk[14] and Syk within the sIg complex are phosphorylated on tyrosine after crosslinking the BCR.[17] Syk, unlike Src-family PTK, has two SH2 domains [Syk.SH2(N) and Syk.SH2(C)] followed by a kinase domain. Various SH2 domains with distinct binding specificities are found in molecules involved in signal transduction pathways such as phospholipase C (PLC) γ1 and γ2 isoforms, the GTPase activating protein (GAP) which binds to and regulates p21ras (Ras) activity, the p85 regulatory subunit of phosphatidylinositol 3' kinase (PI-3 K), certain protein tyrosine phosphatases (PTPase) such as PTP1C, and key "adaptor proteins" such as Grb2/Sem-5, SOS, Shc, Crk, and Nck.[9,19] These domains enable signaling proteins to interact via recognition of motifs containing PY; thus, the two SH2 domains of Syk are likely to play an essential role in Syk activation, localization, and/or phosphorylation of substrates (see below).

In T cells, a 70-kDa PTK, ZAP-70, rapidly associates with the TCR ζ chain after TCR stimulation.[20] Syk and ZAP-70 are closely related and ZAP-70, like Syk, has two SH2 domains.[20,21] How might Syk be activated and function? Given the similarities between the BCR and CD3/TCR complexes,[2,22] Syk could be activated in a manner analogous to ZAP-70 in T cells. Weiss[22] proposed that after CD3/TCR ligation, a Src-family kinase such as Fyn is activated probably via dephosphorylation of the "Y527" Src-family regulatory site[23] by a CD45 PTPase isoform. The activated Fyn would then phosphorylate ARH1 motifs like those in ζ chain or ε chain and these newly phosphorylated receptors, especially ζ chains, would recruit cytosolic ZAP-70 to the cell surface, where it would bind via its SH2 motifs[21] and be activated and contribute to signaling. By analogy, sIg crosslinking would activate myristylated src-family PTK attached to B-cell membranes close to the BCR like Lyn or Blk, perhaps via CD45, which then would phosphorylate Igα and Igβ as well as surface molecules interacting with sIg such as CD19 and CD22 (see below). These surface molecules with PY could then recruit membrane or cytosolic proteins, e.g., Syk, involved in signaling into BCR complexes.

The fact that dense B cells, unlike the Daudi B-cell line, have little or no Syk associated with the Ig receptor[16] is consistent with this model. Alternatively, Syk may be loosely associated with a component of the Ig complex in resting cells via a non-SH2 interaction and may change its association in activated cells. Our recent findings are most consistent with this model; Daudi B cells have Syk associated with the sIgM complex detectable by Western blotting which is not phosphorylated on tyrosine.[17] After sIgM crosslinking, the amount of Syk in the complex does not increase significantly, but tyrosine phosphorylation of Syk does. However, most studies of Syk have been done with cell lines, so further studies are needed to clarify these possibilities.

To test whether Syk via its SH2 domains can bind to proteins in the BCR as ZAP-70 does with the TCR, we constructed glutathione S-transferase (GST) fusion proteins of one Syk SH2 domain [GST-Syk.SH2(N) or GST-Syk.SH2(C)] or both domains [GST-Syk.SH2(N/C)]. Surface IgM complexes were isolated from the Daudi B-cell line, subjected to in vitro kinase assays, released in NP-40, and reprecipitated with GST or one of the GST-Syk.SH2 fusion proteins. Tyrosine phosphorylated proteins in the Ig complex could be precipitated by GST-Syk.SH2(N/C). A subset of these phosphoproteins could also be precipitated by GST-Syk.SH2(N) or GST-Syk.SH2(C). In contrast, the control GST protein did not bind any of these phosphoproteins.[24] Moreover, Syk.SH2(N/C) could immunoprecipitate the Ig complex most effectively after crosslinking the BCR. This may be a result of the increase in tyrosine phosphorylation of the various components of the Ig complex. Individual bands associated with the BCR including Syk itself, Lyn, and Igα could be cut out of gels and eluted into lysis buffer, and they could be immunoprecipitated with Syk.SH2N/C. These results suggest that Syk may bind to more than one component of the BCR and these interactions are mediated most efficiently via both of its SH2 domains. At this juncture, we do not know if Syk.SH2(N/C) associates more strongly with one member of the BCR or another. However, unlike studies with the ZAP-70, we have no strong evidence yet for recruitment of Syk from the cytoplasm into the BCR upon B-cell activation through sIgM crosslinking.

Does Syk interact with any src-family kinases? In a mast cell line Syk and Lyn are associated with β subunit of FcεRI[25] and after crosslinking of BCR on mouse and human B cells phosphorylated p72 coprecipitate with p53/56Lyn.[8] We have found that Syk can interact directly with Lyn in human B cells.[26] In digitonin and NP-40 lysates of mature B cell lines Daudi, B104, T5-1 and CESS Syk was precipitated in the complex with pp53/56 and pp120. The phosphorylated p53/56 doublet was identified to be Lyn by sequential immunoprecipitation after in vitro kinase assay. Another Syk-associated molecule, pp120, could not be reprecipitated with anti-serum against either the p125fak or the substrate for Src family kinases pp120.[26] Because crosslinking the BCR in two Syk-negative cell lines does not induce release of $[Ca^{2+}]i$ (Law et al., unpublished) and crosslinking membrane-bound Syk does induce release of $[Ca^{2+}]i$,[18] our current hypothesis is that interaction with Syk and Lyn and with Syk and pp120 are needed for activation of PLCγ1.

### Surface Molecules That May Associate with the BCR

**CD19.** After crosslinking of sIgM, certain surface molecules may associate with the BCR such as the Fc receptor RII (FcγRII, CD32),[27] CD19, a B-cell-specific member of the Ig supergene family,[28] and perhaps the PTPase CD45[29] (Fig. 1). The findings with CD19 are especially interesting since crosslinking CD19 can augment the proliferation of B cells induced by anti-Ig[30,31] and since crosslinking sIg leads to rapid phosphorylation of CD19 and association of CD19 with PI-3 K.[32,33] CD19 may also inhibit B cells depending on how the B cell is activated[30] and the source of B cell used. The basis of CD19 stimulatory vs. inhibitory signaling is not yet understood. The findings with CD45 are also provocative since $sIgM^+$ $CD45^-$ cells have been reported not to release $[Ca^{2+}]i$ after sIg crosslinking[29] until they are transfected to express CD45. Furthermore, B cells from CD45-defective mice cannot be induced to proliferate by crosslinking sIgM.[34] Thus, as suggested for T cells,[22] the CD45 PTPase may be required for activation of src-family PTK and antigen-induced signaling in B cells.

**CD22.** CD22 is a B-cell-specific member of the Ig supergene family related to cell-cell interaction molecules including myelin-associated glycoprotein (MAG) and carcinoembryonic antigen (CEA) (see Clark[35] for review). An adhesion molecule, CD22 binds to both B and T cell lines.[36-38] Binding of CD22/Ig recombinant globulins (Rg) to T cells blocks anti-CD3 antibodies from inducing the tyrosine phosphorylation of PLC $\gamma 1$,[39] suggesting that interaction of CD22 with a ligand(s) may inhibit T cell activation. A number of cell surface molecules are bound by CD22.Rg including human CD45RO.[39] The ligands for CD22 are not fully defined, but it is clear CD22 is a sialic acid-binding lectin that recognizes a restricted set of glycoproteins containing N-linked oligosaccharides with $\alpha 2,6$-linked sialic acids.[40,41]

In vitro kinase assays on sIgM immunoprecipitates revealed that the human sIgM complex not only has sIgM and Ig$\alpha$ and Ig$\beta$, but also has both ser/thr and tyr kinase activities in the complex and two phosphorylated proteins about 75 and 150 kDa in size.[16,42] The 75-kDa protein has been identified as Syk[16] and the 150-kDa protein as CD22.[16,43] Mouse CD22 and the sIgM complex also associate in some B-cell lines, e.g., BCL-1, but not others (Law et al., in preparation). The association of CD22 with the BCR appears to be functionally significant since (1) $CD22^+$ B cells and not $CD22^-$ B cells can be signaled to release $[Ca^{2+}]i$ after crosslinking sIgM;[44] (2) CD22 has a PTK activity associated with it and is rapidly phosphorylated on tyrosine after crosslinking sIgM;[16,43,45] (3) CD22 has one or more ARH1 motifs in its cytoplasmic tail; and (4) CD22 binds to ligands on lymphocytes. These results suggest that CD22 may function via interactions with its ligands to regulate antigen-specific activation of B cells.

## ACTIVATION OF THE TRANSCRIPTION FACTOR NF-kB BY CROSS-LINKING CD40

CD40, a 45- to 50-kDa transmembrane glycoprotein expressed on B cells, is a member of the NGF receptor family.[46] It plays a critical role in activation and proliferation of normal B cells[47-49] and is essential for survival of germinal center (GC) B cells.[50] Signals delivered via CD40 can greatly influence the cellular responses evoked by the BCR. Valentine and Licciardi[51] found that anti-CD40 could block activation-induced apoptosis by anti-IgM of an immature human B cell line, Ramos. Tsubata et al.[52] then showed that T cells expressing the ligand for CD40 (CD40L) could inhibit anti-IgM-induced apoptosis of an immature mouse B cell line, WEHI 231. Evidence for the importance of CD40-CD40L interaction in the normal immune response has come from studies with X-linked hyper-IgM syndrome (HIM) patients in whom the cooperation between B and T cells has been disturbed owing to a defective CD40L.[53] These patients do not form GC and do not switch from producing IgM to other Ig classes. Recently, a mouse model for HIM has been defined[54]: mice deficient in CD40 have a phenotype very similar to HIM patients, i.e., no GC formation or T-cell-dependent isotype class switching. The fact that CD40L-deficient and CD40-deficient individuals have the same phenotype suggests the CD40L-CD40 interaction is not redundant and is a single receptor-coreceptor system. How the CD40 receptor signals B cells to proliferate or differentiate is not well understood.

Recently Lalmanach-Girard et al.[55] showed that during T-dependent B-cell activation there is an induction of binding activity for an NF-kB site containing oligonucleotide in B cells. This induction of NF-kB was eliminated by interrupting the CD40L-CD40 interaction. These results suggest that CD40 signaling activates NF-kB, although the authors could not rule out the involvement of other cell-surface molecules in this activation process. We decided to take a more reductionistic approach to exclude the participation of additional molecules.[56] Our approach was based on the observation that CD40 crosslinking leads to secretion of IL-6 in normal B cells and in a mouse B-cell line transfected with human

CD40.[57] The promoter of the IL-6 gene is well characterized and contains, among others, binding sites (protoenhancer elements) for transcription factors NF-kB, AP-1, and NF-IL6.[58] Since it is known that multiple copies of protoenhancer elements can support the expression of reporter genes from a heterologous promoter, we used thymidine-kinase/luciferase reporter plasmids driven by polymerized protoenhancer elements in a transient expression system to investigate whether solely crosslinking CD40 leads to the activation of one or more of the above transcription factors.

Our studies have shown that signaling through CD40 functionally activates NF-kB, while AP-1 and NF-IL-6 are affected only marginally or not at all. Furthermore, the increase in luciferase activity is correlated with the number of copies of NF-kB binding sites and depends on the wildtype consensus sequence of NF-kB binding sites in the reporter construct. The increase in enzyme activity after the onset of stimulation is rapid and can be detected within a few hours. Using gel retardation assays, we can detect NF-kB consensus site binding activity in the extracts of nuclei of anti-CD40-stimulated Daudi cells within minutes. The level of NF-kB appears to remain elevated as long as CD40 is crosslinked.

Recently it has been shown that anti-CD40 signaling induces protein tyrosine kinase activity (PTK) in B cells.[59,60] This activity is required for homotypic adhesion among B cells.[61] The aggregation is partially mediated via an increased level of ICAM-1 (CD54) on the cell surface. Intriguingly potential NF-kB binding sites have been reported for the CD54 promoter.[62,63] Thus, we are testing whether NF-kB is involved in the upregulation of CD54 expression, whether PTKs are required for activating NF-kB after CD40 crosslinking, or whether other transcription factors are involved in this process. Experiments are in progress to address those questions.

## REGULATION OF T-CELL CD40L EXPRESSION BY SIGNALS TRANSDUCED THROUGH CD28

Costimulation of T cells with anti-CD28 mAb's or with cells expressing the natural ligand for CD28, B7/BB-1 (CD80), leads to increased T-cell proliferation and IL-2 production in response to anti-CD3.[64] We investigated the effect of anti-CD28 costimulation on T-cell-dependent B-cell growth and Ig secretion. Increases in B-cell proliferation and Ig production were not due solely to increases in soluble lymphokine production, and were correlated with augmented surface expression of CD40L on T cells. Furthermore, anti-CD28 enhanced T-dependent B-cell responses could be inhibited by a soluble, chimeric CD40-Ig fusion protein.

Increases in CD40L surface expression were accompanied by upregulation of steady-state mRNA levels, especially early (2-8 hours) after T-cell activation. In addition, regulation of T-cell CD40L mRNA expression was similar to IL-2 mRNA by a number of criteria. Anti-CD28 synergized with suboptimal concentrations of anti-CD3 to induce CD40L mRNA, and CD40L expression induced by either anti-CD3 alone or in combination with anti-CD28 was inhibited by low concentrations of cyclosporin A.[65] During experiments aimed at determining mRNA half-life, CD40L mRNA was stabilized by costimulation with anti-CD28 (compared with anti-CD3 alone), consistent with the fact that CD28 extends the half-life of mRNAs containing AUUUA motifs in their 3' untranslated region. A major difference between the regulation of IL-2 and CD40L mRNA was observed in their inducibility by anti-CD3 alone, such that IL-2 was poorly induced and difficult to detect in anti-CD3-stimulated T cells. These data underscore similarities and differences between the major soluble and contact-dependent proteins that T cells use to initiate B cell growth and Ig production. Future studies are aimed at determining differences in transcriptional control and comparing promoter motifs between IL-2 and CD40L. Sequencing of a genomic clone containing the CD40L enhancer/promoter is currently underway.

## ACKNOWLEDGMENTS

We thank Karen Chandran, Kevin Draves, and Geraldine Shu for expert technical assistance. This work was supported by NIH grants GM37905, GM42508, RR00166, and DE08229. I. Berberich is supported by an AIDS scholarship of the German Cancer Research Center, S. J. Klaus is supported by American Cancer Society grant PF-3949, and C.-L. Law is a Leukemia Society Special Fellow.

## REFERENCES

1. Moller, G. Ed. 1993. The B-cell antigen receptor complex. *Immunol. Rev. vol. 132*. Munksgaard, Copenhagen. *206pp.*
2. Cambier, J. C. 1992. Signal transduction by T- and B-cell antigen receptors: converging structures and concepts. *Cur. Opin. Immunol. 4:257.*
3. Shaw, A. C., Mitchell, R. N., Weaver, Y. K., Campos-Torres, J., Abbas, A. K., and Leder, P. 1992. Mutations of immunoglobulin transmembrabne and cytoplasmic domains: effect on intracellular signalling and antigen presentation. *Cell 63:381.*
4. Sanchez, M., Misulovin, Z., Burkhardt, A. L., et al. 1993. Signal transduction by immunoglobulin is mediated through Igα and Igβ. *J. Exp. Med. 178:1049.*
5. Kim, K.M., Alber, G., Weiser, P., and Reth, M. 1993. Signaling function of the B-cell antigen receptors. *Immunol. Rev. 132:125.*
6. Clark, M. R., Campbell, K. S., Kazlauskas, A., et al., 1992. The B cell antigen receptor complex: association of Ig-alpha and Ig-beta with distinct cytoplasmic effectors. *Science 258:123.*
7. Matsuuchi, L., Gold, M. R., Travis, A., et al. 1992. The membrane IgM-associated proteins MB-1 and Ig-β are sufficient to promote surface expression of a partially functional B-cell antigen receptor in a non lymphoid cell line. *Proc. Nat. Acad. Sci. USA 89:3404.*
8. Yamamoto, T., Yamanashi, Y., and Toyoshima, K. 1993. Association of src-family kinase Lyn with B-cell antigen receptor. *Immunol. Rev. 132:187.*
9. Songyang, Z., Shoelson, S. E., Chaudhuri, M., et al. 1993. SH2 domains recognize specific phosphopeptide sequences. *Cell 72:767.*
10. Cambier, J. C., Bedzyk, W., Campbell, K., et al. 1993. The B-cell antigen receptor: structure and function of primary, secondary, tertiary and quaternary components. *Immunol. Rev. 132:85.*
11. Taniguchi, T., Kobayashi, T., Kondo, J., et al. 1991. Molecular cloning of a porcine gene Syk that encodes a 72-kDa protein-tyrosine kinase showing high susceptibility to proteolysis. *J. Biol. Chem. 266:15790.*
12. Yamada, T., Taniguchi, T., Yang, C., Yasue, S., Saito, H., and Yamamura, H. 1993. Association with B-cell-antigen receptor with protein-tyrosine kinase p72syk and activation by engagement of membrane IgM. *Eur. J. Biochem. 213:455.*
13. Burg, D. L., Harrison, M. L., and Geahlen, R. L. 1993. Cell cycle-specific activation of the PTK72 protein-tyrosine kinase in B lymphocytes. *J. Biol. Chem. 268:2304.*
14. Hutchcroft, J. E., Harrison, M. L., and Geahlen, R. L. 1991. B lymphocyte activation is accompanied by phosphorylation of a 72-kDa protein-tyrosine kinase. *J. Biol. Chem. 266:14846.*
15. Hutchcroft, J. E., Harrison, M. L., and Geahlen, R. L. 1992a. Association of the 72-kDa protein-tyrosine kinase PTK72 with the B cell antigen receptor. *J. Biol. Chem. 267:8613.*
16. Leprince, C., Draves, K. E., Geahlen, R. L., Ledbetter, J. A., and Clark, E. A. 1993. CD22 associates with the human surface IgM-B cell antigen receptor complex. *Proc. Nat. Acad. Sci. USA 90:3236.*
17. Law, C-L., Sidorenko, S. P., Chandran, K. A., Draves, K. E., Chan, A. C., Weiss, A., Edelhoff, S., Disteche, C. M., and Clark, E. A. 1994a. Molecular cloning of human Syk, a cell protein tyrosine kinase associated with the sIgM/B cell receptor complex. *J. Biol. Chem.*, in press.
18. Kolanus, W., Romeo, C., and B. Seed. 1993. T cell activation by clustered tyrosine kinases. *Cell 74:171.*
19. Mayer, B. and Baltimore, D. 1993. Signaling through SH3 and SH2 domains. *Trends Cell Biol. 3:8.*
20. Chan, A. C., Iwashima, M., Turck, C. W., and Weiss, A. 1992. ZAP-70: a 70 kd protein-tyrosine kinase that associates with the TCR ζ chain. *Cell 71:649.*
21. Wange, R. L., Malek, S. N., Desiderio, S., and Samelson, L. E. 1993. Tandem SH2 domains of ZAP-70 bind to T cell antigen receptor ζ and CD3ε from activated Jurkat T cells. *J. Biol. Chem. 268:19797.*
22. Weiss, A. 1993. T cell antigen receptor signal transduction: A tale of tails and cytoplasmic protein-tyrosine kinases. *Cell 73:209.*
23. Cooper, J. A. and Howell, B. 1993. The when and how of Src regulation. *Cell 73:1051.*
24. Law, C-L., Chandran, K., Sidorenko, S. P., Draves, K. E., and Clark, E. A. 1994b. Both SH2 domains of the spleen tyrosine kinase, Syk, are required for efficient binding to components of the B cell antigen receptor complex. Submitted.

25. Hutchcroft, J. E., Geahlen, R. L., Deanin, G. G., and Oliver, J. M. 1992b. Fc epsilon RI-mediated tyrosine phosphorylation and activation of the 72-kDa protein-tyrosine kinase, PTK72, in RBL-2H3 rat tumor mast cells. *Proc. Nat. Acad. Sci. USA 89:9107.*
26. Sidorenko, S. P., Law, C-L., Chandran, K. A., and Clark, E. A. 1994. The human spleen tyrosine kinase, Syk, associates with p53/56Lyn and a 120 kDa phosphoprotein, pp120. Submitted.
27. Amigorena, S., Bonnerot, C., Drake, J. R., et al. 1992. Cytoplasmic domain heterogeneity and functions of IgG Fc receptors in B lymphocytes. *Science 256:1808.*
28. Pesando, J. M., Bouchard, L. S., and McMaster, B. E. 1989. CD19 is functionally and physically associated with surface immunoglobulin. *J. Exp. Med. 170:2159.*
29. Justement, L. B., Campbell, K. S., Chien, N. C., and Cambier, J. C. 1991. Regulation of B cell antigen receptor signal transduction and phosphorylation by CD45. *Science 252:1839.*
30. Barrett, T. B., Shu, G. L., Draves, K. E., Pezzutto, A., and Clark, E. A. 1990. Signaling through CD19, Fc receptors or transforming growth factor-β: each inhibits the activation of resting human B cells differently. *Eur. J. Immunol. 20:1053.*
31. Carter, R. H. and Fearon, D. T. 1992. CD19: Lowering the threshold for antigen receptor stimulation of B lymphocytes. *Science 256:105.*
32. Tuveson, D. A., Carter, R. H., Soltoff, S. P., and Fearon, D. T . 1993. CD19 of B cells as a surrogate kinase insert region to bind phosphatidylinositol 3 kinase. *Science 260:986.*
33. Chalupny, N. J., Kanner, S. B., Schieven, G. L., et al. 1993. Tyrosine phosphorylation of CD19 in pre-B-cells and mature B-cells. *EMBO J. 12:2691.*
34. Kishihara, K., Penninger, J., Wallace, V. A., et al. 1993. Normal B lymphocyte development but impaired T cell maturation in CD45-exon6 protein tyrosine phosphatase-deficient mice. *Cell 74:143.*
35. Clark, E. A. 1993. CD22, a B-cell-specific receptor, mediates adhesion and signal transduction. *J. Immunol. 150:4715.*
36. Stamenkovic, I., Sgroi, D., Aruffo, A., Sy, M. S., and Anderson, T. 1991. The B lymphocyte adhesion molecule CD22 interacts with leukocyte common antigen CD45RO on T cells and alpha 2-6 sialyltransferase, CD75, on B cells. *Cell 66:1133.*
37. Wilson, G. L., Fox, C. H., Fauci, A. S., and Kehrl, J. H. 1991. cDNA cloning of the B cell membrane protein CD22: a mediator of B-B cell interactions. *J. Exp. Med. 173:137.*
38. Torres, R. M., Law, C-L., Santos-Argumedo, L., et al. 1992. Identification and characterization of the murine homologue of CD22, a lymphocyte restricted adhesion molecule. *J. Immunol. 149:2641.*
39. Aruffo, A., Kanner, S. B., Sgroi, D., Ledbetter, J. A., and Stamenkovic, I. 1992. CD22-mediated stimulation of T cells regulates T-cell receptor/CD3-induced signaling. *Proc. Nat. Acad. Sci. USA 89:10242.*
40. Sgroi, D., Varki, A., Braesch-Anderson, S., and Stamenkovic, I. 1993. CD22, a B cell-specific immunoglobulin superfamily member is a sialic acid-binding lectin, *J. Biol. Chem. 268:7011.*
41. Powell, L. D., Sgroi, D., Sjoberg, E. R., Stamenkovic, I., and Varki, A. 1993. Natural ligands of the B cell adhesion molecule CD22ß carry N-linked oligosaccharides with a2,6 linked sialic acids that are required for recognition. *J. Biol. Chem. 268:7019.*
42. van Noesel, C. J., Brouns, G. S., van Schijndel, G. M., Bende, R. J., Mason, D. Y., Borst, J., and van Lier, R. A. 1992. Comparison of human B cell antigen receptor complexes: membrane-expressed forms of immunoglobulin (Ig)M, IgD, and IgG are associated with structurally related heterodimers. *J. Exp. Med. 175:1511.*
43. Peaker, C. J. G. and Neuberger, M. S. 1993. Association of CD22 with the B cell antigen receptor. *Eur. J. Immunol. 23:1358.*
44. Pezzutto, A., Rabinovitch, P. S., Dörken, B., Moldenhauer, G., and Clark, E. A 1988. Role of CD22 human B cell surface antigen in the regulation of intracellular free calcium responses induced by anti-immunoglobulin. *J. Immunol. 140:1791.*
45. Schulte, R. J., Campbell, M.-A., Fischer, W. H., and Sefton, B. M. 1992. Tyrosine phosphorylation of CD22 during B cell activation. *Science 258:1001.*
46. Clark, E. A. 1990. CD40: A cytokine receptor in search of a ligand. *Tissue Antigens 35:33*
47. Clark, E. A. and Ledbetter, J. A. 1986. Activation of human B cells mediated through two distinct cell surface differentiation antigens, Bp35 and Bp50. *Proc. Nat. Acad. Sci. USA 83:4494*
48. Banchereau, J., de Paoli, P., Valle, A., Garcia, E., and Rousset, F. 1991 Long-term human B cell lines dependent on interleukin 4 and anti-CD40. *Science 251:70*
49. Spriggs, M. K., Armitage, R. J., Strockbine, L., et al. 1992. Recombinant human CD40 ligand stimulates B cell proliferation and immunoglobulin E secretion. *J. Exp. Med. 176:1543.*
50. Liu, Y-J., Joshua, D. E., Williams, G. T., Smith, C. A., Gordon, J., and MacLennan, I. C. M. 1989. Mechanisms of antigen-driven selection in germinal centres. *Nature 342:929.*
51. Valentine, M. A. and Licciardi, K. A. 1992. Rescue from anti-IgM-induced cell death by the B cell surface proteins CD20 and CD40. *Eur. J. Immunol. 22:3141-3146.*
52. Tsubata, T., Wu, J., and Honjo, T. 1993. B-cell apoptosis induced by antigen receptor crosslinking is blocked by a T-cell signal through CD40. *Nature 364:645.*
53. Hill, A. and Chapel, H. 1993. The fruits of cooperation. *Nature 361:494*

54. Kawabe, T., Yoshida, K., Yoshida, N., Kishimoto, T., and Kikutani, H. 1993. Generation and analysis of CD40 deficient mice. *Tis. Antigens 42:309.*
55. Lalmanach-Girard, A. C., Chiles, T. C., Parker, D. C., and Rothstein, T. L. 1993. T cell-dependent induction of NF-kB in B cells. *J. Exp. Med. 177:1215.*
56. Berberich, I., Shu, G., and Clark, E. A. Crosslinking CD40 on B cells rapidly activates the transcription factor NF-kB. Submitted
57. Clark, E. A. and Shu, G. L. 1990. Linkage between IL-6 and CD40 signaling: IL-6 activates the phosphorylation of CD40. *J. Immunol. 145:1400.*
58. Hirano, T., Akira, S., Taga, T., and Kishimoto, T. 1990. Biological and clinical aspects of interleukin 6. *Immunol. Today 11:443*
59. Uckun, F. M., Schieven, G. L., Dibirdik, I., et al.,1991. Stimulation of protein tyrosine phosphorylation, phosphoinositide turnover, and multiple previously unidentified serine/threonine-specific protein kinases by the pan-B-cell receptor CD40/Bp50 at discrete developmental stages of human B-cell ontogeny. *J. Biol. Chem. 266:17478.*
60. Ren, C. L., Morio, T., Fu, S. M., and Geha, R. S. 1994. Signal transduction via CD40 involves activation of *lyn* kinase and phosohatidylinositol-3-kinase, and phosphorylation of phospholipase Cγ2 *J. Exp. Med. 179:673.*
61. Kansas, G. S. and Tedder, T. F. 1991. Transmemberane signals generated through MHC class II, CD19, CD20, CD39 and CD40 antigens induce LFA-1-dependent and independent adhesion in human B cells through a tyrosine kinase-dependent pathway. *J. Immunol. 147:4094.*
62. Stade, B. G., Messer, G., Riethmuller, G., and Johnson, J. P. 1990. Structural characteristics of the 5' region of the human ICAM-1 gene. *Immunobiology 182:79.*
63. Vorarberger, G., Schafer, R., and Stratowa, C. 1991. Cloning of the human gene for intercellular adhesion molecule 1 and analysis of its 5'-regulatory region. *J. Immunol. 147:2777.*
64. Schwartz, R. H. 1992. Costimulation of T lymphocytes: The role of CD28, CTLA-4, and B7/BB-1 in interleukin-2 production and immunotherapy *Cell 71:1055.*
65. Klaus, S. J., Pinchuk, L., Ochs, H. D., Fanslow, W. C., Armitage, R. J., and Clark, E. A. 1994. Costimulation through CD28 enhances T cell-dependent B cell activation via a CD40-CD40L interaction. *J. Immunol. 152:* in press

# ANALYSIS OF THE (YXXL/I)2 SIGNALLING MOTIFS FOUND IN THE CYTOPLASMIC SEGMENT OF THE MOUSE CD3-ζ CHAIN

Anne-Marie K. Wegener and Bernard Malissen

Centre d'Immunologie INSERM-CNRS de Marseille-Luminy
Case 906
13288 Marseille Cedex 9
France

## INTRODUCTION

The specific recognition of antigen by T cells and its ensuing transduction into intracellular signals is accomplished by the T cell antigen receptor (TCR)-CD3 complex. The transducing subunits of this multimolecular complex, termed CD-3-$\gamma$, -$\delta$, -$\epsilon$, -$\zeta$ and -$\eta$, are noncovalently associated with the antigen binding TCR $\alpha\beta$ (or TCR $\gamma\delta$) dimer. They possess large intracytoplasmic segments which are responsible for their signalling properties and made of a recurrent functional domain of ~20 amino acids. Common to each copy of this domain is a pair of Tyr-X-X-Leu/Ile sequences (where X corresponds to a variable residue) separated by seven or eight variable residues. This consensus sequence, hereafter referred to as the $(YXXL/I)_2$ motif, is expressed as a single copy in the cytoplasmic tail of the CD3-$\delta$, CD3-$\gamma$ and CD3-$\epsilon$ subunits.[1] The CD3-$\zeta$ polypeptide appears to be unique among this set of signalling devices in the sense that it displays three concatenated copies of the $(YXXL/I)_2$ motif (denoted as ζa, ζb, and ζc).[2]

Cross-linking of chimeric molecules composed of the extracellular and transmembrane parts of the CD4, CD8$\alpha$, CD16 or CD25 molecules and of a single copy of the $(YXXL/I)_2$ motif suffice to elicit most of the early and late activation events that normally occur when antigen receptors are stimulated.[3] The precise mechanisms by which the oligomerization of the various $(YXXL/I)_2$ motifs generate intracellular signals remain unknown. However, as an immediate result of ligand binding, the tyrosine residues found in each YXXL/I sequence are phosphorylated (pY) and likely to act as "docking" sites which bind with high affinity to the SH2 (src-homology 2) domains found in certain intracellular adaptor and effector molecules. [Adaptors are distinguished from effectors by the fact that they lack any recognizable catalytic sequences]. For instance, a 70 kDa protein tyrosine kinase, termed ZAP-70, has been shown to specifically associate with the tyrosine phosphorylated T cell receptor CD3-$\zeta$ and CD3-$\epsilon$ polypeptides. Further experiments using glutathione S-transferase (GST)/ZAP-70 fusion proteins suggested that the high affinity binding to phospho-$\zeta$ or phospho-$\epsilon$ required that the two SH2 domains found in ZAP-70 bind in a *coordinated fashion* to the pair of pXXXL/I sequences found within each of the CD3-$\zeta$ and $\epsilon$ motifs. The compact structure of the SH2 domain makes it feasible for two concatenated SH2 domains to simultaneously bind to pY residues that

*Mechanisms of Lymphocyte Activation and Immune Regulation V*
Edited by S. Gupta *et al.*, Plenum Press, New York, 1994

are located in the same polypeptide fewer than 10 amino acids apart.[4] Neither of the ZAP-70 SH2 domains expressed singly as GST-fusion proteins could bind phospho-$\zeta$ or phospho-$\epsilon$.[5] In contrast, the single SH2 domain of the Shc adaptor protein appear capable of interacting directly with *one* of the pYXXL/I sequences found within phosphorylated $\zeta$ chains.[6] Therefore, once phosphorylated, the two YXXL/I sequences found within a given motif appear capable of being recognized either in a *concerted* mode by intracellular effector/adaptor proteins containing tandem SH2 domains (e.g., ZAP-70, syk) or *independently* by effector/adaptor proteins containing a single SH2 domain (e.g., Shc).

The $(YXXL/I)_2$ motifs are probably too small to have any intrinsic protein tyrosine kinase (PTK) activity. In the resting state, the unphosphorylated motifs may be constitutively associated with at least one resident PTK [fyn,[7] lck[8]]. Upon receptor cross-linking, these resident PTK appear to be activated and responsible for the phosphorylation of the various $(YXXL/I)_2$ motifs. The association of the resting receptors to the resident PTK appears independent of the YXXL/I sequences, and, most likely, determined by the amino acids which immediately flank them. It had been originally hypothesized that all the phosphorylated $(YXXL/I)_2$ motifs have the same binding specificity (i.e., are capable of docking to the very same set of intracellular effector/adaptor proteins). Accordingly, the presence of several redundant $(YXXL/I)_2$ motifs in a given antigen receptor (e.g., $> 10$ for the TCR, $> 4$ for the BCR, and $> 3$ for the Fc$\epsilon$RI$\gamma$) may have merely evolved as a way of amplifying the intracellular signal(s) resulting from ligand binding. However, recent results suggest that the various motifs are non-redundant, each being capable of association with distinct subsets of SH2-containing proteins, and thus responsible for the coupling of a receptor to a unique activation pathway. Along that line, it should be noted that the specificity of the $(pYXXL/I)_2$:SH2 association may be largely provided by the two residues which are immediately C-terminal to the pY residue. (residues pY+1 and pY+2 in ref. 4).

To refine the analysis of the function of the mouse CD3-$\zeta$ chain, we have isolated the three $(YXXL/I)_2$ motifs it contains and characterized their signalling properties.

## METHODS

### Cells

BW5147$\alpha^-\beta^-$ (hereafter referred to as BW$^-$) is a variant of the BW5147 thymoma lacking functional TCR $\alpha$ and $\beta$ chain genes.[2]

### Chimera Constructions

The CD8/$\zeta$ mutants were constructed by PCR as described previously,[3] sequenced using the dideoxy chain termination method and subsequently subcloned in the pH$\beta$APR-1-neo expression vector.[2]

### Transfection by Protoplast Fusion

Transfection of BW5147$\alpha^-\beta^-$ with protoplasts and selection in the presence of G418-sulphate were performed as described.[2]

### Stimulation of CD8/$\zeta$ Transfectants

In all experiments, 0.25 ml cultures were prepared containing $10^5$ responding cells. For stimulation with antibodies coated on the surface of microtiter wells, wells were precoated with 50 $\mu$l of DME culture medium containing various concentrations of

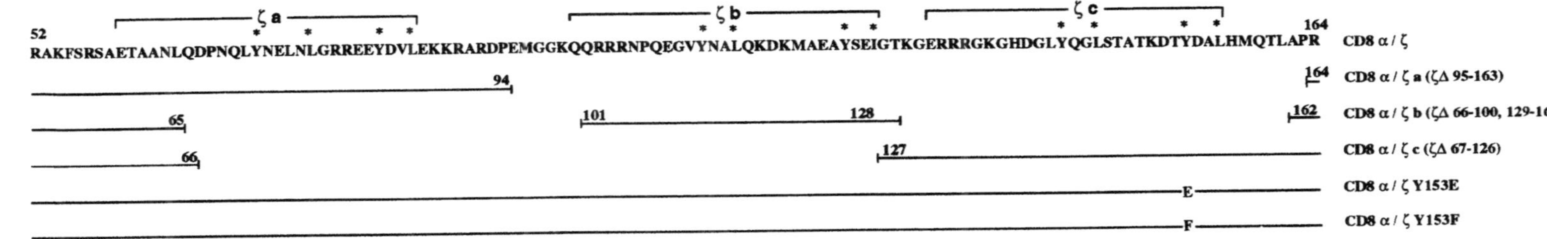


**Figure 1.** Amino acid sequence of the ζ cytoplasmic segment present in the CD8/ζ chimeras containing isolated $(YXXL/I)_2$ motifs (CD8/ζa, CD8/ζb and CD8/ζc). Also shown is the amino acid sequence of two CD8/ζ chimeras containing a substitution at residue Y153 (CD8/ζY153E and CD8/ζY153F). The sequence of the wild-type ζ cytoplasmic tail is shown in the single-letter amino acid code. The tyrosine residues (asterisks) and the localization of the three $(YXXL/I)_2$ motifs (ζa, ζb and ζc) are highlighted. The sequences of the mutated ζ cytoplasmic segments are indicated under the wild-type ζ sequence. The ζ residues are numbered according to reference 9.

antibodies. After 2 hr at room temperature and 1 hr at 4°C, the wells were washed three times with DME and used. At the end of the cultures (18-24 hr), supernatants were harvested and assayed for their level of IL-2.

## RESULTS AND DISCUSSION

To investigate the function of each of the three $(YXXL/I)_2$ motifs present in the mouse CD3-ζ polypeptide, a set of CD8-ζ chimeras was constructed and assessed for their relative transducing abilities independently of CD3-$\gamma$, -$\delta$ and $\epsilon$. As summarized in Figure 1, three truncated ζ cDNA constructs were developed. The first one corresponds to an internal deletion encompassing residues 95-163 and sparing the N-terminal $(YXXL/I)_2$ motif (ζa). The second one lacks both residues 66-100 and 129-161 and preserves the integrity of the central motif denoted as ζb. The third deletion removes residues 67-126 and spares the C-terminal $(YXXL/I)_2$ motif (ζc). These truncated ζ cDNAs were separately fused a few residues upstream of their membrane-spanning domain with the mouse CD8-$\alpha$ extracellular domain, giving the chimeric products CD8/ζa, CD8/ζb and CD8/ζc. When expressed in $BW^-$, the CD8/ζa, CD8/ζb, CD8/ζc and wild-type CD8ζ (CD8ζwt) chimeras were readily detected at the surface of the corresponding transfectants (data not shown). Examination of a number of independent transfectants showed that the levels of surface expression reached by the CD8/ζ chimeras containing internal truncations were lower than those observed with the wild-type CD8/ζ chimera. When tested for their ability to trigger IL-2 production upon antibody-mediated cross-linking (Figure 2), each of the three individualized motifs retained the capacity to induce IL-2 production.

However, as previously documented,[2,8,11] clones expressing truncations or internal deletions of the ζ cytoplasmic segment showed a markedly reduced signalling efficiency.

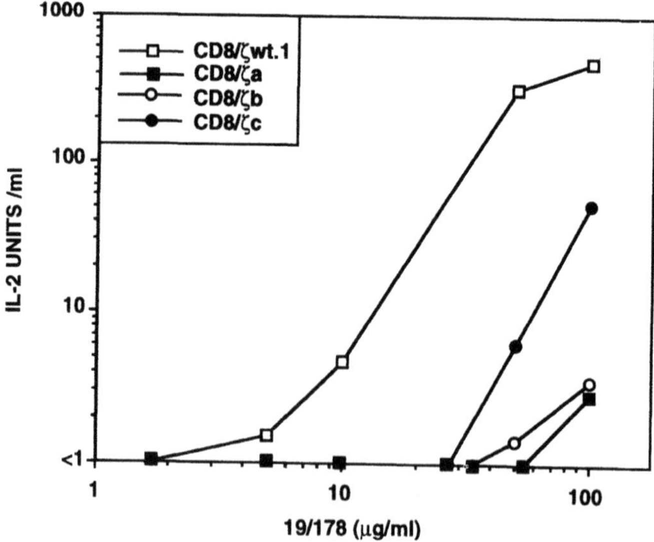


**Figure 2.** IL-2 production of $BW^-$ cells transfected with CD8/ζ chimeras containing isolated $(YXXL/I)_2$ motifs in response to stimulation with anti-CD8$\alpha$ antibodies. All cell lines showed comparable staining with the anti-CD8$\alpha$ antibody 19/178.[10]

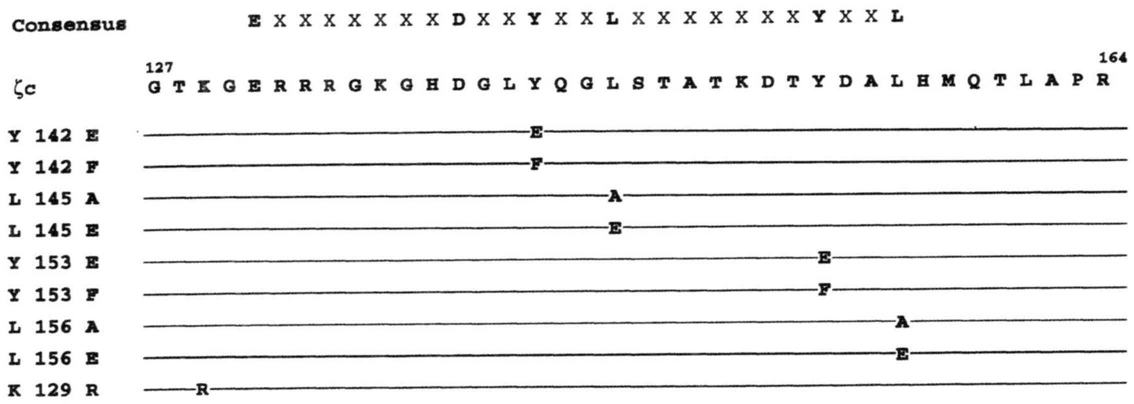

```
Consensus                   E X X X X X X X D X X Y X X L X X X X X X X Y X X L

               127                                                                        164
ζc             G T K G E R R R G K G H D G L Y Q G L S T A T K D T Y D A L H M Q T L A P R

Y 142 E        - - - - - - - - - - - - - - - E - - - - - - - - - - - - - - - - - - - - - -
Y 142 F        - - - - - - - - - - - - - - - F - - - - - - - - - - - - - - - - - - - - - -
L 145 A        - - - - - - - - - - - - - - - - - - A - - - - - - - - - - - - - - - - - - -
L 145 E        - - - - - - - - - - - - - - - - - - E - - - - - - - - - - - - - - - - - - -
Y 153 E        - - - - - - - - - - - - - - - - - - - - - - - - - - E - - - - - - - - - - -
Y 153 F        - - - - - - - - - - - - - - - - - - - - - - - - - - F - - - - - - - - - - -
L 156 A        - - - - - - - - - - - - - - - - - - - - - - - - - - - - - A - - - - - - - -
L 156 E        - - - - - - - - - - - - - - - - - - - - - - - - - - - - - E - - - - - - - -
K 129 R        - - R - - - - - - - - - - - - - - - - - - - - - - - - - - - - - - - - - - -
```

**Figure 3.** Mutational analysis of the C-terminal $(YXXL/I)_2$ motif ($\zeta c$) of the $\zeta$ polypeptide. The sequences of the mutated $\zeta c$ motifs used in this study are indicated under the corresponding wild-type sequence. Amino acid identities with the latter are indicated with dashes.

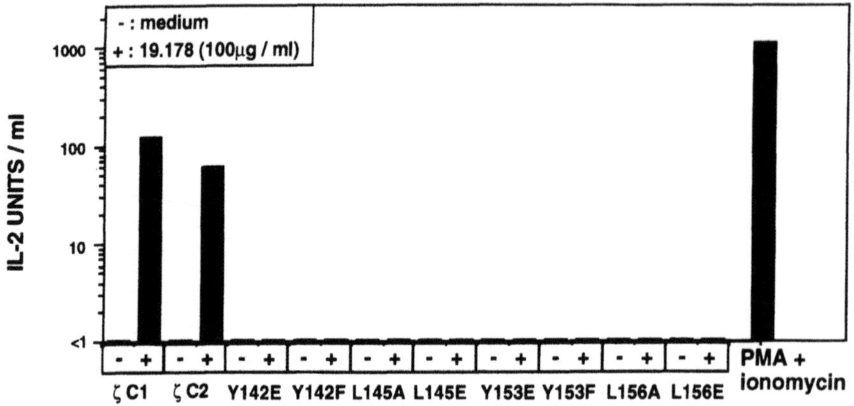


**Figure 4.** IL-2 secretion of BW⁻ cells transfected with mutated CD8/ζc genes, in response to stimulation with 19/178, an anti-CD8α antibody (+), or medium alone (-). All cells were capable of IL-2 production upon addition of phorbol 12-myristate 13-acetate (PMA) and ionomycin.

It should be noted that the two YXXL/I sequences found in the functional ζb motif are spaced by eight amino acids (Figure 1) and thus differ from the 7-amino acid spaced-YXXL/I sequences found in most of the $(YXXL/I)_2$ motifs identified to date.

Mutational analysis of the $(YXXL/I)_2$ motifs present in the CD3-ε and N-terminal segment of CD3-ζ (ζa) has shown that substitution of either of the two tyrosine or leucine/isoleucine residues abrogates their signalling capacity.[11,12] The introduction of similar mutations in the ζc motif should therefore allow analysis of their relative contribution to its signalling properties. Accordingly, the tyrosines 142 and 153 were individually mutated to either phenylalanine (F) or glutamic acid (E) (Figure 3). Flow cytometry analysis showed that these mutated CD8/ζc chimeras were properly expressed at the surface of BW⁻ (data not shown). Despite levels of surface expression comparable to those observed with two wild-type CD8/ζc transfectants (ζc1 and ζc2 in Figure 4), the transfectants expressing the Y142F, Y142E, Y153F or Y153E mutations did not respond to anti-CD8 antibody (Figure 4). Substitution of residues L145 or L156 resulted in the complete inactivation of the corresponding CD8/ζc chimeras (Figure 4). Because substitutions of Y153 and L156 abolishes the signalling function of the CD8/ζc chimera, it appears likely that the signalling ability of the CD3-η polypeptide[11] can be fully attributed to the ζa and ζb motifs.

Our data extend to the mouse CD3-ζ subunit the findings of Romeo et al.[11] and of Irving et al.[13] and confirm that the three individualized motifs it contains are indeed capable of inducing a late T cell activation event. Finally, mutational analysis of the ζc motif indicated that its two YXXL/I sequences constitute functionally important components.

## ACKNOWLEDGEMENTS

This work was supported by institutional grants from CNRS, INSERM, and specific grant from ARC. A.-M. K.W. is supported by a fellowship from the Carlsberg Foundation.

## REFERENCES


1. M. Reth, Antigen receptor tail clue, *Nature* 338:383 (1989).
2. A.-M.K. Wegener, F. Letourneur, A. Hoeveler, T. Brocker, F. Luton, and B. Malissen, The T cell receptor/CD3 complex is composed of at least two autonomous transduction modules, *Cell* 68:83 (1992).
3. B. Malissen and A.-M. Schmitt-Verhulst, Transmembrane signalling through the T cell receptor-CD3 complex, *Curr. Op. Immunol.* 5:324 (1993).
4. S. Zhou, S.E. Shoelson, M. Chaudhuri, G. Gish, T. Pawson, W.G. Haser, F. King, T. Roberts, S. Ratnofsy, R.J. Lechleider, *et al.*, SH2 domains recognize specific phosphopeptide sequences, *Cell* 72:767 (1993).
5. R.L. Wange, S.N. Malek, S. Desiderio, and L.E. Samelson, Tandem SH2 domains of ZAP-70 bind to T cell antigen receptor $\zeta$ and CD3-$\epsilon$ from activated Jurkat cells, *J. Biol. Chem.* 268:19797 (1993).
6. K.S. Ravichandran, K.K. Lee, Z. Songyang, L.C. Cantley, P. Burn, and S.J. Burakoff, Interaction of Shc with the $\zeta$ chain of the T cell receptor upon T cell activation, *Science* 262:902 (1993).
7. L.K. Timson-Gauen, A.N.T. Kong, L.E. Samelson, and A.S. Shaw, $p59^{fyn}$ tyrosine-kinase associates with multiple T-cell receptor subunits through its unique amino-terminal domain, *Mol. Cell. Biol.* 12:5438 (1992).
8. D.B. Straus and A. Weiss, The CD3 chains of the T cell antigen receptor associate with ZAP-70 tyrosine kinase and are tyrosine phosphorylated after receptor stimulation, *J. Exp. Med.* 178:1523 (1993).
9. A.M. Weissman, M. Baniyash, D. Hou, L.E. Samelson, W.H. Burgess, and R.D. Klausner, Molecular cloning of the $\zeta$ chain of the T cell antigen receptor, *Science* 239:1018 (1988).
10. D. Blanc, C. Bron, J. Gabert, F. Letourneur, H.R. MacDonald, and B. Malissen, Gene transfer of the Ly3 chain of the mouse CD8 molecular complex, *Eur. J. Immunol.* 18:613 (1988).
11. C. Romeo, A. Amiot, and B. Seed, Sequence requirements for induction of cytolysis by the T cell antigen/Fc receptor $\zeta$ chain, *Cell* 68:889 (1992).
12. F. Letourneur and R.D. Klausner, Activation of T cells by a tyrosine kinase domain in the cytoplasmic tail of CD3-$\epsilon$, *Science* 255:79 (1992).
13. B.A. Irving, A.C. Chan, and A. Weiss, Functional characterization of a signal transducing motif present in the T cell antigen receptor $\zeta$ chain, *J. Exp. Med.* 177:1093 (1993).

# MOLECULAR AND GENETIC INSIGHTS INTO T CELL ANTIGEN RECEPTOR SIGNAL TRANSDUCTION

Arthur Weiss[1], Makio Iwashima[1], Bryan Irving[1], Nicolai S. C. van Oers[1], Theresa A. Kadlecek[1], David Straus[1], and Andrew Chan[2]

[1]Departments of Medicine and of Microbiology and Immunology
Howard Hughes Medical Institute
University of California, San Francisco
San Francisco, Ca 94143

[2]Department of Medicine
Howard Hughes Medical Institute
Washington University School of Medicine
St. Louis, Mo. 63110

## INTRODUCTION

T lymphocytes are relatively quiescent cells, poised in the $G_0$ stage of the cell cycle, until they encounter antigen. The initiation of an immune response by T cells requires the recognition of antigen by the T cell antigen receptor (TCR) and the conversion of this recognition event into biochemical signals that can induce a cellular response. This cellular response leads to the clonal expansion of antigen-specific differentiated effector T cells.

Stimulation of the TCR results in the induced tyrosine phosphorylation of a variety of cellular proteins, including subunits of the TCR, phospholipase C γ1 (PLC γ1), the product of the protooncogene vav, mitogen-activated protein kinase and many others (1, 2). The phosphorylation of these proteins results in a cascade of biochemical events which contribute to the ultimate cellular response. For instance, phosphorylation of PLC γ1 results in an increase in its catalytic activity (3). The activation of PLC γ1 is responsible for the increase in cytoplasmic free calcium and activation of protein kinase C, events that have been causally linked to the induction of interleukin 2 gene transcription (2). Thus, the regulation of protein tyrosine phosphorylation is a critical mechanism that the TCR uses to initiate cellular responses.

The mechanism by which the TCR initiates protein tyrosine phosphorylation has been of great interest. Unlike many growth factor receptors,

*Mechanisms of Lymphocyte Activation and Immune Regulation V*
Edited by S. Gupta *et al.*, Plenum Press, New York, 1994

the TCR does not have intrinsic protein tyrosine kinase (PTK) or protein tyrosine phosphatase (PTPases) activity. Instead, the TCR uses recently recognized structural domains to interact with intracellular PTKs. This interaction perturbs a pre-existing dynamic equilibrium between cellular PTKs and PTPases leading to an increase in net cellular protein tyrosine phosphorylation. Recent progress has been made in understanding the molecular mechanisms by which the TCR transduces signals.

## STRUCTURAL DOMAINS IN THE TCR CD3 AND ζ CHAINS THAT ARE INVOLVED IN SIGNAL TRANSDUCTION

The TCR on most mature T cells is an oligomeric complex consisting of ligand (antigen)-binding subunits and subunits involved in signal transduction (Figure 1). The Ti αβ heterodimer contains all of the information necessary to recognize peptide antigens bound to major histocompatibility complex molecules. The CD3 and ζ chains are involved in signal transduction. The signaling function of the CD3 and ζ chains have been most clearly established by studies employing chimeric receptors. In these chimeras, the cytoplasmic domains of the CD3 ε or ζ chains were freed from their interactions with other chains of the TCR by linking

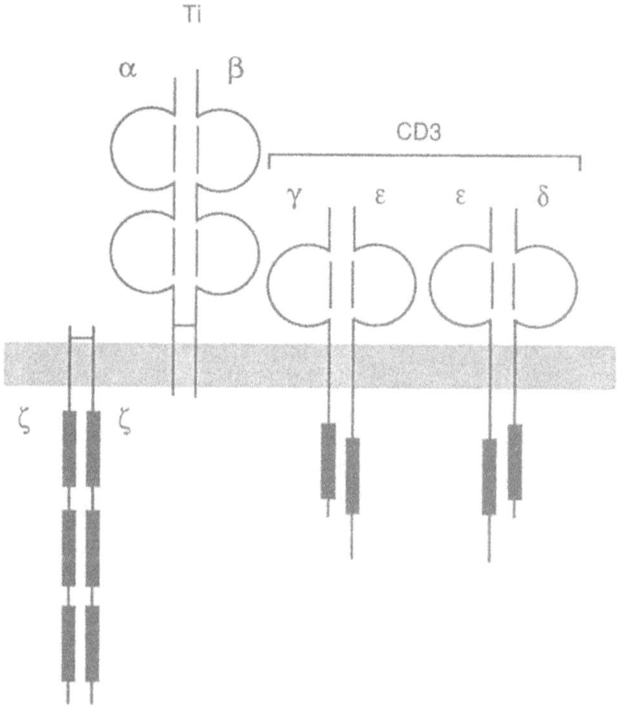


Figure 1. Schematic representation of the oligomeric TCR. The black rectangles represent the ARAMs present in the cytoplasmic domains of the ζ and CD3 chains.

| | | | | | | | | | | | | | | | | | |
|---|---|---|---|---|---|---|---|---|---|---|---|---|---|---|---|---|---|
| hζ1 | N | Q | L | **Y** | N | E | **L** | N | L | G | R | R | E | E | - | **Y** | D | V | **L** |
| hζ2 | E | G | L | **Y** | N | E | **L** | Q | K | D | K | M | A | E | A | **Y** | S | E | **I** |
| hζ3 | D | G | L | **Y** | Q | G | **L** | S | T | A | T | K | D | T | - | **Y** | D | A | **L** |
| hCD3γ | E | Q | L | **Y** | Q | P | **L** | K | D | R | E | D | D | Q | - | **Y** | S | H | **L** |
| hCD3ε | N | P | D | **Y** | E | P | **I** | R | K | G | Q | R | D | L | - | **Y** | S | G | **L** |
| hCD3δ | E | Q | V | **Y** | Q | P | **L** | R | D | R | D | D | A | Q | - | **Y** | S | H | **L** |
| rIgE FcRγ | D | A | V | **Y** | T | G | **L** | N | T | R | N | Q | E | T | - | **Y** | E | T | **L** |
| rIgE FcRβ | D | R | L | **Y** | E | E | **L** | - | H | V | Y | S | P | I | - | **Y** | S | A | **L** |
| mIgα | E | N | L | **Y** | E | G | **L** | N | L | D | D | C | S | M | - | **Y** | E | D | **I** |
| mIgβ | D | H | T | **Y** | E | G | **L** | N | I | D | Q | T | A | T | - | **Y** | E | D | **I** |
| BLV gp30 | D | S | D | **Y** | Q | A | **L** | L | P | S | A | P | E | I | - | **Y** | S | H | **L** |
| EBV LMP2A | H | S | D | **Y** | Q | P | **L** | G | T | Q | D | Q | S | L | - | **Y** | L | G | **L** |

Figure 2. Sequences of antigen recognition activation motifs (ARAMs) in hematopoietic antigen receptors and viral proteins.

them to the extracellular and transmembrane domains of other transmembrane molecules (4-6). Quite remarkably, these cytoplasmic domains conferred upon these chimeric receptors the signal transduction functions characteristic of the intact multisubunit TCR complex. For instance, stimulation of a CD8/ζ chimera with an anti-CD8 monoclonal antibody induced protein tyrosine phosphorylation, activation of the inositol phospholipid pathway, increases in cytoplasmic free calcium, expression of activation antigens (such as CD69) and even induction of interleukin-2 production (4). Such studies indicated that the cytoplasmic domains of both the CD3 ε and ζ chains contain the sequence information necessary to couple the TCR to intracellular signal transduction machinery.

This redundancy of function was explained in subsequent experiments when the functional domains of these chains were mapped by using chimeric receptors containing ζ and CD3 ε cytoplasmic sequences (6-8). These studies revealed that a common sequence motif, termed herein as ARAM (for antigen recognition activation motif) which was triplicated within the cytoplasmic domain of ζ and contained as a single copy in each of the CD3 chains (Figure 1), is responsible for coupling chimeric receptors to the TCR-regulated signal transduction mechanisms. The ARAM sequences, depicted in Figure 2, alone are sufficient to confer signal transduction function upon chimeric receptors.

This motif, first noted by Reth (9), contains the consensus sequence of $YXXL(X)_{6\text{-}8}YXXL$. Functional ARAMs are found in the non-ligand binding subunits of other antigen receptors including the B cell antigen receptor and the mast cell IgE Fc receptor. The ARAMs within these antigen receptors that are expressed on distinct cells of the hematopoietic lineage appear to have arisen from a common evolutionary precursor, as their exon-intron organization is similar (10). Moreover, two membrane proteins, gp30 and LMP2, encoded by the bovine leukemia virus and the Epstein Barr virus which are involved in B cell transformation, also contain ARAM sequences. Thus, these two viruses may have usurped the ARAM as a mechanism to communicate with host cell signal transduction mechanisms that could regulate cell growth. It is not yet clear whether the ARAMs are required for cellular transformation by these viruses.

Each of the antigen receptors that contain ARAMs contain multiple copies. This raises questions regarding the functional significance of the incorporation of multiple copies of the ARAM into an individual oligomeric antigen receptor with a single ligand binding subunit. It is possible that the ARAMs contain sufficient sequence disparity to interact with distinct intracellular proteins involved in signal transduction. Some evidence in support of this hypothesis has been reported (6, 11). An alternative, though not necessarily exclusive, possibility is that the presence of multiple ARAMs within a receptor with a single binding subunit may represent a means of signal amplification. We recently provided evidence to support this latter hypothesis by linking a single copy or three copies of a ζ ARAM to the transmembrane and extracellular domains of CD8 and compared the signal transduction capabilities of these chimeric molecules (8). The chimera with three copies of the ζ motif induced greater increases in tyrosine phosphoproteins, cytoplasmic free calcium and transcriptional activity of a reporter construct that incorporated an interleukin 2 transcriptional element than the receptor incorporating one copy of the motif. Moreover, the response of the chimeric receptor with three copies of a single ζ motif approached that of the-wild type ζ sequence with three distinct ARAMs. Thus, the presence of multiple ARAMs within a single receptor is likely to represent a means of signal amplification. Effectively, this serves to increase the sensitivity of the T cell response to antigen.

## ARAMS INTERACT SEQUENTIALLY WITH TWO FAMILIES OF PTKS

Biochemical and genetic studies have suggested the involvement of at least two families of PTKs in TCR signal transduction. Two Src family members, Fyn and Lck, have been implicated. Fyn has been coimmunoprecipitated with the TCR from some cells, albeit at low stoichiometry (12, 13). In mice in which the *fyn* gene has been disrupted, T cells develop normally (14, 15). However, the most mature single positive ($CD4^+$ or $CD8^+$) thymocytes have markedly diminished responses to TCR stimuli, though peripheral T cells have a much milder defect and can respond to antigen. Lck is associated with the coreceptors CD4 and CD8 (16). Cell lines, which are deficient in Lck kinase function, have a marked impairment in TCR signal transduction function (17, 18). Disruption of the *lck* gene results in a pronounced thymocyte developmental arrest (19). These observations provide strong evidence for the involvement of both of these Src family members in TCR signal transduction.

The Syk and ZAP-70 PTKs are recently identified PTKs with common structural features (20, 21). Unlike the Src family PTKs, Syk and ZAP-70 lack an N-terminal myristilation site, SH3 domain or C-terminal negative regulatory tyrosine phosphorylation site. Instead, Syk and ZAP-70 contain tandem N-terminal SH2 domains and a more C-terminal kinase domain. ZAP-70 is exclusively expressed in T cells and natural killer cells (21). In contrast, Syk is expressed more broadly within cells of the hematopoietic lineage (20). However, Syk is expressed at relatively low levels in thymocytes and even lower levels in peripheral T cells when compared to B cells (22). In the Jurkat T cell line, ZAP-70 and Syk are recruited from the cytoplasm and associate with the stimulated TCR (22, 23). Studies with ZAP-70 indicate that it associates exclusively with the tyrosine phosphorylated CD3 and ζ chains (21). Moreover, this association has been mapped to the ARAMs within ζ (8). ZAP-70 and Syk are also inducibly tyrosine phosphorylated following TCR stimulation (22, 23).

Recent studies have provided insights into the molecular events involved in the interactions of these two PTK families with the TCR ARAMs. The sequential interactions of these PTKs with an ARAM-containing receptor are depicted in Figure 3 and have been recently reported elsewhere (24). Such a model may apply to the molecular events involved in B cell antigen receptor and IgE Fc receptor signal transduction. During the initial event shown, stimulation of the TCR induces the tyrosine phosphorylation of ARAMs by a Src family member. Genetic and biochemical evidence support this model. We previously isolated a signal transduction mutant from the Jurkat T cell leukemic line, J.CaM1, in which no increase in cytoplasmic free calcium or protein tyrosine phosphorylation is observed following TCR stimulation (25, 26). J.CaM1 does not express a functional Lck PTK. Stimulation of the TCR on J.CaM1 fails to induce the phosphorylation of ζ or the CD3 chains. Thus, Lck is required for this initial event in the Jurkat line. In some T cells, this function may be provided by Fyn, though Fyn is expressed in J.CaM1. In J.CaM1, in addition to failure of the ARAMs to be phosphorylated, ZAP-70 is neither recruited to the TCR nor is it tyrosine phosphorylated (24). These observations place Lck upstream of ZAP-70 and required for ζ and CD3 chain phosphorylation.

We have used a heterologous system to study these interactions in further detail. We used the Cos 18 cell line stably transfected with a CD8/ζ chimera, as a TCR surrogate, to study the interaction of the ζ ARAMs with Src family members and Syk/ZAP-70 members which were transfected into the cell transiently (21, 24). In this system, transfection of the Cos 18 with either Lck or Fyn can induce some low level phosphorylation of the CD8/ζ chimera. The association of ZAP-70 or Syk with the CD8/ζ chimera as well as the phosphorylation of ZAP-70 or Syk requires Lck or Fyn. This is also associated with an increase in CD8/ζ phosphorylation. Using mutants of Lck, we have shown that the kinase activity of Lck is required for the phosphorylation of CD8/ζ as well as the recruitment and phosphorylation of ZAP-70. In contrast, the ability of the SH2 domain of Lck is not required for any of these functions in this heterologous system. These observations on J.CaM1 and in the Cos 18 cells strongly support the notion that Lck (or in some T cells, Fyn) plays a role which is upstream of ZAP-70 in initiating signal transduction by the TCR.

The presence of both tyrosines within an ARAM appears to be critically important in signal transduction and in the recruitment of ZAP-70. Mutation of either tyrosine within an ARAM eliminates the functional capability of the ARAM to confer signal transduction function upon chimeric receptors (6, 7). This suggested to us that the phosphorylation of both tyrosines is required for the function of the ARAM. This requisite function could represent the interaction of the ARAM with ZAP-70 or Syk. To test this hypothesis, biotinylated synthetic peptides based on the 28 residues surrounding the N-terminal ζ ARAM were prepared in which either, neither or both tyrosines were phosphorylated. These peptides were mixed with lysates from the Jurkat T cell line and were assessed for their ability to form stable complexes with ZAP-70. Only the doubly phosphorylated ARAM was able to form a stable complex with ZAP-70 (24). Moreover, a mixture of the singly phosphorylated peptides could not form a complex with ZAP-70 suggesting that ZAP-70 could not bridge ARAMs. Similar observations have been made with other doubly phosphorylated ARAMs with both ZAP-70 and Syk (B. Irving and A. Weiss, unpublished observations). This

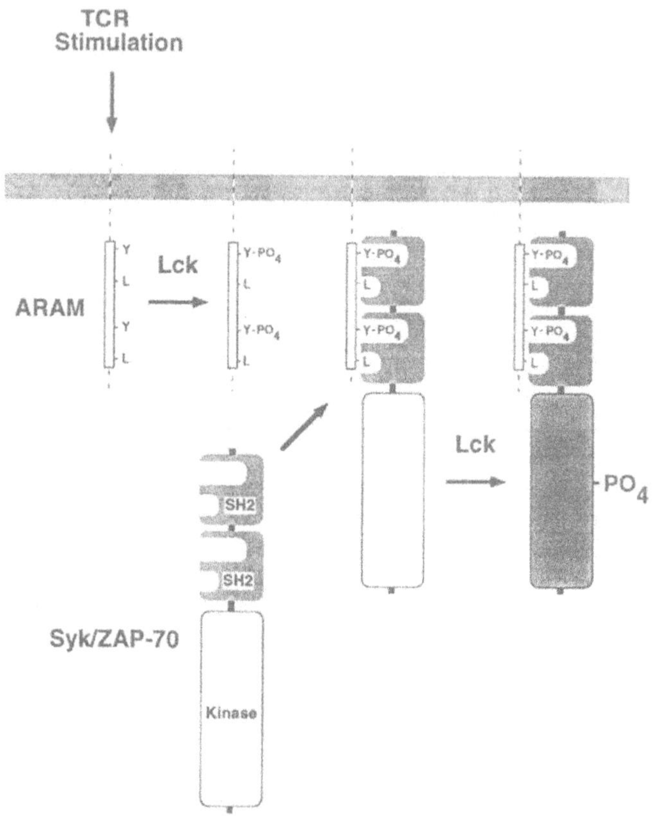


Figure 3. Model for the sequential interactions of a TCR ARAM with the Lck and ZAP-70 PTKs.

supports the notion that only the doubly phosphorylated ARAM will be functionally able to interact with ZAP-70 and may offer an explanation for the required presence of the two tyrosines within an ARAM for signal transduction function.

The observation that ZAP-70 interacts only with the phosphorylated form of the ζ chain (21) suggested that its SH2 domains were involved. SH2 domains form stable complexes with tyrosine phosphorylated residues and specificity is conferred upon the interaction by sequence differences in the SH2 domain and differences in the sequences surrounding the phosphotyrosine residue (27). The requirement for a doubly phosphorylated ARAM in forming a stable interaction with ZAP-70 or Syk suggested that both of the ZAP-70 SH2 domains contribute to forming the stable complex. To test this hypothesis, point mutations were introduced into either of the two ZAP-70 SH2 domains which changed a critical arginine (of the FLVR sequence) in the phosphotyrosine binding site to lysine; this abolishes phosphotyrosine binding function of SH2 domains. Neither of these mutants, transfected into the CD8/ζ-expressing Cos 18 cells together with Lck, could form a stable complex with the CD8/ζ chain chimera (24). These results suggest that both SH2 domains are required to form a complex with the doubly phosphorylated ARAM.

Finally, the kinase activity of ZAP-70 is not required for the phosphorylation of the ζ chain, its recruitment to the ζ chain, or for its own phosphorylation. A mutant of ZAP-70, deficient in kinase function, associated with the CD8/ζ chimera and became tyrosine phosphorylated if cotransfected with Lck (24). This suggests that Lck may be responsible for ZAP-70 phosphorylation, though it is not clear whether the sites phosphorylated within the kinase deficient form are identical to wild-type form ZAP-70. The importance of ZAP-70 tyrosine phosphorylation is not clear. No increase in the catalytic activity of ZAP-70 has been observed, though this may reflect the lack of identifiable ZAP-70 substrates at the present time. Interestingly, even with the kinase deficient form of ZAP-70, an increase in phosphorylation of the CD8/ζ chimera was observed. This may result from the protection of the tyrosine phosphorylated residues by the bound ZAP-70 SH2 domains. A similar role has been proposed for the SH2 domains of PLC γ1 interacting with the EGF receptor (28). This may result in potentiation of signal transduction.

The sequential interactions of the TCR ARAMs with these two families of PTKs is an attractive one, as it incorporates the function of the coreceptors. Thus, CD4 or CD8 can serve to enhance the sensitivity of the system by serving to colocalize Lck to the TCR ARAMs to help initiate the signal transduction events depicted in Figure 3. Thus, this would imply that coreceptors not only function to bind to MHC molecules and contribute to the binding energetics of the TCR/major histocompatibility complex molecule/peptide, but would also serve to deliver and concentrate Lck to the ARAMs.

A critical functional role for ZAP-70 is suggested by several lines of evidence. First, in the Cos 18 cell system described above, a very large increase in cellular tyrosine phosphorylation is observed if ZAP-70 is cotransfected with either Lck or Fyn (21, 24). This is only observed with the kinase active form of ZAP-70. Second, the sequence motifs responsible for ζ chains signal transduction, the ARAMs, are identical to sequence requirements for ZAP-70 interactions (8). Finally, recently a form of severe combined immunodeficiency has been described which results from ZAP-70 deficiency and is characterized by a functional defect in TCR signal transduction (29, 30). These observations provide strong support for a role of ZAP-70 in TCR signal transduction.

## SUMMARY

The T cell response to antigen is initiated by the TCR through its ability to regulate the phosphorylation of tyrosine residues of critical intracellular molecules. The TCR uses a modular system in which sequence motifs, ARAMs, within the ζ and CD3 chains to communicate with intracellular PTKs. These PTK interact with the TCR in a sequential and coordinated manner. This rather complex modular system may have evolved to insure that all requirement of the signalling mechanism are met before a response is initiated. Mistakes could result in autoimmunity. It is likely that the general features of this system will reflect the events involved in B cell antigen receptor and IgE Fc receptor signal transduction since ARAMs as well as Src and Syk/ZAP-70 family members are used in these systems as well.

## ACKNOWLEDGMENTS

This work was supported in part by grants from the National Institutes of Health (to A.W. and to A.C.C.), the Arthritis Foundation (to A.C.C.) and by a postdoctoral training award from the American Federation for Clinical Research/Merck, Sharp and Dohme (to A.C.C).

## REFERENCES

1. L.E. Samelson, and R.D. Klausner, Tyrosine kinases and tyrosine-based activation motifs. Current research on activation via the T cell antigen receptor, *J. Biol. Chem.* 267:24913 (1992).
2. A. Weiss, and D.R. Littman, Signal transduction by lymphocyte antigen receptors, *Cell* 76:263 (1994).
3. S. Nishibe, M.I. Wahl, S.M. Hernandez-Sotomayor, N.K. Tonk, S.G. Rhee, and G. Carpenter, Increase of the catalytic activity of phospholipase C-γ1 by tyrosine phosphorylation, *Science* 250:1253 (1990).
4. B. Irving, and A. Weiss, The cytoplasmic domain of the T cell receptor ζ chain is sufficient to couple to receptor-associated signal transduction pathways, *Cell* 64:891 (1991).
5. C. Romeo, and B. Seed, Cellular immunity to HIV activated by CD4 fused to T cell or Fc receptor polypeptides, *Cell* 64:1037 (1991).
6. F. Letourneur, and R.D. Klausner, Activation of T cells by a tyrosine kinase activation domain in the cytoplasmic tail of CD3ε, *Science* 255:79 (1992).
7. C. Romeo, M. Amiot, and B. Seed, Sequence requirements for induction of cytolysis by the T cell antigen/Fc receptor ζ chain, *Cell* 68:889 (1992).
8. B.A. Irving, A.C. Chan, and A. Weiss, Functional characterization of a signal transducing motif present in the T cell receptor ζ chain, *J. Exp. Med.* 177:1093 (1993).
9. M. Reth, Antigen receptor tail clue, *Nature* 338:383 (1989).
10. A.-M.K. Wegener, F. Letourneur, A. Hoeveler, T. Brocker, F. Luton, and B. Malissen, The T cell receptor/CD3 complex is composed of at least two autonomous transduction modules, *Cell* 68:83 (1992).
11. M.R. Clark, K.S. Campbell, A. Kazlauskas, S.A. Johnson, M. Hertz, T.A. Potter, C. Pleiman, and J.C. Cambier, The B cell antigen receptor complex association of Ig-α and Ig-β with distinct cytoplasmic effectors, *Science* 258:123 (1992).

12. L.E. Samelson, A.F. Phillips, E.T. Luong, and R.D. Klausner, Association of the fyn protein-tyrosine kinase with the T-cell antigen receptor, *Proc. Natl. Acad. Sci. USA* 87:4358 (1990).
13. G.P. Sarosi, P.M. Thomas, M. Egerton, A.F. Phillips, K.W. Kim, E. Bonvini, and L.E. Samelson, Characterization of the T cell antigen receptor-p60$^{fyn}$ protein tyrosine kinase association by chemical cross-linking, *Int. Immunol.* 4:1211 (1992).
14. M.W. Appleby, J.A. Gross, M.P. Cooke, S.D. Levin, X. Qian, and R.M. Perlmutter, Defective T cell receptor signaling in mice lacking the thymic isoform of p59$^{fyn}$, *Cell* 70:751 (1992).
15. P.L. Stein, H.-M. Lee, S. Rich, and P. Soriano, pp59$^{fyn}$ mutant mice display differential signaling in thymocytes and peripheral T cells, *Cell* 70:741 (1992).
16. A. Veillette, N. Abraham, L. Caron, and D. Davidson, The lymphocyte-specific tyrosine kinase p56$^{lck}$, *Sem. Immunol.* 3:143 (1991).
17. D. Straus, and A. Weiss, Genetic evidence for the involvement of the Lck tyrosine kinase in signal transduction throught the T cell antigen receptor, *Cell* 70:585 (1992).
18. L. Karnitz, S.L. Sutor, T. Torigoe, J.C. Reed, M.P. Bell, D.J. McKean, P.J. Leibson, and R.T. Abraham, Effects of p56$^{lck}$ on the growth and cytolytic effector function of an interleukin-2-dependent cytotoxic T-cell line, *Molecular and Cellular Biology* 12:4521 (1992).
19. T.J. Molina, K. Kishihara, D.P. Siderovski, W. van Ewijk, A. Narendran, E. Timms, A. Wakeham, C.J. Paige, K.-U. Hartmann, A. Veillette, D. Davidson, and T.W. Mak, Profound block in thymocyte development in mice lacking p56$^{lck}$, *Nature* 357:161 (1992).
20. T. Taniguchi, T. Kobayashi, J. Kondo, K. Takahashi, H. Nakamura, J. Suzuki, K. Nagai, T. Yamada, S. Nakamura, and H. Yamaamura, Molecular cloning of a porcine gene syk that encodes a 72-kDa protein-tyrosine kinase showing high susceptibility to proteolysis, *J. Biol. Chem.* 266:15790 (1991).
21. A.C. Chan, M. Iwashima, C.W. Turck, and A. Weiss, ZAP-70: A 70kD protein tyrosine kinase that associates with the TCR $\zeta$ chain, *Cell* 71:649 (1992).
22. A.C. Chan, N.S.C. van Oers, A. Tran, L. Turka, C.-L. Law, J.C. Ryan, E.A. Clark, and A. Weiss, Differential expression of ZAP-70 and Syk protein tyrosine kinases, and the role of this family of protein tyrosine kinases in T cell antigen receptor signaling, *J. Immunol.* In Press, (1994).
23. A.C. Chan, B.A. Irving, J.D. Fraser, and A. Weiss, The $\zeta$-chain is associated with a tyrosine kinase and upon T cell antigen receptor stimulation associates with ZAP-70, a 70 kilodalton tyrosine phosphoprotein, *Proc. Natl. Acad. Sci. USA* 88:9166 (1991).
24. M. Iwashima, B.A. Irving, N.S.C. van Oers, A.C. Chan, and A. Weiss, Sequential interactions of the TCR with two distinct cytoplasmic tyrosine kinases, *Science* 263:1136 (1994).
25. M.A. Goldsmith, and A. Weiss, Isolation and characterization of a T-lymphocyte somatic mutant with altered signal transduction by the antigen receptor, *Proc. Natl. Acad. Sci. USA* 84:6879 (1987).
26. D.B. Straus, and A. Weiss, The CD3 chains of the T cell antigen receptor associate with the ZAP-70 tyrosine kinase and are tyrosine phosphorylated after receptor stimulation, *J. Exp. Med.* 178:1523 (1993).

27. T. Pawson, and G.D. Gish, SH2 and SH3 domains: From structure to function, *Cell* 71:359 (1992).
28. D. Rotin, B. Margolis, M. Mohammadi, R.J. Daly, G. Daum, N. Li, E.H. Fischer, W.H. Burgess, A. Ullrich, and J. Schlessinger, SH2 domains prevent tyrosine dephosphorylation of the EGF receptor, identification of Tyr992 as the high affinity binding site for SH2 domains of phospholipase Cγ, *EMBO J.* 11:559 (1992).
29. A.C. Chan, T.A. Kadlecek, M.E. Elder, A.H. Filipovich, W.-L. Kuo, M. Iwashima, T.G. Parslow, and A. Weiss, ZAP-70 deficiency in an autosomal recessive form of severe combined immunodeficiency, *Science* In Press, (1994).
30. M.E. Elder, D. Lin, J. Clever, A.C. Cahn, T.J. Hope, A. Weiss, and T.G. Parslow, Human severe combined immunodeficiency due to a defect in ZAP-70 -- A T-cell tyrosine receptor-linked tyrosine kinase, *Science* In Press, (1994).

# SIGNAL TRANSDUCTION DURING T CELL DEVELOPMENT

Dan R. Littman[1,2], Craig B. Davis, Nigel Killeen, and Hua Xu[2]

Departments of Microbiology and Immunology and of [1]Biochemistry and Biophysics and [2]Howard Hughes Medical Institute, University of California at San Francisco, San Francisco, CA 94143-0414

## INTRODUCTION

The CD4 molecule has important roles in signaling during both T cell development and peripheral helper T cell activation[1,2]. The ectodomain of CD4 binds to a membrane-proximal domain of MHC class II molecules on antigen presenting cells[3], while its cytoplasmic tail interacts with the cytoplasmic protein tyrosine kinase (PTK) p56*lck* (lck), a member of the src family of protein tyrosine kinases (reviewed in reference 4). These properties of CD4 contribute to the simultaneous interaction of the T cell antigen receptor (TCR) and CD4 with the same MHC class II molecule and to the localization of lck to the TCR complex. It has been thought that the function of CD4 (and of the class I-specific coreceptor CD8) in signal transduction is to facilitate localization of lck within the TCR complex, thus initiating a PTK signaling cascade. However, recent studies suggest that the primary function of CD4 may be to physically stabilize the interaction of the TCR with MHC, and that a direct signaling role of the CD4-associated lck molecule may not be required[4,5]. In this chapter, we discuss a non-catalytic role of the CD4-associated lck molecule in facilitating T cell activation and developmental signaling.

## ROLE OF CD4-ASSOCIATED LCK IN AN ANTIGEN-SPECIFIC T CELL HYBRIDOMA

The role of the CD4-lck interaction has been analyzed in the T cell hybridoma 171.3, which lacks expression of endogenous TCR-$\alpha$ and TCR-$\beta$ chains, as well as of CD4 and CD8. This cell line has been reconstituted with genes encoding a TCR that recognizes a peptide of hen egg lysozyme (HEL) in the context of the class II molecule $A^b$. The response

*Mechanisms of Lymphocyte Activation and Immune Regulation V*
Edited by S. Gupta *et al.*, Plenum Press, New York, 1994

of this hybridoma to antigen requires that it express CD4. In earlier studies, we showed that only CD4 molecules that can bind lck permit IL-2 secretion in response to antigen. Point mutations in either of two cysteine residues required for binding of lck resulted in complete loss (more than 1,000-fold) of the antigen-specific response[6].

## Studies on CD4-Lck Chimeric Proteins

To specifically focus our study on the role of the CD4-associated lck molecule, we have prepared fusion constructs consisting of the coding regions for the CD4 extracellular and transmembrane domains linked to lck. Since the chimeric products of such constructs were expressed well on the surface of the hybridoma and also reconstituted the antigen-specific response, we were able to initiate a structure/function analysis of the associated lck molecule. In the initial study, we were surprised to find that point mutations of lysine-227 within the ATP binding site to either alanine or arginine, which resulted in loss of phosphotransferase activity, had only a minor effect (5-fold) on the antigen-specific response[4]. When the entire kinase domain was deleted from the chimeric protein, the response to antigen actually increased in sensitivity. These findings indicated that the kinase domain of the CD4-associated lck is not required for its enzymatic activity (at least in this T cell hybridoma), but may instead regulate accessibility of other regions of the molecule to various protein-protein interactions.

During the past several years, “adaptor” molecules, consisting of SH2 and SH3 domains (such as Sem5/GRB2, Nck, and Crk) have been shown to mediate interactions between various cytoplasmic signaling molecules[7]. Since the kinase domain of lck is preceded by the myristylated N-terminal segment that binds to CD4 as well as by SH3 and SH2 domains, it is attractive to postulate that these regions may mediate a similar adaptor function for CD4-associated lck. It has been proposed that phosphorylation of the C-terminal tyrosine-505 of lck triggers its intramolecular interaction with the SH2 domain, thus essentially keeping the molecule in an "off" configuration[8]. The CD45 protein tyrosine phosphatase (PTPase) is thought to act on this residue to activate lck. Our results support this model and suggest that the interaction of phosphorylated tyrosine-505 with SH2 not only reduces kinase activity, but, more importantly, hinders access of other phophotyrosine-containing proteins to the SH2 (plus SH3) domain of lck.

## Role of the Lck SH2 and SH3 Domains

To determine whether the SH2 and/or SH3 domains of CD4-associated lck have an important function in the absence of kinase activity, we introduced several point mutations or deletions in the chimeric constructs. A point mutation within the chimeric protein's SH2 domain (arginine 154 changed to lysine, destroying P-tyr binding activity) decreased

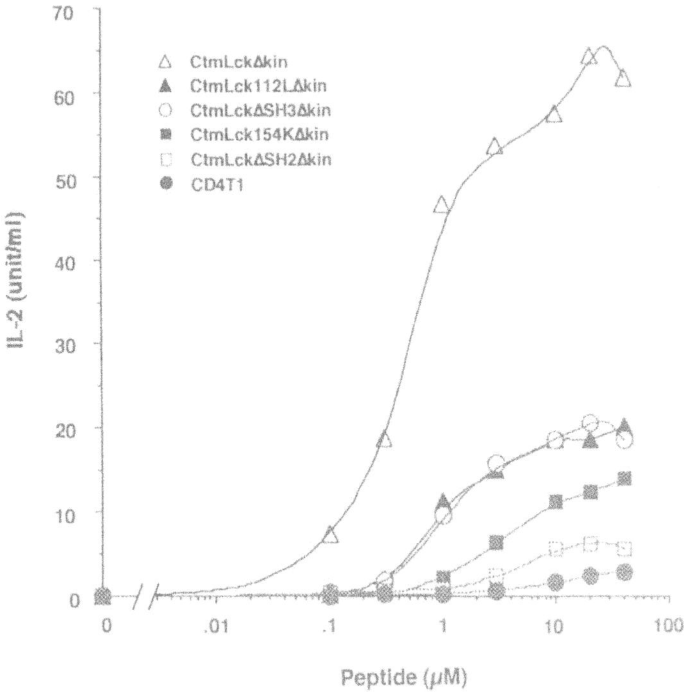


**Figure 1.** Role of the lck SH2 and SH3 domains in signaling by the CD4/lck chimeric proteins. The various chimeric constructs with a deletion of the entire kinase region (Δkin) and point mutations or deletions (Δ) of the SH2 or SH3 domains were introduced by retroviral transduction into 171.3 cells. Cell populations sorted to express similar levels of surface CD4 were stimulated with HEL peptide in the presence of antigen presenting cells, as described in reference 6, and secreted IL-2 levels were measured using the CTLL assay[6]. The CD4T1 control contains a tailless form of CD4 that cannot bind to lck.

response to antigen about 5-fold relative to the wild-type chimera[4]. However, when this mutation was coupled to the point mutation in the kinase domain, there was virtual elimination of the activity of the CD4 chimeric protein[4]. These results indicate that the SH2 domain has a critical role in lck function, but that the kinase activity of lck can still contribute to T cell activation in the absence of SH2 function. This is consistent with the finding that when the kinase domain is present but is defective in enzymatic activity there is a slightly diminished response due to the loss of the kinase activity.

The increased response upon complete removal of the kinase domain is consistent with the notion that this region normally blocks access to domains within the rest of the protein. To test this hypothesis in greater detail, we combined mutations or deletions of the SH2 and SH3 domains with the deletion of the kinase domain. The mutant constructs were introduced into the 171.3 cells by retroviral infection, and cell populations were sorted for expression of equivalent surface levels of CD4. The R154K mutation resulted in almost complete loss of antigen response, as did the deletion of the SH2 domain (Figure 1). A proline to leucine

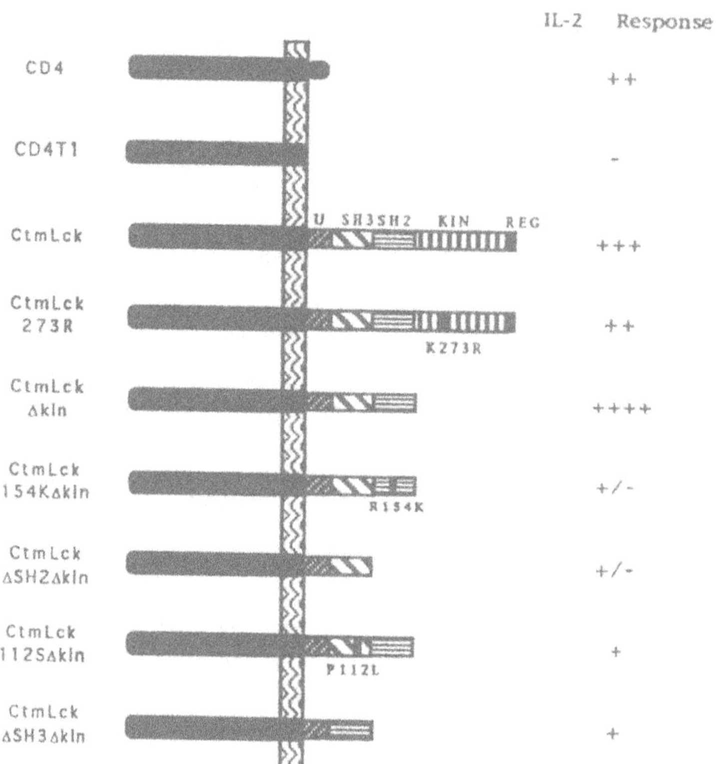


**Figure 2.** Relative coreceptor function of mutant CD4 and CD4/lck chimeras.

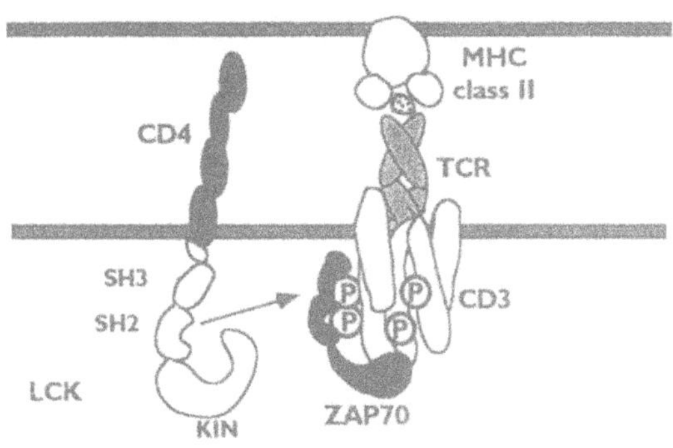


**Figure 3.** Model for the role of CD4-associated lck and its SH2/SH3 domains in TCR-mediated signal transduction. Following binding of TCR to MHC/peptide, tyrosine phosphorylation of chains within the CD3 complex results in recruitment of ZAP70 which, in turn, increases overall tyrosine phosphorylation of associated proteins. The SH2 domain of lck may then directly interact with these proteins, thus recruiting CD4.

mutation at position 112 of the SH3 domain, which is equivalent to one of the *sem5* alleles in *C. elegans* and results in loss of binding to proline-containing peptides, also resulted in a significant decrease in the function of the chimeric kinase-deleted chimera. A molecule with a deletion of the entire SH3 domain had a similar phenotype (Figure 1). A summary of the various mutations in the chimeric CD4/lck molecules and of the relative antigen-specific responses is shown in Figure 2.

It should be noted that the mutations in SH3 had less severe effects on antigen response than the mutations in SH2. Recent studies have suggested that, in src family kinases, SH3 domains can contribute to the function of the adjacent SH2 domains[9]. It is therefore possible that the effects that we observe upon mutation of the SH3 domain of lck reflects decreased function of the SH2 domain. Alternatively, the SH3 domain may itself have a specific function, such as linking CD4 to the cytoskeleton, and may therefore contribute to signal transduction.

These results further support the notion that a major function of CD4-associated lck is to regulate the location of CD4 within the plasma membrane, as shown in the model in Figure 3. This model proposes that CD4 concentration is generally too low to permit sufficient contact with MHC class II molecules that are simultaneously recognized by the TCR. The Lck molecule may direct CD4 to the relevant class II molecule through its SH2 interaction with newly-phosphorylated proteins associated with the TCR complex. Such interactions would enable CD4 to stabilize the TCR-MHC interaction, permitting signaling to proceed.

## FUNCTION OF CD4 IN T CELL DEVELOPMENT

### Role of CD4-Lck Interaction in Positive Selection

The avidity model for CD4 function predicts that the need for a CD4-lck interaction can be circumvented by increasing the surface concentration of CD4. This prediction is borne out in studies on T cell development, using CD4-knockout mice reconstituted with wild-type or mutant CD4 transgenes[5]. In the absence of CD4 gene expression, the level of mature peripheral helper T cells is reduced 5-7 fold, due to impaired positive selection in the thymus. Normal level expression of transgene-derived wild-type CD4, regulated by CD3δ or Lck promoters, restored the proportions of T cell subsets to wild-type levels. Surprisingly, expression of a tailless CD4 also restored helper lineage development, but this required high level transgene expression. Analysis of multiple transgenic lines expressing tailless CD4 showed a direct correlation between level of transgene expression and extent of lineage rescue[5]. Effective replacement of wild-type CD4 with a tailless variant argues against an absolute requirement for CD4-lck association in T cell development and is consistent with an avidity-enhancing function for CD4. High CD4 levels on developing thymocytes thus appear to increase the probability of CD4 interactions with relevant MHC class II molecules, resulting in TCR-mediated signaling and positive selection.

## Lineage Commitment in the Presence and Absence of CD4

Although CD4 function is required for efficient positive selection of developing thymocytes, a sizeable proportion of mature T cells in CD4-null mice have helper activity. These cells, which express neither CD4 nor CD8, have normal MHC class II-restricted helper responses to Leishmania[10]. Therefore, the requirement for CD4 in the development of helper T cells is not absolute. It is not yet clear if helper cells that can differentiate in the absence of CD4 have special properties; for example, these cells may constitute a separate thymocyte lineage, they may have higher affinity T cell receptors, or they may express molecules other than CD4 that can compensate for the function of this glycoprotein. However, a reporter transgene whose expression is directed by the CD4 transcriptional regulatory sequences was expressed in these $CD4^-CD8^-$ helper cells in CD4-null mice, strongly suggesting that these cells are committed to the CD4 lineage[11].

The mechanism involved in the differentiation of double positive ($CD4^+CD8^+$) thymocytes into single positive mature cells has been the subject of a lively debate during the past year[12]. Some have proposed that instructional signals, potentially involving the CD4 and CD8 coreceptors, result in CD4 or CD8 lineage commitment[13]. The development of CD4-lineage helper cells in the CD4-null mice[10] and the finding that tailless CD4 rescues positive selection of this lineage[5] argue strongly against instructional signals dependent on co-recognition of MHC class II by CD4 and TCR. These and other results favor a model of stochastic down-regulation of CD4 or CD8 and subsequent selection of cells bearing the appropriate TCR-coreceptor combination. Selection of $CD4^+$ cells would hence require co-engagement of CD4 and TCR on the same MHC class II molecule, and class II-reactive cells that commit to the CD8 lineage would die.

Further evidence for a stochastic mechanism was obtained by analyzing development of class II-restricted thymocytes in mice that express a CD4 transgene in all T cell lineages[14].

The instructional model predicts that, in these mice, all cells bearing a class II-specific TCR should shut off CD8 and commit to the CD4 lineage. In contrast, if single positive cells arise stochastically independent of TCR specificity, $CD8^+$ cells with class II-specific TCR's should be detected due to their rescue in the presence of transgenic CD4. Using mice whose T cells were heavily skewed for class II specificity, we found that expression of a CD4 transgene in all developing thymocytes resulted in the rescue of CD8 lineage cells bearing the class II-specific TCRs[14]. Moreover, these rescued cells exhibited the properties of cytotoxic T lymphocytes and not of T helper cells. These results indicate that lineage commitment can occur independently of the MHC specificity of a particular TCR, but that completion of positive selection generally requires expression of the appropriate coreceptor, which participates in setting the requisite avidity between TCR and MHC.

### Contribution of CD4 to TCR-MHC Avidity in Development

Whereas expression of normal levels of a CD4 transgene rescued development of the CD8 lineage very efficiently in mice that have a pigeon cytochrome c (PCC)/MHC class II-specific TCR, overexpression of CD4 resulted in a phenotype strongly suggestive of intrathymic deletion: approximately a ten-fold drop in the number of thymocytes, a pronounced reduction in the immature DP subpopulation, and the appearance in the periphery of mature T cells that expressed the PCC-specific TCR and the CD4 transgene, but not endogenous CD4 or CD8[15]. This result suggests that increasing the avidity of thymocyte:class II interactions up to a certain point increases the efficiency of CD8 lineage rescue, but increasing avidity beyond that point adversely affects generation of both CD4 and CD8 lineages. These results are consistent with models of thymocyte development in which a threshold TCR avidity for self MHC is required for positive selection and a higher avidity threshold, when surpassed, results in negative selection[16].

## CONCLUSIONS

Several different approaches now provide strong evidence that the CD4 coreceptor functions as an avidity regulator during T cell development and activation. First, the associated lck molecule appears to act as a molecular bridge between the TCR complex and CD4, most likely via the SH2/SH3 domains that probably interact with proteins within the TCR complex. Second, high level expression of CD4, in the absence of lck binding, can provide the necessary signals for positive selection. Third, positive selection of helper T cells with TCRs presumed to have high affinity for MHC can occur even in the absence of CD4. Finally, high level expression of CD4 results in negative selection of cells bearing certain TCRs that would normally mediate positive selection.

These results raise the question of why the CD4 and CD8 coreceptors are necessary. Since thymocytes with TCRs of sufficiently high affinity for MHC can mature in the absence of coreceptor, expression of coreceptor may only influence which of the many precursor thymocytes will undergo positive or negative selection. An answer to this paradox may lie in the correlation of the T cell's MHC specificity with its functional program. Thus, MHC class I-reactive cells are generally cytotoxic, while class II-specific cells are of the helper phenotype. Down-regulation of either CD4 or CD8 may be mechanistically coupled to the commitment to each of these phenotypes, thus ensuring that the great majority of cells that complete intrathymic development also have a match between MHC specificity and function. Our observation that shutoff of CD4 gene expression in CD8 lineage cells is correlated with commitment to a program of CTL differentiation supports this model[14]. Our results thus indicate that CD4 acts primarily to confirm thymocyte lineage commitment by stabilizing TCR:MHC interactions, rather than by generating signals that force commitment to the CD4 lineage.

To determine if a common signal regulates shutoff of CD4 and activation of CTL-specific genes, we have begun to dissect the mechanism of transcriptional regulation of the CD4 gene. A sequence 13 kb upstream of the transcription initiation site was found to have T cell-specific enhancer activity[17]. This sequence directed expression of reporter genes independent of T cell subset specificity. We have now identified a second regulatory sequence, within the first intron of the CD4 gene, that specifically silences gene expression in CD8 lineage T cells[18]. This silencing element functions independently of position and orientation, and is also active on heterologous regulatory elements. Studies aimed at identifying trans-acting factors that bind to the silencer sequence and shut off CD4 expression in CD8 single positive thymocytes may provide the tools that will unravel the mechanism of the stochastic process that governs T cell lineage commitment.

## Acknowledgments

This work was supported by grants from the National Institutes of Health and by the Howard Hughes Medical Institute. The transgenic facility at UCSF was supported in part by a grant from the Markey Foundation.

## REFERENCES

1. C.A. Janeway, Jr., The role of CD4 in T-cell activation: accessory molecule or co-receptor? *Immunol. Today* 10:234 (1989).
2. M. Julius, C.R. Maroun, and L. Haughn, Distinct roles for CD4 and CD8 as co-receptors in antigen receptor signalling, *Immunol. Today* 14:177 (1993).
3. R. König, L.-Y. Huang, and R.N. Germain, MHC class II interaction with CD4 mediated by a region analogous to the MHC class I binding site for CD8, *Nature, Lond.* 356:796 (1992).

4. H. Xu and D.R. Littman, A kinase-independent function of Lck in potentiating antigen-specific T cell activation, *Cell*, 74:633 (1993).

5. N. Killeen and D.R. Littman, Helper T cell development in the absence of CD4-p56$^{lck}$ association, *Nature, Lond.*, 364:729 (1993).

6. N. Glaichenhaus, N. Shastri, D.R. Littman, and J.M. Turner, Requirement for association of p56lck with CD4 in antigen-specific signal transduction in T cells, *Cell* 64:511 (1991).

7. F. McCormick, How receptors turn Ras on, *Nature, Lond.* 363:15 (1993).

8. M. Sieh, J.B. Bolen, and A. Weiss, CD45 specifically modulates binding of Lck to a phosphopeptide encompassing the negative regulatory tyrosine of Lck, *EMBO J.* 12:315 (1993).

9. S.M. Murphy, M. Bergman, and D.O. Morgan, Suppression of c-src activity by C-terminal src kinase involves the c-src SH2 and SH3 domains: analysis with Saccaromyces cerevisiae, *Mol. Cell. Biol.* 13:5290 (1993)

10. R.M. Locksley, S.L. Reiner, F. Hatam, D.R. Littman,and N. Killeen, Helper T cells without CD4: control of leishmaniasis in CD4-deficient mice, *Science* 261:1448 (1993).

11. N. Killeen, S. Sawada, and D.R. Littman, unpublished results.

12. C.B.Davis and D.R. Littman, Thymocyte lineage commitment: is it instructed or stochastic? *Curr. Opin. Immunol.* 6:266 (1994).

13. H. von Boehmer and P. Kisielow, Lymphocyte lineage commitment: instruction versus selection, *Cell* 73:207 (1993).

14. C.B. Davis, N. Killeen, M.E.C. Crooks, D. Raulet, and D.R. Littman, Evidence for a stochastic mechanism in the differentiation of mature subsets of T lymphocytes, *Cell* 73:237 (1993).

15. C.B. Davis, N. Killeen, S. Sawada, and D.R. Littman, unpublished results.

16. P. Travers, T cells. Immunological agnosia, *Nature, Lond.* 363:117 (1993).

17. S. Sawada and D.R. Littman, Identification and characterization of a T cell-specific enhancer adjacent to the murine CD4 gene, *Molec. Cell. Biol.* 11:5506 (1991).

18. S. Sawada, J.D. Scarborough, N. Killeen, and D.R. Littman, A lineage-specific transcriptional silencer regulates CD4 gene expression during T lymphocyte development, *Cell*, in press (1994).

# THE REGULATION AND FUNCTION OF p21RAS IN SIGNAL TRANSDUCTION BY THE T CELL ANTIGEN RECEPTOR

Doreen Cantrell, M. Izquierdo Pastor, and M. Woodrow

Lymphocyte Activation Laboratory
Imperial Cancer Research Fund
44 Lincolns Inn Field
London WC2A 3PX

## INTRODUCTION

T lymphocyte activation is initiated by triggering of the T cell antigen receptor (TCR). The TCR is a multichain complex comprising a disulfide-linked heterodimer of the idiotypic $\alpha\beta$ chains noncovalently associated with a signal transduction complex composed of the invariant CD3 $\gamma$, $\delta$, $\epsilon$ chains and the $\zeta$ (16 kDa) and $\eta$ (22 kDa) subunits.[1-3] The cytoplasmic domains of the CD3 and $\zeta$ subunits are crucial for TCR coupling to intracellular PTKs which is absolutely required for all subsequent T cell responses.[4-8] PTKs are known to couple the TCR to an inositol lipid specific phospholipase C, PLC$\gamma$1[9] thus enabling the TCR to regulate the hydrolysis of membrane phosphoinositides (PtdIns), in particular phosphatidylinositol (4,5)-biphosphate (PtdIns(4,5$P_2$), liberating inositol (1,4,5)-triphosphate (Ins(1,4,5)$P_3$) and 1,2-diacylglycerol (DAG). Ins(1,4,5)$P_3$ releases $Ca^{2+}$ from the endoplasmic reticulum which results in an initial rapid rise in the concentration of intracellular $Ca^{2+}$; DAG is known to activate the serine/threonin kinase, protein kinase C (PKC).[10-12]

A second PTK controlled signalling pathway to originate from the TCR involves the guanine nucleotide binding proteins p21ras.[13,14] The ras proteins bind GTP and have an intrinsic GTPase activity that catalyses the hydrolysis of GTP to GDP. A consensus from ras studies is that the GTP bound form of the protein is the biologically active form.[15] In T cells stimulated via the TCR the ras proteins rapidly accumulate in the active GTP bound state. Activation of PKC with phorbol esters or diacylglycerols also induces the accumulation of ras-GTP complexes. However, the functional relevance of PKC for receptor coupling to ras is not clear. In particular, PKC does not couple the TCR to p21ras; instead, a PKC independent pathway involving PTKs plays a critical role in TCR regulation of ras.[16] The realisation that [$Ca^{2+}$] and PKC do not couple the TCR to ras is significant because it indicates that at least two signalling pathways can originate from the TCR: one mediated by PLC$\gamma$1 and the other by ras.

*Mechanisms of Lymphocyte Activation and Immune Regulation V*
Edited by S. Gupta *et al.*, Plenum Press, New York, 1994

## THE FUNCTION OF p21RAS IN T CELL ACTIVATION

It has been known for several years that calcium and PKC are important intracellular signals in T cell activation[17,18] but more recently, it was recognised that p21ras also has a crucial function in T cells. This conclusion was derived from transient transfection protocols that examined the consequences of expressing mutated constitutively active or dominant inhibitory ras mutants on T cell activation. The ras oncogene p21-v-Ha-ras is mutated at codon 12 (Ser > Val) and 59 (Ala > Thr). These mutations render the ras protein insensitive to negative regulation by ras- GAPs such that it accumulates in cells in an "active" GTP bound state. A second ras mutant, N17 ras, is able to block cellular ras function by competing for ras guanine nucleotide exchange proteins, thereby preventing endogenous ras activation by stopping exchange of GTP onto cellular ras. p21-v-Ha ras, when expressed in T cells, can synergise with a calcium signal to activate the IL-2 gene.[19,20] Contrastingly, expression of N17 ras inhibits TCR induction of the IL-2 gene which indicates that ras function is essential for TCR signal transduction.[20]

Ras regulation of the IL-2 gene is mediated by ras control of the transcriptional factor NFAT.[21] Current knowledge indicates that NFAT is a heterodimer comprising an AP-1 component that is found only in the nucleus of activated T cells and a constitutively expressed cytoplasmic component that translocates to the nucleus during T cell activation.[22-25] The calcium signalling requirement for NFAT and hence IL-2 expression reflects that a calcium sensitive phosphotase calcineurin regulates translocation of the NFAT cytoplasmic component to the nucleus.[22,26-28] The expression by transfection of the truncated "activated" calcineurin generates a powerful signal that substitutes for calcium and synergises with p21ras to activate NFAT.[29] The role of ras in NFAT induction is probably explained because p21ras regulates a kinase cascade that stimulates the expression and function of AP-1.[30-32]

The transmission of signals from p21ras to the nucleus is proposed to involve the regulation of the activity of Map kinases or ERKs (mitogen activated protein kinases or extracellular signal regulated kinases).[32-36] T lymphocytes express at least two MAP kinases, ERK1 and ERK2, that are stimulated in response to TCR triggering.[37,38] Two intracellular pathways for ERK2 regulation co-exist in T cells: one mediated by ras and the other by PKC.[39] The TCR stimulates both p21ras and PKC but data obtained from transfection studies show that expression of the inhibitory ras mutant, N17 ras which prevents endogenous ras activation suppresses TCR induction of ERK2.[39] Accordingly it appears that p21 ras and not PKC couples the TCR to the regulation of MAP kinases. Map kinases translocate to the nucleus when activated and their known substrates include transcriptional factors such as ELK-1, c-jun and c-myc.[40-42] Thus one mechanism whereby p21ras could couple the TCR to the nucleus is via stimulation of the ERK kinase cascade.

## THE MECHANISMS THAT COUPLE THE TCR TO p21RAS

The level of active p21ras-GTP complexes is determined by a balance of the rate of hydrolysis of bound GTP and the rate of exchange of bound GDP for cytosolic GTP.[43] The GTPase activity of p21ras is controlled by GTPase activating proteins of which two mammalian proteins are known: p120-GAP and neurofibromin.[44,45] Proteins that regulate guanine nucleotide exchange on ras have also been characterised and in mammalian cells they include the homologues of the yeast CDC25 protein and the *Drosophilia* "son of sevenless" gene product, SOS.[46,47] It has also been described that the protooncogene Vav is an exchange protein for p21ras in T cells even though from its structure Vav is a more likely candidate for a *rho/rac* exchange protein.[48,49]

The alternative mechanisms that could couple the TCR to p21ras are TCR stimulation of guanine nucleotide exchange or TCR inhibition of ras-GAP proteins. Experiments to explore ras regulatory mechanisms have shown that TCR triggering inhibits the activity of ras-GAPs.[13,50] It is not yet known which GAP protein, i.e. p120-GAP or neurofibromin, is the target for TCR regulation nor is the mechanism that allows the TCR to inhibit ras-GAP proteins understood, although clearly PTKs are involved in some way.[16] It has also been described that PTKs may allow the TCR to regulate a putative ras exchange protein, Vav. Vav is a tyrosine phosphorylated in response to TCR triggering and this tyrosine phosphorylation appears to stimulate the guanine nucleotide exchange capacity of the protein.[48] Any analysis of ras regulatory mechanisms in T cells must provide a model to explain PKC regulation of p21ras since this appears to be a lymphocyte specific. Notably, PKC regulates accumulation of ras-GTP complexes in T cells and B cells but not in fibroblast or epithelial cells.[14] Vav is a potential ras regulatory protein that is expressed only in haematopoietic cells. However, Vav cannot explain the apparent lymphoid specificity of the PKC-p21ras link because Vav is expressed in all haematopoietic cells[49] and mast cells or myeloid cells that do express Vav do not express a PKC mediated route for p21ras regulation.[15,51]

The relative contribution of ras-Gaps or exchange proteins to ras regulation *in vivo* is not easy to determine. The most direct strategy is to examine the kinetics of guanine nucleotide metabolism by p21ras. In T cells such studies have been carried out in permeabilised cells. These experiments indicate that in T cells, guanine nucleotide exchange onto p21ras is rapid and unchanged by cell activation. Instead the rate of GTP hydrolysis by ras decreases upon TCR triggering allowing ras to accumulate in the GTP bound state.[50] One interpretation of these data is that TCR coupling to p21ras is mediated by ras-GAP proteins and not by guanine nucleotide exchange proteins.

The rapid kinetics of ras guanine nucleotide exchange and its apparent uncoupling from receptor activation is a fundamental difference between T cells and other cell types such as fibroblasts. In fibroblasts, growth factors such as epidermal growth factor (EGF), platelet derived growth factor (PDGF), and insulin regulate p21ras by stimulating ras guanine nucleotide exchange.[52-54] The ras exchange protein in fibroblasts is the mammalian homologue of the Drosophila SOS gene.[55,56] SOS is recruited to the cell membrane by the adapter protein Grb-2/Sem5.[54] Grb-2 is composed of one SH2 domain and two SH3 domains.[57,58] The SH3 domains of Grb2 bind to the carboxy terminal proline rich domain of SOS[55] and the SH2 domain binds to tyrosine phosphorylation sites in a number of molecules including the Y1068 autophosphorylation site of the EGF receptor, Shc, IRS-1, and the cytoplasmic tyrosine phosphatase Syp which in turn binds to Y1009 in the PDGF receptor.[54,59-63] The Grb-2 SH2 domain can thus recruit SOS to the cell membrane by interacting with these tyrosine phosphorylated molecules. T cells express Grp-2 and SOS, and given the conservation of this signalling pathway it seems reasonable to assume that Grb-2 and SOS play some role in p21ras regulation in T cells. Interestingly, T cells express a high ratio of SOS to p21ras compared to fibroblasts, which could explain why basal rates of guanine nucleotide exchange on p21ras are relatively high in these cells (unpublished observations). It has also been observed that the SOS/Grb-2 complex can associate with 2 proteins that are tyrosine phosphorylated in TCR activated T cells: Shc and a membrane bound tyrosine phosphoprotein of 36kDa molecular weight.[64,65] The predominant complex that exists in TCR stimulated cells is one comprising SOS/ Grb-2/ p36. Nevertheless, there is as yet no direct evidence that this complex functions to regulate ras guanine nucleotide exchange in T cells. Indeed, it is very difficult to assess whether changes in the *in vitro* activity/cell localisation of ras regulatory proteins like the Ras-GAPs, Vav or SOS have a major regulatory consequence for p21ras *in vivo*. However, given the importance of p21ras in T cell activation, the mechanisms that couple the TCR to p21ras will be essential for TCR signal transduction.

## REFERENCES


1. M. Reth, Antigen receptor tail clue, *Nature* 338:383 (1989).
2. A.M. Weissman, J.S. Bonifacino, R.D. Klausner, L.E. Samelson, and J.J. O'Shea, T cell antigen receptor: structure, assembly and function. *Year Immunol.* 4:74 (1989).
3. A. Weiss, T cell antigen receptor signal transduction: a tale of tails and cytoplasmic protein-tyrosine kinases, *Cell* 73:209 (1993).
4. C.H. June, M.C. Fletcher, J.A. Ledbetter, G.L. Schieven, J.N. Siegel, A.F. Phillips, and L.E. Samelson, Inhibition of tyrosine phosphorylation prevents T cell receptor-mediated signal transduction, *Proc. Natl. Acad. Sci. U.S.A.* 87:7722 (1990).
5. R.D. Klauser and L.E. Samelson, T cell antigen receptor activation pathways: the tyrosine kinase connection, *Cell* 64:875 (1991).
6. C.E. Rudd, CD4, CD8 and the TCR-CD3 complex: a novel class of protein-tyrosine kinase receptors, *Immunol. Today* 11:400 (1990).
7. L.E. Samelson, A.F. Phillips, E.T. Luong, and R.D. Klausner, Association of the fyn protein-tyrosine kinase with the T-cell antigen receptor, *Proc. Natl. Acad. Sci. U.S.A.* 87:4358 (1990).
8. A.C. Chan, M. Iwashima, C.W. Turck, and A. Weiss, Zap 70: A 70kd protein tyrosine that associates with the TCR zeta chain, *Cell* 71:649 (1992).
9. A. Weiss, G. Koretzky, R.C. Schatzmann, and T. Kadlecec, Functional activation of the T cell antigen receptor induces tyrosine phosphorylation of phospholipase C gamma 1, *Proc. Natl. Acad. Sci. U.S.A.* 88:5484 (1991).
10. M.J. Berridge and R.F. Irvine, Inositol phosphates and cell signalling, *Nature* 341: 197 (1989).
11. Y. Nishizuka, The molecular heterogeneity of protein kinase C and its implication for cellular regulation, *Nature* 334:661 (1988).
12. N. Berry and Y. Nishizuka, Protein kinase C and T cell activation, *Eur. J. Biochem.* 89:205 (1989).
13. J. Downward, J.D. Graves, P.H. Warne, S. Rayter, and D.A. Cantrell, Stimulation of p21ras upon T-cell activation, *Nature* 346:719 (1990).
14. J. Downward, J.D. Graves, and D.A. Cantrell, The regulation and function of p21ras in T lymphocytes, *Immunol. Today* 13:92 (1992).
15. T. Satoh, M. Nafakuku, and Y. Kaziro, Function of Ras as a molecular switch in signal transduction, *J. Biol. Chem.* 267:24149 (1992).
16. M. Izquierdo, J. Downward, J.D. Graves, and D.A. Cantrell, Role of protein kinase C in T-cell antigen receptor regulation of p21ras: evidence that two p21ras regulatory pathways coexist in T cells, *Mol. Cell. Biol.* 12:3305 (1992).
17. D.A. Cantrell, M.K. Collins, and M.J. Crumpton, Autocrine regulation of T-lymphocyte proliferation: differential induction of IL-2 and IL-2 receptor, *Immunology* 65:343 (1988).
18. A. Weiss and J.B. Imboden, Cell surface molecules and early events involved in human T lymphocyte activation, *Adv. Immunol.* 41:1 (1987).
19. C.T. Baldari, G. Macchia, and J.L. Telford, Interleukin-2 promoter activation in T cells expressing activated Ha-ras, *J. Biol. Chem.* 267:4289 (1992).
20. S. Rayter, M. Woodrow, S.C. Lucas, D. Cantrell, and J. Downward, p21ras mediates control of IL2 gene promoter function in T cell activation, *EMBO J.* 11:4549 (1992).
21. M. Woodrow, S. Rayter, J. Downward, and D.A. Cantrell, p21ras function is important for T cell antigen receptor and protein kinase C regulation of nuclear factor of activated cells, *J. Immunol.* 150:1 (1993).

22. W.M. Flanagan, B. Corthésy, R.J. Bram, and G.R. Crabtree, Nuclear association of a T-cell transcription factor blocked by FK-506 and cyclosporin A, *Nature* 352:803 (1991).
23. J. Jain, P.G. McCaffrey, V.E. Valge-Archer, and A. Rao, Nuclear factor of activated T cells contains Fos and Jun, *Nature* 356:801 (1992).
24. J.P. Northrop, K.S. Ullman, and G.R. Crabtree, Characterisation of the nuclear and cytoplasmic components of the lymphoid-specific nuclear factor of activated T cells (NFAT) complex, *J. Biol. Chem.* 268:22917 (1993).
25. L.H. Boise, B. Petryniak, X. Mao, C.H. June, C.Y. Wang, T. Lindsten, R. Bravo, K. Kovary, J.M. Leiden, and C.B. Thompson, The NFAT-1 DNA binding complex in activated T cells contains Fra-1 and Jun-B, *Mol. Cell. Biol.* 113:1911 (1993).
26. S.L. Schreiber and G.R. Crabtree, The mechanism of action of cyclosporin A and FK506, *Immunol. Today* 13:136 (1992).
27. N.A. Clipstone and G.R. Crabtree, Identification of calcineurin as a key signalling enzyme in T-lymphocyte activation, *Nature* 357:695 (1992).
28. S.J. O'Keefe, J. Tamura, R.L. Kincaid, M.J. Tocci, and E.A. O'Neill, FK-506 and Cs-A-sensitive activation of the interleukin-2 promoter by calcineurin, *Nature* 357:692 (1992).
29. M. Woodrow, N. Clipstone, and D.A. Cantrell, p21ras and calcineurin synergise to regulate NFAT, *J. Exp. Med.* 178: 1517 (1993).
30. B. Binetruy, T. Smeal, and M. Karin, Ha-ras augments c-jun activation and stimulates phosphorylation of its activation domains, *Nature* 251:122 (1991).
31. T. Smeal, B. Binetruy, D.A. Mercola, M. Birrer, and M. Karin, Oncogenic and transcriptional cooperation with Ha-Ras requires phosphorylation of c-Jun on serines 63 and 73, *Nature* 354:494 (1991).
32. T. Hunter and M. Karin, The regulation of transcription by phosphorylation, *Cell* 70: 375 (1992).
33. S.J. Leevers and C.J. Marshall, MAP kinase regulation--the oncogene connection, *Trends Cell. Biol.* 2:283 (1992).
34. D.J. Robbins, M. Cheng, E. Zhen, C.A. Vanderbilt, L.A. Feig, and M.H. Cobb, Evidence for a Ras-dependent extracellular signal-regulated protein kinase (ERK) cascade, *Proc. Natl. Acad. Sci. U.S.A.* 89:6924 (1992).
35. S.M. Thomas, M. DeMarco, G. D'Arcangelo, S. Halegoua, and J.S. Brugge, Ras is essential for nerve growth factor and phorbol ester-induced tyrosine phosphorylation of MAP kinases, *Cell* 68:1031 (1992).
36. S.L. Pelech and J.S. Sanghera, MAP kinases: charting the regulatory pathways, *Science* 257:1355 (1992).
37. A.E. Nel, C. Hanekon, and L. Hultin, Protein kinase C plays a role in the induction of tyrosine phosphorylation of lymphoid microtubule-associated protein-2 kinase. Evidence for a CD3-associated cascade that includes p56lck and that is defective in HPB-ALL, *J. Immunol.* 147:1933 (1991).
38. C.E. Whitehurst, T.G. Boulton, M.H. Cobb, and T.G. Geppert, Extracellular signal-related kinases in T cells. Anti-CD3 and 4-beta-phorbol 12-myristate 13-acetate induced phosphorylation and activation, *J. Immunol.* 148:3230 (1992).
39. M. Izquierdo, S.J. Leevers, C.J. Marshall, and D.A. Cantrell, p21ras couples the T cell antigen receptor to extracellular signal-related kinase 2 in T lymphocytes, *J. Exp. Med.* 178:1199 (1993).
40. B.J. Pulverer, J.M. Kyriakis, J. Avruch, E. Nikolakaki, and J.R. Woodgett, Phosphorylation of c-jun mediated by MAP kinases, *Nature* 353:670 (1991).
41. R. Marais, J. Wynne, and R. Treisman, The SRF accessory protein ELK-1 contains

a growth factor-regulated transcriptional activation domain, *Cell* 73:381 (1993).

42. A. Seth, F.A. Gonzalez, S. Gupta, D.L. Raden, and R.J. Davis, Signal transduction within the nucleus by mitogen-activated protein kinase, *J. Biol. Chem.* 34:24796 (1992).
43. J. Downward, Regulation of p21ras by GAPs and guanine nucleotide exchange proteins in normal and oncogenic cells, *Curr. Opin. Genet. Dev.* 2:13 (1992).
44. F. McCormick, ras GTPase activating protein: signal transmitter and signal terminator, *Cell* 56:5 (1989).
45. R. Ballester, D. Marchuk, M. Boguski, A. Saulino, R. Letcher, M. Wigler, and F. Collins, The NF1 locus encodes a protein functionally related to mammalian GAP and yeast IRA proteins, *Cell* 63: 851 (1990).
46. C. Shou, C.L. Farnsworth, B.G. Neel, and L.A. Feig, Molecular cloning of cDNAs encoding a guanine-nucleotide-releasing factor for Ras p21, *Nature* 358:351 (1992).
47. D. Bowtell, P. Fu, M. Simon, and P. Senior, Identification of murine homologues of the *Drosophilia* Son of Sevenless gene: potential activators of ras, *Proc. Natl. Acad. Sci. U.S.A.* 89:6511 (1992).
48. E. Gulbins, K.M. Coggeshall, G. Baier, S. Katzav, P. Burn, and A. Altman, Tyrosine kinase stimulated guanine nucleotide exchange of Vav in T cell activation, *Science* 260:822 (1993).
49. J.M. Adams, H. Houston, J. Allen, T. Lints, and R. Harvey, The hematopoietically expressed vav proto-oncogene shares homology with the dbl GDP-GTP exchange factor, the bcr gene and a yeast gene (CDC24) involved in cytoskeletal organization, *Oncogene* 7:611 (1992).
50. J.D. Graves, J. Downward, S. Rayter, P. Warne, A.L. Tutt, M. Glennie, and D.A. Cantrell, CD2 antigen mediated activation of the guanine nucleotide binding proteins p21ras in human T lymphocytes, *J. Immunol.* 146:3709 (1991).
51. M. Izquierdo, J. Downward, W.J. Leonard, H. Otani, and D.A. Cantrell, IL-2 activation of p21ras in murine myeloid cells transfected with human IL-2 receptor beta chain, *Eur. J. Immunol.* 22:817 (1992).
52. R.H. Medema, A.M.M. Vries-Smits, G.C.M. van der Zon, J.A. Maassen, and J.L. Bos, Ras activation by insulin and epidermal growth factor through enhanced exchange of guanine nucleotides on p21ras, *Mol. Cell. Biol.* 13:155 (1993).
53. L. Buday and J. Downward, Epidermal growth factor regulates the exchange rate of guanine nucleotides on p21ras in fibroblasts, *Mol. Cell. Biol.* 13:1903 (1993).
54. L. Buday and J. Downward, Epidermal grwoth factor regulates p21ras through the formation of a complex of receptor, Grb2 adapter protein and Sos nucleotide exchange factor, *Cell* 73:611 (1993).
55. S.E. Egan, B.W. Giddings, M.W. Brooks, L. Buday, A.M. Sizeland, and R.E. Weinberg, Association of SOS Ras exchange protein with GRB2 is implicated in tyrosine kinase signal transduction and transformation, *Nature* 363:45 (1993).
56. F. McCormick, How receptors turn Ras on, *Nature* 363:15 (1993).
57. K. Matuoka, M. Shibata, A. Yamakawa, and T. Takenawa, Cloning of ASH, a ubiquitous protein composed of one src homology region (SH)2 and two SH3 domains from human and rat cDNA libraries, *Proc. Natl. Acad. Sci. U.S.A.* 89:9015 (1992).
58. E.J. Lowenstein, R.J. Daly, A.G. Batzer, W. Li, B. Margolis, R. Lammers, A.

Ullrich, and J. Schlessinger, The SH2 and SH3 containing protein Grb-2 links receptor tyrosine kinases to ras signaling, *Cell* 70:431 (1992).
59. M. Rozakis-Adcock, R. Fernley, S. Wade, T. Pawson, and D. Bowtell, The SH2 and SH3 domains of mammalian Grb-2 couple the EGF receptor to the Ras activator mSOS, *Nature* 363:83 (1993).
60. M. Rozakis-Adcock, J. McGlade, G. Mbamalu, G. Pelicci, R. Daley, W. Li, A. Batzer, S. Thomas, J. Brugge, P.G. Pelicci, *et al.*, Association of the Shc and Grb2/Sem5-containing proteins is implicated in activation of the Ras pathway by tyrosine kinases, *Nature* 360:689 (1992).
61. G. Pelicci, L. Lanfrancone, F. Grignani, J. McGlade, F. Cavallo, G. Forni, I. Nicoletti, F. Grignani, T. Pawson, and Pelicci PG, A novel transforming protein (SHC) with an SH2 domain is implicated in mitogenic signal transduction, *Cell* 70:93 (1992).
62. E.Y. Skolnik, C.H. Lee, A. Batzer, L.M. Vicentini, M. Zhou, R. Daly, M.J. Myers Jr., J.M. Backer, A. Ullrich, M.F. White, *et al.*, The SH2/SH3 domain-containing protein GRB2 interacts with tyrosine-phosphorylated IRs1 and Shc: implications for insulin control of ras signalling, *EMBO J.* 12:1929 (1993).
63. W. Li, R. Nishimura, A. Kashishian, A.G. Batzger, W.J.H. Kim, J. Cooper, and J. Schlessinger, A new function for a phosphotyrosine phosphatase: Linking Grb-2 SOS to a receptor tyrosine kinase, *Mol. Cell. Biol.* 14:509 (1994).
64. K.S. Ravichandran, K.K. Lee, Z. Songyang, L.C. Cantley, P. Burn, and S.J. Burakoff, Interaction of Shc with the z chain of the T cell receptor upon T cell activation, *Science* 262:902 (1993).
65. L. Buday, S.E. Egan, P. Rodriguez-Viciana, D.A. Cantrell, and J. Downward, A complex of Grb-2 adaptor protein, SOS exchange factor and a 36kDa membrane bound tyrosine phosphoprotein is implicated in Ras activation in T cells, *J. Biol. Chem.* 269:9019 (1994).

# IMMUNOLOGICAL TOLERANCE BY ANTIGEN-INDUCED APOPTOSIS OF MATURE T LYMPHOCYTES

Lixin Zheng, Stefen A. Boehme, Jeffrey M. Critchfield, Juan Carlos Zuniga-Pflucker, Matthew Freedman and Michael J. Lenardo#

Laboratory of Immunology, National Institute of Allergy and Infectious Diseases, National Institutes of Health, Bethesda, Maryland 20892

## INTRODUCTION

Although the immune system has evolved to mount defensive lymphocyte activation responses to pathogens, it has long been known that antigen can specifically abrogate an immune response.[1] This form of acquired immunological tolerance can occur in adult animals and in certain circumstances has been shown to involve T lymphocytes. We have found conditions under which antigen causes the programmed death of mature T lymphocytes. We first determined that susceptibility of T lymphocytes to programmed death results from the exposure to growth lymphokines such as interleukin-2 (IL-2) followed by cell cycle progression. Susceptible T lymphocytes can then be triggered to die by T cell receptor (TCR) engagement. We have found that this mechanism leads to T cell death when T cells are exposed to large concentrations of antigen under fully activating conditions. Our results suggest that the induction of death by antigen is a feedback regulatory process that controls the intensity of immune responses involving T lymphocytes. We have called this mechanism "propriocidal regulation" and propose that it represents one means by which antigen-induces tolerance. We have recently shown that autoreactive T lymphocytes are susceptible to death by the propriocidal mechanism. This allowed us to use antigen to delete pathogenic T lymphocytes resulting in clinical and pathological amelioration of experimental allergic encephalomyelitis in mice. In this review we shall describe the results mentioned above, which are contained in several recent publications,[1,3,4,6,12] that support the concept that antigen-induced apoptosis of mature T lymphocytes is a potent form of immunological tolerance.

*Mechanisms of Lymphocyte Activation and Immune Regulation V*
Edited by S. Gupta *et al.*, Plenum Press, New York, 1994

## RESULTS

### Antigen-Induced Death - The Propriocidal Mechanism

Previous investigations had suggested the paradigm that thymocytes, but not mature peripheral T lymphocytes were susceptible to TCR-induced death.[2] In part, this notion stemmed from the finding that the typical response of a mature T lymphocyte to antigen stimulation was activation and proliferation. However, we have found conditions under which mature T cells respond to antigen by an initial activation process followed by programmed cell death. The first evidence came from the study of a mature CD4+ T lymphocyte clone, A.E7, that recognizes to the 81-104 amino acid fragment of pigeon cytochrome c in the context of the $E^k$ major histocompatibility complex (MHC) class II molecule. When A.E7 cells are exposed to IL-2 and begin active cell cycling, they become susceptible to TCR-mediated programmed cell death.[3] A.E7 cells induced to die by TCR stimulation had the microscopic morphology as well as a DNA fragmentation pattern consistent with apoptosis. The level of induced death was directly proportional both to the level of IL-2-stimulated cell cycling as well as to the intensity of subsequent antigen-receptor stimulation. This implied that T lymphocytes that were exposed to growth lymphokine prior to TCR occupancy, a reversal of the normal sequence of events that occurs when a T cell encounters antigen, it would undergo programmed death instead of proliferating.

To explore this phenomenon in T lymphocytes that had not been propagated for long periods of time in vitro, fresh lymph node cells were examined. In this series of experiments, several questions were addressed: 1) would T lymphocytes taken directly ex vivo exhibit a sensitivity to death caused by IL-2 followed by antigen receptor stimulation?; 2) are both CD4+ and CD8+ cells susceptible?; and 3) in a population of susceptible T lymphocytes, do only those T cells that are directly challenged through the TCR undergo death and not bystander cells? This latter point is important because if only directly challenged T lymphocytes die, then the mechanism becomes a potent selective pathway to ablate specific classes of antigen-reactive mature T cells much as negative selection acts upon immature thymocytes. Lymph node T cells taken directly ex vivo were first stimulated with concanavalin A in order to cause the expression of high affinity IL-2 receptors which are not normally present on most resting T lymphocytes. Cells were then stimulated with IL-2 either in high concentrations (100-140 I.U. per ml) or low concentrations (2-5 I.U. per ml). Because lymph node T cells cannot survive for extended periods of time in culture without IL-2, such conditions could not be tested. Nonetheless, low concentrations of IL-2 had the effect of maintaining cells in a healthy state in vitro with relatively little proliferation whereas high concentrations of IL-2 caused a marked degree of proliferation. To selectively stimulate distinct subpopulations of the IL-2-treated T lymphocytes, TCR binding antibodies that were fixed to plastic dishes were used. A comparison was made between a monoclonal antibody (mAb) against TCRs bearing a Vß8 chain and a mAb against TCRs bearing a Vß6 chain. The number of viable T cells was quantitated after 2 days of culture either with no antibody, anti-Vß8, and anti-Vß6. In this experiment, the appropriate concentration of IL-2 was added during the re-stimulation on anti-TCR mAbs so an to prevent any lymphokine withdrawal apoptosis. For T cells that had been treated with high concentrations of IL-2, either anti-Vß8 or anti-Vß6 caused the induction of apoptosis in a fraction of the population and reduced cell numbers compared to cultures in which no antibody cross-linking was carried out. This effect depended on the dose of IL-2 because in T cells propagated in low concentrations of IL-2 actually showed a modest increase in cells numbers and very little, if any, death following TCR cross-linking. Moreover, both $CD4^+$ and $CD8^+$ cells were susceptible to TCR-induced death. Similar results were obtained using cells from a transgenic mouse line expressing a TCR that recognizes a peptide antigen (Ag) derived from pigeon cytochrome c. Transgenic lymph node T cells carried in 50 I.U. per ml and then exposed to peptide antigen on fully co-stimulatory antigen-presenting cells (APCs) underwent progressively higher levels of cell death at increasing concentrations of Ag (Fig. 1).

This showed that antigen presented in fully activating form could induce the death of T cells that were cycling in IL-2. Further studies have shown that if T cells are cycling in the IL-2, death caused by TCR signals is neither worsened or improved by co-stimulatory signals.[4] Taken together, these results show that TCR stimulation can trigger an intrinsic pathway that leads to mature T cell apoptosis. A number of other studies have confirmed these observations and show that the mechanism leading to death of mature T cells after lymphokine and TCR stimulation is a characteristic of all types of T lymphocytes including αß and γδ, $CD4^+$ and $CD8^+$, as well as $T_H1$ and $T_H2$ subpopulations.[1]

Based on the in vitro analysis, it was hypothesized that antigen-induced death of mature T lymphocytes represented a feedback regulatory mechanism. It was possible that during a strong antigenic challenge large amounts of lymphokine would be produced that would drive a large number of the specifically-reactive T lymphocytes into the cell cycle. The continued presence or reintroduction of antigen could then cause the programmed death of a significant fraction of cycling T cells. This would serve to limit or reverse T cell expansion and the production of lymphokines. This would be autoregulatory because the level of antigen and growth lymphokine would determine the degree of apoptosis. Thus, the more intense the stimulation, the greater the level of proliferative suppression due to antigen provoked T lymphocyte death. This feedback mechanism was called "propriocidal regulation" to draw an analogy to other feedback regulatory systems that are common in various organ systems in multicellular organisms.[5] In particular, the proprioceptive neurons in the central nervous system that control limb movement by stimulating opposing muscle groups. We envisioned that this form of informational feedback was crucial to governing the strength of antigen reactions in T lymphocytes.

## Antigen-Induced T Cell Apoptosis Depends On Cell Cycling

Further studies addressed the role of cell cycle progression in the susceptibility to death. Two pieces of evidence suggested that cell cycle progression might be a key

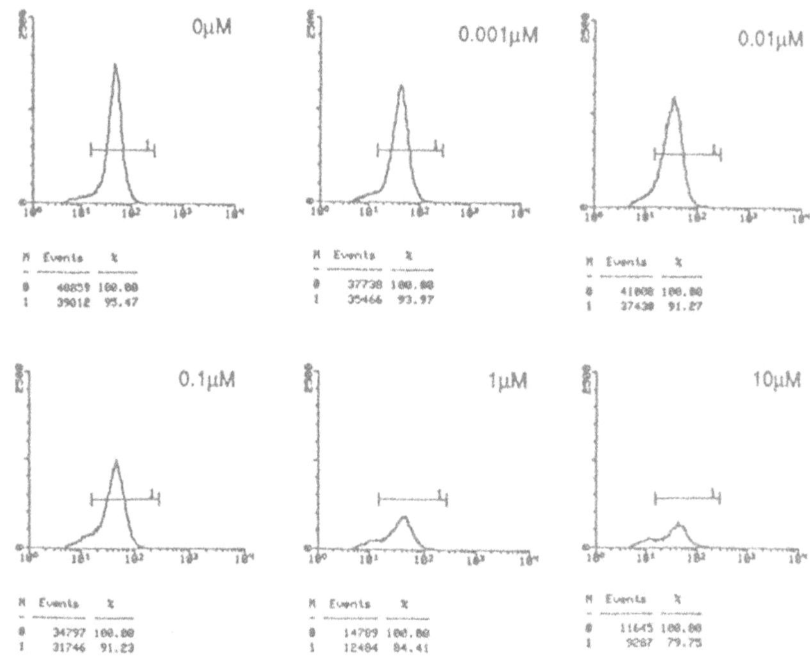


**Figure 1.** Induction of apoptosis in transgenic lymph node cells treated with IL-2 and peptide antigen. Lymph node cells that contain transgenes for the alpha and beta chains of a T cell receptor that recognizes an epitope (amino acids 81-104) of pigeon cytochrome c in the context of the $E^k$ molecule were first stimulated with 3 ug per ml concanavalin A for 48 hrs and washed extensively with alpha-methyl-mannoside. The cells were then incubated with 50 I.U. per ml IL-2 for 48 hrs and then incubated with the indicated concentrations of antigen with irradiated splenocyte antigen-presenting cells (APCs) from B10.A mice. Simulations went for 48 hrs and included 50,000 T cells, 200,000 APCs, and 50 I.U. per ml IL-2. Cultures were then analyzed for viable transgene bearing T cells by staining for the Vα11 transgene with a FITC-labelled mAb and then carrying out flow cytometry as previously described.[4, 12]

determinant of the sensitivity of T cells to death: 1) The degree of death caused by any given level of antigen can be directly correlated by the level of cell cycling as indicated by the incorporation of $^3$H-thymidine, and 2) removal of lymphokine does not immediately protect from TCR-induced death. In the latter case, the loss of susceptibility to death occurs gradually over a period of days as the cells exit the cell cycle and become "resting." If cell cycling was the critical switch to apoptosis susceptibility, then other T cell growth lymphokines should predispose T cells to death. Indeed, IL-4 treatment of T cell clones or lymph node cells will lead to a vulnerability to antigen-induced death.[6] This susceptibility to death is also proportion to the level of cell cycling. To further elucidate the relationship between cell cycling and lymphokine addition, experiments were carried out with various cell cycle blocking agents.[6] It was found that agents that block in $G_1$ phase of the cell cycle such as deferoxamine, mimosine, or dibutyryl cyclic AMP would completely protect from TCR-induced death. By contrast, in cells that were blocked in S phase or at the $G_1$-S boundary using either excess thymidine or aphidocholin, there is no protection from TCR-induced death. Taken together, these studies demonstrate that lymphokines predispose to apoptosis by driving T cells from a resting state into S phase or later stages of the cell cycle.

## Superantigen-Mediated Deletion In Vivo

Based on the parameters of antigen-induced apoptosis determined from in vitro studies, it was conjectured that shortly after immunization, responding T cells in vivo might become transiently sensitive to apoptosis. This may occur 2-3 days after antigen encounter when the production of growth lymphokine and mitogenic response would be maximal. Re-administration of the antigenic substance at this time could then lead to the death of a significant fraction of the activated and cycling cells. This possibility was tested by immunizing BALB/c mice with the superantigen, *Staphylococcal* enterotoxin B (SEB). SEB will activate selected subpopulations of T lymphocytes, for example, those bearing a Vß8 chain in the TCR, but does not activate other subpopulations such as those that bear a Vß6 chain. We used a protocol in which a large dose of SEB was given on day 1 followed by re-administration of SEB on days 3 and 5. At eight days, the animals were sacrificed and peripheral T cells were examined for the fractions that bear either Vß6 or Vß8 chains in their TCRs. It was found that repetitive doses of SEB caused a marked deletion of Vß8$^+$ T cells but not Vß6$^+$ T cells.[3] To determine the role of IL-2 to in the deletion observed, groups of SEB-treated mice were also given antibodies that block either IL-2 or IL-4 responses. Antibodies that block IL-2 were shown to largely abrogate the deletion whereas IL-4 blocking antibodies had no such effect. This indicates that under these experimental conditions, IL-2 appeared to contribute to the T cell deletion caused by SEB. Furthermore, these results go against the alternative possibility that lymphokine withdrawal accounted for the cell loss.[7] For example, if cells had become strongly activated and then growth lymphokine was consumed, it might be expected that repeated doses of SEB would rescue the activated T cells. In fact, further analyses showed that repeated injections always caused greater deletion than single injections (A. Kuang and M. Lenardo, unpublished observations). Moreover, antibodies against IL-2 that would potentially exacerbate withdrawal of this lymphokine instead protected against cell loss.

## Propriocidal Apoptosis As A Mechanism of Tolerance

To explore the potential role of the propriocidal mechanism in immunological tolerance, we next investigated the phenomenon of "high dose suppression" or "high zone tolerance." This antigen effect has been characterized both in antigen responses in vivo as well as in T lymphocyte responses in vitro.[1,8,9] It is typified by three characteristics that occur at high doses of antigen: 1) suppressed apparent proliferation of the cells as measured by $^{3}$H-thymidine incorporation, 2) production of large amounts of IL-2, and 3) expression of high affinity IL-2 receptors (Fig. 2). By contrast, lower, non-suppressive, doses of antigen manifest higher $^{3}$H-thymidine incorporation despite the fact that the level of IL-2 accumulation may be actually less than at suppressive doses of antigen. Thus, high dose suppression may be defined as the loss of $^{3}$H-thymidine incorporation despite the presence of IL-2 and high affinity receptors for this growth lymphokine. We reasoned that at high doses of antigen, all of the requirements of the propriocidal mechanism would be met. The production of large amounts of IL-2 could stimulate a majority of the T cells into cycle. Then, high concentrations of antigen could precipitate apoptosis in the cycling cells. To test this idea, we analyzed cultures of T cells stimulated at high doses of antigen for evidence of programmed cell death.[12] We found little death in cultures exposed to no antigen or nonsuppressive doses of antigen at which $^{3}$H-thymidine incorporation was maximal. Cell counts revealed that an increase in cell number quantitatively accounted for the increased $^{3}$H-thymidine incorporation at nonsuppressive doses of antigen. Also, the loss of $^{3}$H-thymidine uptake at suppressive doses of antigen was directly related to a decreased number of cells. Cell cycle analysis showed that there was comparable numbers of cells entering S phase at both low and high doses of antigen suggesting that there was no cell cycle blockade at high doses of antigen. Further analysis showed that ladders of DNA fragmentation indicating apoptosis were present at high antigen doses. Also, it was found that antibodies that antagonize either IL-2 or its high affinity receptor were able to prevent T cell death at high antigen doses suggesting the death is lymphokine dependent. Taken together these results indicated that programmed cell death of activated, cycling cells caused a suppression of proliferation at high antigen doses.

## Antigen-Induced Deletion Ameliorates Autoimmune Encephalomyelitis

The mechanism of T cell death that we had elucidated suggested to us the possibility that conventional peptide antigens, under the appropriate conditions, could be used to program the specifically reactive T cells to die. For cases in which the T lymphocytes were responsible for disease processes, their eradication could lead to clinical improvement. Based on the mechanism of apoptosis, T cell deletion would result from a strong activation response in which cell cycling was followed by TCR re-engagement. Thus, unlike previous treatment strategies based on the impairment of activation by TCR stimulation - such as the attempt to induce T cell clonal anergy - this strategy aimed at full activation of T cells.[11] Indeed, apoptosis would result from overstimulation by antigen.

To test this idea, we studied the effects of antigen treatment in experimental allergic encephalomyelitis (EAE), a mouse disease model that reflects clinical and pathological features of the human disorder, multiple sclerosis.[12] In this disease model, pathogenic $CD4^+$ T lymphocytes that react against various protein components of the myelin such as myelin basic protein (MBP) or proteolipid protein (PLP), cause destruction of myelin sheaths in the central nervous system. This leads to severe paralysis and eventually death. We employed the adoptive transfer model of this disease which cause reproducibly severe disease that has a relapsing and remitting course as is the case for many human autoimmune diseases (Fig. 3). The model is initiated by priming susceptible strains of mice with MBP in complete Freund's adjuvant. Ten days later, the draining lymph nodes are removed and re-stimulated in vitro with MBP. After four days of in vitro stimulation, the activated T cells are injected intravenously into naive recipients. The disease is then scored as follows: .... We found that animals that received no antigen treatment rapidly developed severe paralysis on days 7-9. By contrast, animals that had received injections of MBP twice daily on days 1, 3, and 5 following the transfer of encephalitogenic cells exhibited very little disease. A similar course of treatment together with 60,000 units of IL-2 given each day for days 1-5 after cell transfer also had a dramatic protective effect from paralysis. This suggests that clonal anergy could not account for the protective effects of antigen treatment since IL-2 would be expected to reverse anergy.[11,13] In each case, the mean clinical severity for a group of 5 mice was reduced from approximately grade 4 to less than grade 1 following treatment. The pathological picture of successfully treated mice showed dramatically reduced level of demyelination and destruction of the normal architecture of the spinal cord. Further experiments have revealed that treatments that are administered even beyond the onset of symptoms have a significant ameliorative effect on the clinic course of disease. Thus, repeated high dose antigen treatments create a state of tolerance in which an autoimmune process is improved.

To determine whether T cell deletion had occurred by the mechanisms described above, several experiments were undertaken.[12] First it was shown that T lymphocytes expressing a transgenic TCR that recognizes an encephalitogenic epitope of MBP are profoundly deleted when repeated antigen treatments are administered. Paralytic disease caused by adoptive transfer of the encephalitogenic transgenic T cells is improved by the deletion. Also non-transgenic encephalitogenic T cells that are marked with a fluorescent dye are also deleted by repeated antigen treatments leading to the amelioration of disease. These findings strongly support the concept that deletion of mature T lymphocytes is an means by which antigen-specific treatment of immune diseases can be carried out. Other experiments indicated that antigen induces only the deletion of specifically reactive T cells and not bystander T cells. Also, in cases where antigen was effective in causing deletion, the T cells were activated and cycling when adoptively transferred. Repeated antigen treatments were not effective in inducing the deletion of resting T lymphocytes. This is consistent with in vitro experiments that show that antigen can only induce the death of mature T cells that are activated and cycling. Thus, a deletional form of tolerance that closely resembles antigen-induced mature T cell apoptosis can lead to remarkable improvement of an immune-mediated disease.

## CONCLUSIONS

Several years ago we made the observation that mature T lymphocytes that were cycling in IL-2 would undergo programmed cell death upon ligation of the antigen receptor. This observation surprised us and suggested, in contrast to previous thinking, that IL-2 was not simply a growth factor for T lymphocytes. Rather, IL-2 can control T cell number both by inducing proliferation and by predisposing to apoptosis in the continued presence of antigen. This finding also suggested that antigen-induced apoptosis was a feature not only of thymocytes but also of mature, self-tolerant, and "educated" T lymphocytes. This has led us to propose that programmed death of mature T cells serves a fundamentally different purpose that the apoptosis that underlies the selective processes that occur during T cell maturation. We have proposed that programmed death of mature T cells allows tolerance to develop against exogenous antigens. We have further proposed that tolerance to exoantigens by apoptosis is an autoregulatory feedback mechanism that allows a productive activation response to be controlled under conditions in which antigenic stimulation persists. We have called this feedback regulatory mechanism, "propriocial regulation." Thus, antigen-induced apoptosis of mature T cells is selection process in a cybernetic sense. In addition to governing the proliferation of mature T cells, it is possible that antigen-induced apoptosis may contribute to self-tolerance in the periphery. Further experiments will be aimed at testing this possibility. Further experiments will also be directed at determining the molecular processes that underly antigen-stimulated apoptosis.

Perhaps one of the most exciting findings to stem from the initial observation of antigen-induced mature T cell apoptosis is the fact that antigen can be used to eradicate disease-causing T cells in an autoimmune disease. This suggests the possibility that just as the iatrogenic administration of antigen can increase clonal frequency to achieve vaccination, antigen might also be used to eliminate clones of pathogenic T cells. This would achieve a long-sought goal of providing an antigen-specific therapy for diseases caused by T cells. We are now carrying out further tests in various animal models of immunological diseases with the view of developing this approach as a clinical therapy.

## REFERENCES


1. J.M. Critchfield, S.A. Boehme, and M.J. Lenardo, Regulation of antigen-induced mature T lymphocyte apoptosis, *in* : "Apoptosis and the Immune Response," C. Gregory, ed., Wiley and Sons, New York (1972).
2. D.J. McConkey, S. Orrenius, and M. Jondal, Cellular signalling in programmed cell death (apoptosis), *Immunol. Today* 11:120 (1990).
3. M. Lenardo, Interleukin-2 programs mouse alpha beta T lymphocytes for apoptosis, *Nature* 3535:858 (1991).
4. S.A. Boehme, L. Zheng, and M.J. Lenardo, Distinct antigen receptor signal requirements for programmed cell death compared to lymphokine expression in mature T lymphocytes, Manuscript submitted.

5. N. Wiener, Cybernetics; or, Control and Communication in the Animal and the Machine. M.I.T. Press, New York (1961).
6. S.A. Boehme and M.J. Lenardo, Propriocidal apoptosis of mature T lymphocytes occurs at S phase of the cell cycle, *Eur. J. Immunol.* 23:1552 (1993).
7. R.C. Duke and J.J. Cohen, IL-2 addiction: withdrawal of growth factor activates a suicide program in dependent T cells, *Lymphokine Research* 5:289 (1986).
8. N.A. Mitchison, Induction of immunological paralysis in two zones of dosage, *Proc. Roy. Soc. (Lond)* 161:275 (1964).
9. G. Suzuki, Y. Kawase, S. Koyasu, I. Yahara, Y. Kobayashi, and R.H. Schwartz, Antigen-induced suppression of the proliferative response of T cell clones, *J. Immunol.* 140:1359 (1988).
10. D. Moskophidis, F. Lechner, H. Pircher, and R.M. Zinkernagel, Virus persistence in acutely infected immunocompetent mice by exhaustion of antiviral cytotoxic effector T cells, *Nature* 362:758 (1993).
11. R.H. Schwartz, A cell culture model for T lymphocyte clonal anergy, *Science* 248:1349 (1990).
12. J.M. Critchfield, M.K. Racke, J.C. Zuniga-Pflucker, B. Cannella, C.S. Raine, J. Goverman, and M.J. Lenardo, T cell deletion in high antigen dose therapy of autoimmune encephalomyelitis, *Science* in press (1994).

# PROPERTIES OF CA CURRENTS ACTIVATED BY T CELL RECEPTOR SIGNALING

Brett A. Premack and Phyllis Gardner

Departments of Molecular Pharmacology and Medicine
Stanford University Medical Center
Stanford, CA 94305

## INTRODUCTION

Influx of extracellular Ca is one of the early biochemical events associated with signaling through the T cell receptor (TCR) complex[1-4]. The influx of Ca is a critical component of the signaling pathway because it helps to drive the cytoplasmic Ca from resting levels of around 100 nM up to several μM, the concentration required for activation of many intracellular Ca binding proteins and enzymes[5]. Recently there has been a great deal of progress in understanding the molecular basis of the very earliest events in TCR signaling which precede Ca mobilization as well as those later Ca-dependent processes leading to transcription of the IL-2 gene[6-9]. Unfortunately, the molecules involved in initiation, maintenance, and termination of Ca influx remain elusive despite the combined efforts of researchers using molecular, biochemical and electrophysiological approaches. However, during the last year a number of important discoveries have generated an unprecedented interest in the nature of the Ca influx pathway[10,11].

The TCR, like many other surface receptors for growth factors and hormones, mobilizes stored intracellular Ca through the action of the second messenger inositol 1,4,5-trisphosphate ($IP_3$, reviewed in ref. 12). In the case of the T cells the transduction pathway couples through receptor-associated tyrosine kinases, which phosphorylate PLCγ, increasing its catalytic activity toward the membrane-bound substrate $PIP_2$, producing cytoplasmic $IP_3$. $IP_3$ then binds in a highly cooperative fashion to receptors which are mainly localized to a Ca sequestering fraction of the endoplasmic reticulum (ER). The $IP_3$ receptor itself forms a functional intracellular Ca channel, which in the presence of μM $IP_3$, allows Ca to flow down its concentration gradient, rapidly discharging the stored Ca. The release of stored Ca begins within about 30 seconds after crosslinking of the TCR, but is relatively short-lived, and in a few minutes most if the stored Ca has been depleted. During this time, there is a shift in the source of Ca from the storage organelle to influx across the plasma membrane. For a number of years it has been accepted that a second messenger generated by signaling through the receptor was responsible for switching on Ca entry,

*Mechanisms of Lymphocyte Activation and Immune Regulation V*
Edited by S. Gupta *et al.*, Plenum Press, New York, 1994

although the mechanism remained controversial[13-20]. However, more recent evidence suggests that in some cell types, including lymphocytes, Ca influx may be under the direct control of the Ca storage organelle in a process often termed "capacitative Ca entry"[21-23]. In this report we describe some of our recent results using Ca indicator dyes and whole-cell patch-clamp to study the properties of the Ca influx pathway in the Jurkat human T cell line.

## MATERIALS AND METHODS

The general methods for cell culture and ratiometric single cell Ca measurement using indo-1 have been described in detail elsewhere[24,25]. Indo-1/AM, indo-1 K salt and $Cs_4BAPTA$ were obtained from Molecular Probes. The anti-human mouse IgG1 anti-CD3 monoclonal antibody (clone UCHT1) was from Sigma. Thapsigargin was obtained from either Gibco, Calbiochem, or Sigma, no differences in potency were noted.

Conventional whole-cell patch-clamp[26] was carried out using a List EPC-7 patch-clamp amplifier with the current output low-pass filtered at 1.5 KHz using an external 8 pole Bessel filter. During whole-cell voltage-clamp the membrane currents were digitized at a sample rate of 1 KHz. with 12 bit resolution using the Axon Instruments TL-1 interface and pClamp analysis software. During voltage-clamp the holding potential was -60 mV, step current-voltage (I-V) curves used 400 msec steps to -100, -60, -20, +20, +60 mV. Ramp I-V curves were generated with ramp commands from -100 to +60 mV, at a depolarization rate of 0.5 mV/msec.

The pipette solution contained (in mM) 140 CsAsp, 10 CsCl, 2 $MgCl_2$, 0.5 MgATP, 1 $CaCl_2$, 10 $Cs_4BAPTA$, 10 Hepes, pH 7.3. The calculated free Ca concentration of this solution is 25 nM. The substitution of Asp for Cl gives a calculated Cl reversal of -60 mV when used with the extracellular solutions described below. Na and K free NMDG (n-methyl-d-glucamine) external solutions (160 NMDGCl, 2 $CaCl_2$, 1 $MgCl_2$, 5 d-glucose, 10 Hepes, pH 7.3) were used in most patch-clamp experiments to minimize currents flowing through voltage-gated Na and K channels.

## RESULTS AND DISCUSSION

We initially examined the hypothesis that depletion of the Ca store controls Ca entry in Jurkat cells by studying Ca mobilization in intact single cells loaded with the Ca indicator indo-1/AM (Fig. 1). The pattern of Ca mobilization in most individual Jurkat cells stimulated by with monoclonal antibody against CD3 (1 μg/ml) is biphasic. The peak occurs between 1 and 2 minutes, and represents mainly Ca released from intracellular stores by $IP_3$. Over the following 5 minutes there is a decrease to a plateau level which represents mainly influx of extracellular Ca because this phase is abolished when extracellular Ca is chelated with EGTA (Fig. 1A). The patterns of Ca mobilization elicited by anti-CD3 in single cells are remarkably similar to those obtained either by averaging records from single cells or in populations of cells suspended in cuvettes. We see little evidence that out of phase Ca oscillations in single cells account for the peak and plateau responses observed in populations of cells, at least during the first 10 to 20 minutes following TCR stimulation.

Thapsigargin (TG), a potent and specific inhibitor of SERCA type ER Ca-ATPases induces Ca mobilization in many cell types without production of $IP_3$ [27-30]. Application of 500 nM TG initiates Ca responses which are quite similar to those observed following crosslinking of the T cell receptor (Fig. 1B). In particular, it is clear that TG induces a sustained Ca influx which outlasts the release of Ca produced by inhibition of the ER Ca-ATPase.

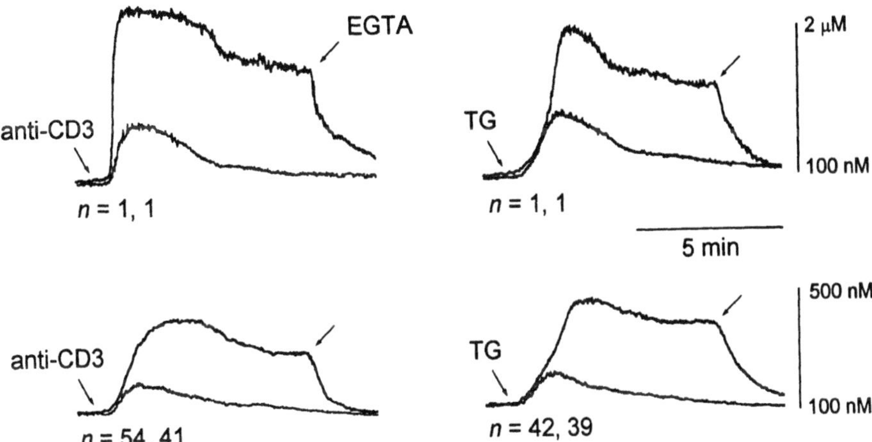


**Figure 1.** Intracellular Ca changes in Jurkat cells stimulated with anti-CD3 antibody (1 μg/ml) or thapsigargin (TG, 500 nM) recorded using indo-1. The top panels show representative single cells stimulated either in the presence of 2 mM Ca containing HBSS (upper traces), or in the same solution with Ca chelated by addition of 3 mM EGTA (lower traces). The bottom panels show averages for similar experiments. The number of cells (*n*) is shown for each record.

We have confirmed that TG releases stored Ca without significant production of $IP_3$ in Jurkat cells using an $IP_3$-specific radioligand binding assay[25]. In other experiments, we have observed that two other structurally unrelated ER Ca-ATPase inhibitors, butylhydroquinone and cyclopiazonic acid, also release stored Ca and induce Ca influx in single Jurkat cells (not shown). Taken together these results suggest that Ca entry occurs independently of $IP_3$ production, and provide strong evidence that release of stored Ca is the trigger for initiation of Ca entry in lymphocytes, as has been suggested previously[29,30-33].

In our laboratory we have concentrated on using whole-cell patch-clamp to more fully characterize the biophysical properties of the receptor-activated Ca influx pathway. These experiments have examined three main questions which are critical to understanding the mechanism of Ca influx. 1) Are the Ca entry pathways activated by receptor signaling and depletion of stored Ca actually the same? 2) Do the electrical properties of the entry pathway allow us to distinguish whether the Ca influx is mediated by an ionic channel, a plasma membrane Ca pump or electrogenic exchanger? 3) What biochemical signals initiate and terminate the influx process?

Figure 2 A and B compare the membrane currents elicited in voltage-clamped Jurkat cells during stimulation with either anti-CD3 antibody or TG. The ionic conditions have been optimized to facilitate recording of currents carried by Ca ions. The only permeant cation in the bathing solution is Ca (2 mM); in this experiment the Na and K have been replaced by NMDG and Hepes buffer. The pipette solution, which perfuses the interior of the cell, contains Cs, which is relatively impermeant, and about 25 nM free Ca with 10 mM CsBAPTA as the Ca buffer. Under these conditions Ca entry is detected as an inward

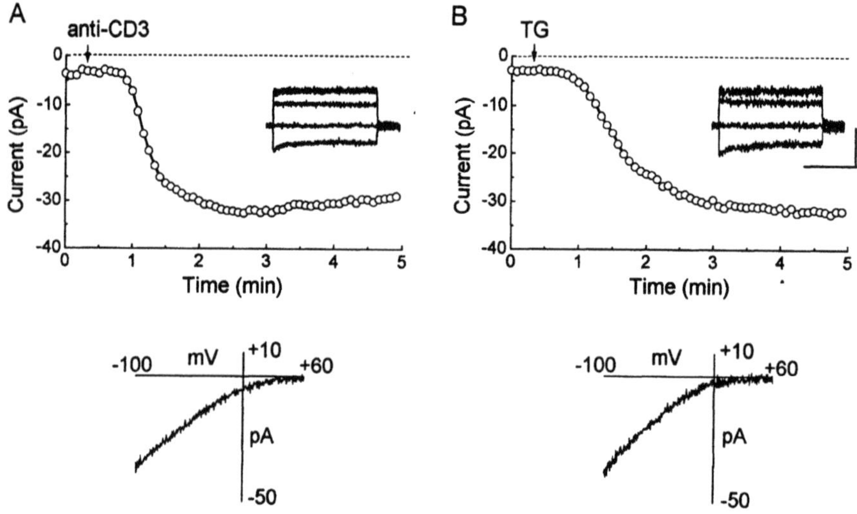


**Figure 2.** Membrane currents recorded from whole-cell voltage-clamped Jurkat cells stimulated with either anti-CD3 antibody (A) or TG (B). Parts A and B (top panel) show the changes in membrane current recorded at the holding potential (-60 mV), sampled every 5 seconds, plotted as a function of time. Stimulation with anti-CD3 or TG causes an increase in inward current which develops slowly over 1 to 2 minutes. The insets show superimposed individual traces recorded for voltage steps to -100, -60, -20, 20 and 60 mV. Leakage and capacitance currents have been removed by subtracting control records from voltage steps obtained prior to stimulation. The inset scale bar represents 25 pA, 200 msec. The lower panels show ramp I-V curves obtained in the same experiment by subtracting control ramps from those taken 5 minutes after application of anti-CD3 or TG.

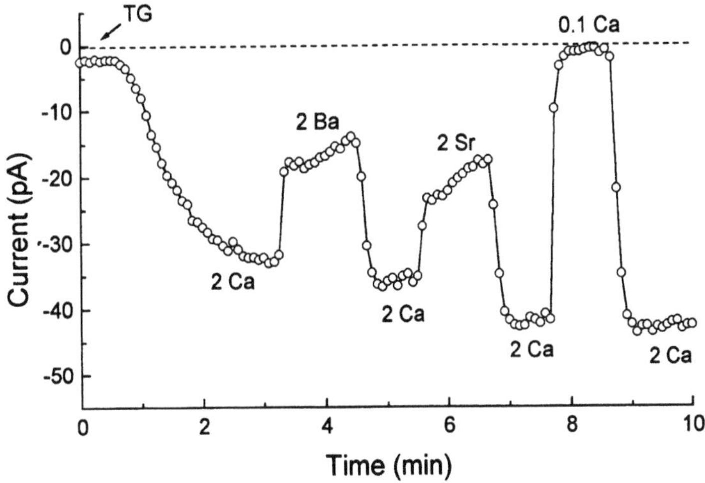


**Figure 3.** Comparison of membrane currents carried by equimolar concentrations of Ca, Ba and Sr. The open circles show the average steady-state inward currents recorded at the holding potential (-60 mV) over time. Ca current induced by 500 nM TG was initially recorded in 2 mM Ca, Na-free (NMDG) external solution, the entire bath was then perfused with test solutions and returned to the original solution. Ramp I-V curves were also generated every 5 seconds and confirm the identity of the inward current.

(shown downward) current deflection. Both anti-CD3 and TG induce an inward current, carried by Ca, which develops slowly over several minutes following an initial delay of up to 1 minute. Once the current has fully developed, there is very little time-dependent activation or inactivation during 400 msec long voltage pulses when the BAPTA is used as the intracellular Ca buffer (Fig. 2 insets). There is more pronounced Ca-dependent inactivation during voltage pulses when the weaker Ca buffer EGTA is used (Premack and Gardner, in preparation; see also ref. 33). The Ca currents generated by receptor stimulation and ER Ca-ATPase inhibition show essentially identical kinetics and a characteristic inwardly rectifying current-voltage (I-V) relation. Other biophysical properties such as the average current density, ionic selectivity and block measured for divalent metal cations (see below and ref. 25) are also the same for receptor stimulation and Ca-ATPase inhibition. These results all suggest that the same Ca current can be activated by either treatment.

In order to address the question of whether the whole-cell membrane current is mediated by an ion channel or a plasma membrane pump or exchanger we have examined the selectivity for various divalent and monovalent cations. As shown in Figure 3, the current activated by TG (or anti-CD3, not shown) is largest when carried by Ca ions, replacement of external Ca with equimolar Ba or Sr reduces the amplitude by 50 to 60% but there is still a significant inward current. Reducing the external Ca 20-fold (to 0.1 mM) abolishes nearly all of the measurable inward current, suggesting that under normal ionic conditions the Ca appears to be the main charge carrier.

The ionic current activated by depleting Ca stores with TG is not measurably permeant to Na in the presence of physiological levels of external Ca. With 2 mM Ca outside of the cell, replacement of all the external Na with Tris or NMDG has very little effect on the amplitude of inward current (not shown, see ref. 24). However, maintenance of this high degree of selectivity for Ca over Na, requires the presence of external Ca. Removal of the external Ca allows a very drastic increase in the permeability to Na. This result is demonstrated in Figure 4. Superfusing an unstimulated cell with a Na-containing solution in which the external Ca has been removed with EGTA (<100 nM free Ca) has only a small effect on baseline current. However, once the current is fully activated by depleting Ca stores with TG, removal of the external Ca produces a large, reversible increase in inward current. It is important to note that under these conditions the large inward current is carried by Na, because replacement of the external Na with Tris or NMDG, which are impermeant, eliminates the current.

These results at first seem somewhat difficult to interpret, but are more easily explained when taken in the context of what is known about monovalent cation selectivity in voltage-gated Ca channels. In L-type Ca channels there is also a remarkable selectivity for Ca over Na ions, but these channels can support a large Na conductance when the external free Ca is lowered below about 1 μM[35-37]. This is thought to occur because Ca ions bind transiently to glutamate residues within the pore of the channel creating a charge barrier which helps to reject Na and other monovalent cations [38-40]. Our results showing Na permeation only in the absence of Ca, significant conductance to Ba and Sr, block by Cd, Co, and Ni[25] are all properties expected for an ionic channel, and not consistent with a Na/Ca exchanger or other Ca pumping mechanism.

To examine the relationship between intracellular Ca and regulation of the Ca current we have measured Ca current and cytoplasmic Ca levels simultaneously during voltage-clamp experiments. In the example shown in Figure 5 the Ca current was initiated by application of carbachol to a Jurkat cell that has been stably transfected with the human M1 muscarinic receptor[41,42,24]. In the M1 transfectants $IP_3$ levels are similar to cells stimulated through the TCR, but the M1 receptor is coupled to a PLCβ isoform via a heterotrimeric G-protein rather than by tyrosine kinases. In these cells, the onset of Ca current is rapid and clearly precedes the rise in cytoplasmic Ca, providing strong evidence that the current causes the Ca increase rather than the reverse (see also ref. 18). In fact, high levels of cytoplasmic Ca tend to lead to spontaneous deactivation of the Ca current. These relations are most readily examined when the Ca current amplitude is rapidly increased by stepping the extracellular Ca from 2 to 10 mM (inset to Fig. 3). Under these conditions the Ca current maxima precedes the cytoplasmic Ca maxima by about 30

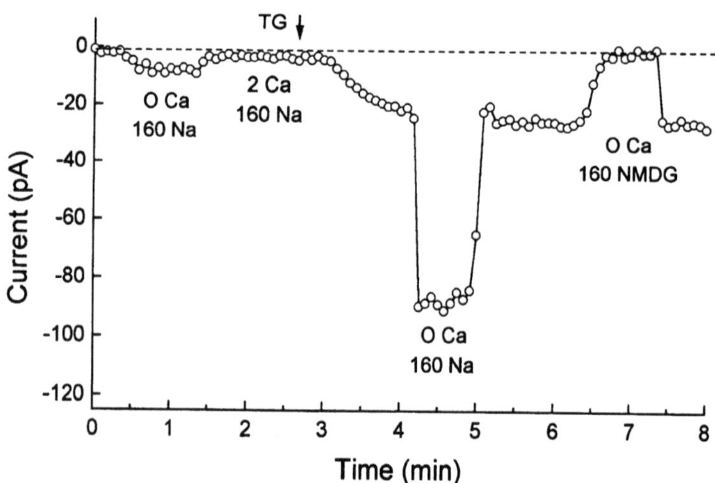


**Figure 4.** The TG-induced Ca current becomes highly permeable to Na in the absence of external Ca. In an unstimulated Jurkat cell, extracellular perfusion with a solution containing Na and less than 100 nM free Ca ("0 Ca") causes only a modest increase in membrane current. Once the depletion-activated current is fully induced in the presence of Ca , removal of external Ca causes a very large, reversible increase in inward current carried by Na. This effect is not observed if Na as well as Ca is eliminated from the bathing solution by substituting NMDG. Holding potential -60 mV.

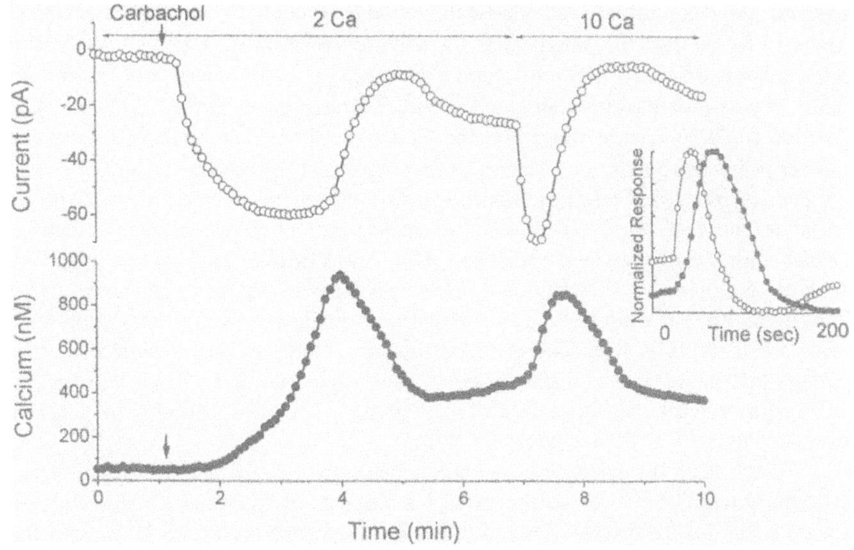


**Figure 5.** The Ca influx current raises cytoplasmic Ca levels in a voltage-clamped Jurkat cell. Simultaneous recordings of membrane current (open circles) and average cytoplasmic Ca concentration (filled circles) calculated from the ratio of indo-1 fluorescence at 400 and 480 nm. Activation of the Ca current by carbachol in muscarinic receptor transfectants (see text), precedes the rise in cytoplasmic Ca (the pipette contains 10 mM BAPTA). Note that spontaneous deactivation of the Ca current is associated with high cytoplasmic Ca levels. The inset shows normalized current and cytoplasmic Ca compared. Holding potential -60 mV. Adapted from reference 24.

seconds. The high intracellular Ca levels are associated with a deactivation of Ca current which appears to take about one minute to fully develop. Note that the Ca current turns off almost completely even in the maintained presence of high levels of external Ca. There appears to be a homeostatic mechanism operating that regulates the amount of Ca entry.

Up to this point we have outlined some of our experimental results which suggest that Ca influx in lymphocytes occurs via voltage-insensitive, highly Ca-selective ionic channels, now we would like to turn to the topic of physiological regulation of these channels. Figure 6 summarizes several mechanisms which have been hypothesized to be involved in the activation of receptor-operated Ca channels. Part A depicts one of the earliest and simplest ideas for regulation of Ca influx; Ca released from the Ca store could traverse the cytoplasm to bind to and open Ca entry channels. In this case the Ca channels could be envisaged to either have integral Ca binding sites analogous to those of Ca-activated K channels, or to be activated by Ca-dependent enzymes. Several lines of evidence argue against this mechanism working in Jurkat cells. As shown above, the activation of the Ca current precedes the rise in cytoplasmic Ca when 10 mM BAPTA is included in the patch pipette. Increasing the Ca buffering capacity of the intracellular solution actually prolongs the duration of the Ca current and raising intracellular Ca by flash photolysis of caged Ca promotes deactivation of the Ca current[24].

It also appears unlikely that inositol phosphates such as $IP_3$ or $IP_4$ directly activate Ca entry channels in intact cells (Fig. 6B), even though Ca-permeable channels are opened by $IP_3$ in excised patches from Jurkat cells[15]. In whole cells, flash photolysis of caged $IP_3$ activates the same Ca current as receptor stimulation, but the delayed onset kinetics are not consistent with direct channel gating by $IP_3$. Intracellular perfusion with $IP_4$ neither activates the Ca current nor augments currents elicited by flash release of $IP_3$[24]. Furthermore, the fact that microsomal Ca-ATPase inhibitors such as TG can activate Ca influx without detectable $IP_3$ production argues against a direct action of $IP_3$ on channel gating. One possible interpretation of the earlier excised patch data is that portions of the Ca store could be excised along with the plasma membrane and in this configuration $IP_3$ may act indirectly to open Ca channels by first depleting the associated Ca store. This idea is supported by excised patch experiments in which Ca permeable channels with properties similar to those of the 7 pS $IP_3$-induced channel are observed following treatment of patches with TG or butylhydroquinone (Kerr and Gardner, unpublished observations). However, a problem with this notion is that recent experiments using non-stationary current fluctuation analysis estimate the single channel conductance from depletion-activated whole-cell currents in lymphocytes to be roughly 24 fS[43]. This conductance estimate suggests that single channel openings would generate current fluxes much too small to be discerned in conventional single channel recordings, and suggest that the 7 pS channels are not related to the depletion-activated Ca current.

The bulk of the evidence supports the idea that in electrically inexcitable cells such as lymphocytes, which lack voltage-gated Ca channels, depletion of the Ca store is the primary signal for Ca entry[22,23,44]. The key question then becomes: How does the Ca storage organelle actually signal to the plasma membrane Ca channels? There are indications that in some cell types one or more diffusible messengers may be involved in this process (Fig. 6C). In pancreatic acinar cells extracellular application of a permeable analogue of cGMP (8-Br-cGMP) could activate Ca current, and cGMP has been proposed to function as an endogenous messenger which couples the stores to Ca influx[45]. We have tested this hypothesis in Jurkat cells and found no evidence of Ca current activation with 1 mM 8-Br-cGMP, even though TG could induce normal Ca currents in the same cells (Fig. 7). Therefore, we have shifted our attention to other possible signaling pathways which may be involved in the generation of a Ca entry stimulus.

Randriamampita and Tsien[46] have recently demonstrated that a small, diffusible messenger, isolated from lectin-stimulated Jurkat cells, can induce sustained Ca entry when

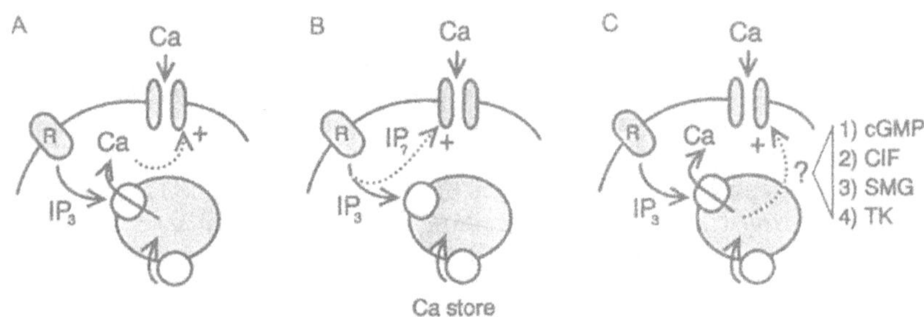


**Figure 6.** Hypothetical mechanisms which have been proposed to underlie receptor-activated Ca entry. The diagrams represent a section of plasma membrane with receptor (R), Ca channel, and the Ca store. The membrane of the Ca storage organelle is assumed to contain both the $IP_3$ receptor/Ca channel, and the TG-sensitive Ca-ATPase. Part A shows released Ca itself as the signal for Ca channel activation. Part B summarizes hypotheses involving inositol phosphate metabolites. Part C shows the central role of the Ca store itself in stimulating Ca entry, as well as possible signals which may couple depletion of the Ca store to Ca channel activation. See text for further explanation.

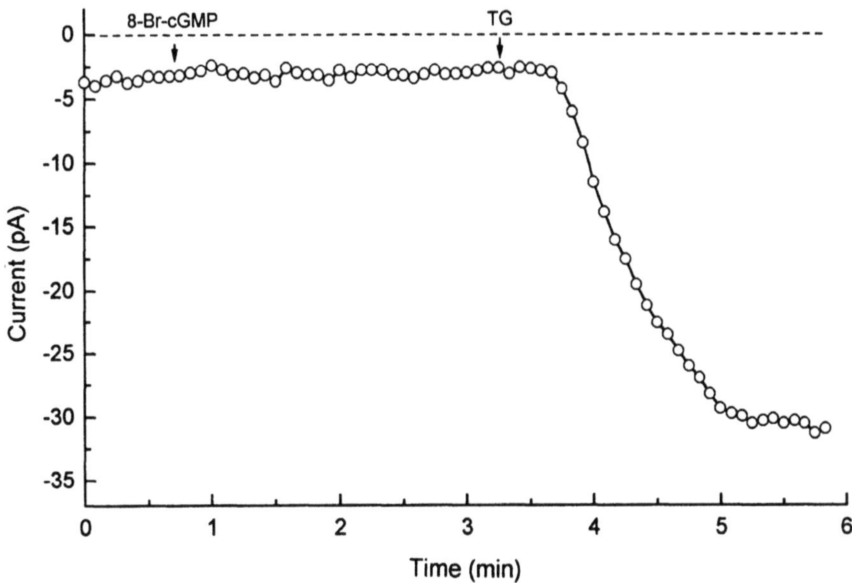


**Figure 7.** Lack of stimulation of Ca current by 8-Br-cGMP in Jurkat cells. The amplitude of Ca current at -60 mV is sampled and plotted every 5 seconds. External application of 1 mM 8-Br-cGMP has no stimulatory effect on Ca current recorded over 3 minutes, but subsequently 500 nM TG induces a characteristic inward Ca current. Ramp I-V curves were also generated every 5 seconds and confirm the identity of the inward current.

applied to macrophage and fibroblast cell lines. The molecule, termed CIF (calcium influx factor) has not been purified, but it does not appear to correspond to known second messengers. CIF seems to be released from storage sites following depletion of the Ca stores, rather than synthesized *de novo*. The messenger apparently then induces opening of plasma membrane Ca channels, either directly or through other molecules. CIF has not yet been shown to activate the characteristic Ca entry current, so it remains to be demonstrated that the CIF-stimulated pathway and stores depletion-activated pathways are actually identical. In another recent development, two laboratories have suggested that there is GTP-dependent step which regulates Ca entry in mast cells and acinar cells[47,48]. Several lines of suggest that a GTP hydrolysis is required; possibly an unidentified small GTP-hydrolyzing protein (SMG) may be involved. Both CIF and a SMG are attractive, and certainly not mutually exclusive, hypotheses which remain to be fully tested in T cells.

There is also evidence implicating a specialized tyrosine kinase signalling pathway in Ca entry in T cells and fibroblasts. In T cell hybridomas, introduction of a constitutively active form of pp60 v-*src*, causes spontaneous elevation of Ca influx without significant effects on $IP_3$ production[49]. Application of the tyrosine kinase inhibitors genistein and tyrphostin blocks bradykinin or TG stimulated Ca entry in HWSP fibroblasts[50]. We are currently examining the effects of genistein and other more potent and specific tyrosine kinase inhibitors such as lavendustin A on TG-induced Ca influx currents (which bypass TCR tyrosine kinases) in Jurkat cells. The preliminary results suggest that these compounds do indeed inhibit activation of the Ca current (Chung and Gardner, in preparation) while more general serine/threonine kinase inhibitors such as H-7 and staurosporine are without pronounced effects (Premack, unpublished observations).

The abundance of new hypotheses suggests that there are rich possibilities for biochemical regulation of Ca entry currents by diffusible messengers and enzymes. In the short term, the critical molecules involved in T cell receptor-activated Ca entry have yet to be defined and therefore remain an intriguing problem.

**Acknowledgements**

This work was supported by NIH training grant GM07065 (to B.A.P) and American Cancer Society grant DB-26E (to P.G.). P.G. is a recipient of the Burroughs Wellcome Faculty Scholar Award.

## REFERENCES

1. R.Y. Tsien, R. Pozzan and T.J. Rink, T-cell mitogens cause early changes in cytoplasmic free Ca and membrane potential in lymphocytes, *Nature* 295: 68 (1982).
2. T.R. Hesketh, G.A. Smith, J.P. Moore, M.V. Taylor and J.C. Metcalfe, Free cytoplasmic calcium concentration and the mitogenic stimulation of lymphocytes, *J. Biol. Chem.* 258: 4876 (1983).
3. J. Imboden and A. Weiss, The T-cell antigen receptor regulates sustained increases in cytoplasmic free Ca through extracellular Ca mobilization, *Biochem. J.* 247: 695 (1987).
4. E.W. Gelfand, R.K. Cheung, G.B. Mills and S. Grinstein, Uptake of extracellular Ca and not recruitment from internal stores is essential for T-lymphocyte proliferation, *Eur. J. Immunol.* 18: 917 (1988).
5. B.A. Premack and P. Gardner, Signal transduction by T cell receptors: Mobilization of Ca and regulation of Ca-dependent effector molecules, *Amer. J. Physiol.* 263: C1119 (1992).
6. N.A. Clipstone and G.R. Crabtree, Identification of calcineurin as a key signalling enzyme in T-lymphocyte activation, *Nature* 357: 695 (1992).
7. J. Jain, P.G. McCaffrey, Z. Miner, T.K. Kerppola, J.N. Lambert, G.L. Verdine, T. Curran and A. Rao, The T cell transcription factor NFATp is a substrate for calcineurin and interacts with Fos and Jun, *Nature* 365: 352 (1993).
8. J. Liu, FK506 and cyclosporin, molecular probes for studying intracellular signal transduction, *Immunol. Today* 14: 290 (1993).
9. J.D. Fraser, D. Straus and A. Weiss, Signal transduction events leading to T-cell lymphokine gene expression, *Immunol. Today* 14: 357 (1993).
10. D.E. Clapham, A mysterious new influx factor?, *Nature* 364: 763 (1993).
11. J.W. Putney Jr., Excitement about Ca signalling in nonexcitable cells, *Science* 262: 676 (1993).
12. M.J. Berridge, Inositol trisphosphate and calcium signalling, *Nature* 361: 315 (1993).
13. V. VonTscharner, B. Prod'hom, M. Baggiolini and H. Reuter, Ion channels in human neutrophils are activated by a rise in the free cytosolic calcium concentration, *Nature* 324: 369 (1986).
14. M. Kuno, J. Goronzy, C.M. Weyand and P. Gardner, Single-channel and whole-cell recordings of mitogen-regulated inward currents in human cloned helper T lymphocytes, *Nature* 323: 269 (1986).
15. M. Kuno and P. Gardner, Ion channels activated by inositol 1,4,5-trisphosphate in plasma membrane of human T lymphocytes, *Nature* 326: 301 (1987).
16. R. Penner, G. Matthews and E. Neher,Regulation of calcium influx by second messengers in rat mast cells, *Nature* 334: 499 (1988).
17. G. Matthews, E. Neher and R. Penner, Second messenger activated calcium influx in rat peritoneal mast cells, *J. Physiol.* 418: 105 (1989).
18. R.S. Lewis and M.D. Cahalan, Mitogen-induced oscillations of cytosolic Ca and transmembrane Ca current in human leukemic T cells, *Cell Regul.* 1: 99 (1989).
19. M. Hoth and R. Penner, Depletion of intracellular calcium stores activates a calcium current in mast cells, *Nature* 355: 353 (1992).
20. A. Luckhoff and D.E. Clapham, Inositol 1,3,4,5-tetrakisphosphate activates an endothelial Ca-permeable channel, *Nature* 355: 356 (1992).

21. J.W. Putney Jr., A model for receptor-regulated calcium entry, *Cell Calcium* 7: 1 (1986).
22. J.W. Putney Jr., Capacitative calcium entry revisited, *Cell Calcium* 11: 611 (1990).
23. J.W. Putney Jr., and G.S. Bird, The signal for capacitative Ca entry, *Cell* 75: 199 (1993).
24. T.V. McDonald, B.A. Premack and P. Gardner, Flash photolysis of caged inositol 1,4,5-trisphosphate activates plasma membrane calcium current in Human T cells, *J. Biol. Chem.* 268: 3889 (1993).
25. B.A. Premack, T.V. McDonald and P. Gardner, Activation of Ca currents in a Jurkat T cells following the depletion of Ca stores by microsomal Ca-ATPase inhibitors, *In press: J. Immunol.* (1994).
26. O.P. Hamill, A. Marty, E. Neher, B. Sakmann and F.J. Sigworth, Improved patch-clamp techniques for high resolution current recording from cells and cell-free membrane patches, *Pflugers Arch.* 391: 85 (1981).
27. T.R. Jackson, S.I. Patterson, O. Thastrup and M.R. Hanley, A novel tumour promoter, thapsigargin, transiently increases cytoplasmic free Ca without generation of inositol phosphates in NG115-401L neuronal cells, *Biochem. J.* 253: 81 (1988).
28. H. Takemura, A.R. Hughes, O. Thastrup and J.J. Putney, Activation of calcium entry by the tumor promoter thapsigargin in parotid acinar cells. Evidence that an intracellular calcium pool and not an inositol phosphate regulates calcium fluxes at the plasma membrane, *J. Biol. Chem.* 264: 12266 (1989).
29. H. Gouy, D. Cefai, S.B. Christensen, P. Debre and G. Bismuth, Ca influx in human T lymphocytes is induced independently of inositol phosphate production by mobilization of intracellular Ca stores. A study with the Ca endoplasmic reticulum-ATPase inhibitor thapsigargin, *Eur. J. Immunol.* 20: 2269 (1990).
30. N. Demaurex, D.P. Lew and K.H. Krause, Cyclopiazonic acid depletes intracellular Ca stores and activates an influx pathway for divalent cations in HL-60 cells, *J. Biol. Chem.* 267: 2318 (1992).
31. M.J. Mason, S.M. Mahaut and S. Grinstein, The role of intracellular Ca in the regulation of the plasma membrane Ca permeability of unstimulated rat lymphocytes, *J. Biol. Chem.* 266: 10872 (1991).
32. B. Sarkadi, A. Tordai, L. Homolya, O. Scharff and G. Gardos, Calcium influx and intracellular calcium release in anti-CD3 antibody-stimulated and thapsigargin-treated human T lymphoblasts, *J. Membr. Biol.* 123: 9 (1991).
33. J. Llopis, S.B. Chow, G.E. Kass, A. Gahm and S. Orrenius, Comparison between the effects of the microsomal Ca-translocase inhibitors thapsigargin and 2,5-di-(tert-butyl)-1,4-benzohydroquinone on cellular calcium fluxes, *Biochem. J.* 553 (1991).
34. M. Hoth and R. Penner, Calcium release-activated calcium current in mast cells, *J. Physiol.* 465: 359 (1993).
35. P. Hess and R.W. Tsien, Mechanism of ion permeation through calcium channels, *Nature* 309: 453 (1984).
36. W. Almers and E.W. McCleskey, Nonselective conductance in calcium channels of frog muscle: Calcium selectivity in a single-file pore, *J. Physiol.* 353: 585 (1984).
37. P. Hess, J.B. Lansman and R.W. Tsien, Calcium channel selectivity for monovalent and divalent cations. Voltage and concentration dependence of single channel current in ventricular heart cells, *J. Gen. Physiol.* 88: 293 (1986).
38. R.W. Tsien, P. Hess, E.W. McCleskey and R.L. Rosenberg, Calcium channels: mechanisms of selectivity, permeation and block, *Ann. Rev. Biophys. Biophys. Chem.* 16: 265 (1987).
39. S. Tang, G. Mikala, A. Bahinski, A. Yatani, G. Varadi and A. Schwartz, Molecular localization of ion selectivity sites within the pore of a human L-type cardiac calcium channel, *J. Biol. Chem.* 268: 13026 (1993).
40. J. Yang, P.T. Ellinor, W.A. Sather, J.F. Zang and R.W. Tsien, Molecular determinants of Ca selectivity and ion permeation in L-type Ca channels, *Nature* 366: 158 (1993).
41. M.A. Goldsmith, D.M. Desai, T. Schultz and A. Weiss, Function of a heterologous muscarinic receptor in T cell antigen receptor signal transduction mutants, *J. Biol. Chem.* 264: 17190 (1989).
42. D.M. Desai, M.E. Newton, T. Kadlecek and A. Weiss, Stimulation of the phosphatidylinositol pathway can induce T-cell activation, *Nature* 348: 66 (1990).
43. A. Zweifach and R.S. Lewis, Mitogen-regulated Ca current of T lymphocytes is activated by depletion of intracellular stores, *P.N.A.S* 90: 6295 (1993).

44. E. Neher, Controls on calcium influx, *Nature* 355: 298 (1992).
45. T.D. Bahnson, S.J. Pandol and V.E. Dionne, Cyclic GMP mediates depletion-activated Ca entry in pancreatic acinar cells, *J. Biol. Chem.* 268: 10808 (1993).
46. C. Randriamampita and R.Y. Tsien, Emptying of intracellular Ca stores releases a novel small messenger that stimulates Ca influx, *Nature* 364: 809 (1993).
47. C. Fasolato, M. Hoth and R. Penner, A GTP-dependent step in the activation of capacitative calcium influx, *J. Biol. Chem.* 268: 20737 (1993).
48. G.S. Bird and J.W. Putney, Inhibition of thapsigargin-induced Ca entry by microinjected Guanine nucleotide analogues, *J. Biol. Chem.* 268: 21486 (1993).
49. B.B. Niklinska, H. Yamada, J.J. O'Shea and J.D. Ashwell, Tyrosine kinase-regulated and inositol phosphate-independent Ca elevation and mobilization in T cells, *J. Biol. Chem.* 267: 7154 (1992).
50. K.M. Lee, K. Toscas and M.L. Villereal, Inhibition of bradykinin- and thapsigargin-induced Ca entry by tyrosine kinase inhibitors, *J. Biol. Chem.* 268: 9945 (1993).

# THE ROLE OF SYK IN CELL SIGNALING

Robert L. Geahlen and Debra L. Burg

Department of Medicinal Chemistry
Purdue University
West Lafayette, IN 47907

## INTRODUCTION

Many multichain antigen receptors on the surfaces of immune cells signal by stimulating the phosphorylation of key intracellular proteins on tyrosine. This occurs even though the transmembrane components of the receptors themselves lack intrinsic enzymic activity. Instead, these receptors associate with and signal through cytoplasmic protein-tyrosine kinases. These kinases are now known to include members of a new family of proteins for which the prototype is a recently discovered enzyme known as $p72^{syk}$.

## CHARACTERIZATION OF THE SYK KINASE

### Identification of PTK72 and $p72^{syk}$

The $p72^{syk}$ protein-tyrosine kinase (Syk), or at least its catalytic domain, was originally identified the old-fashioned way: through protein purification. Interest in the enzyme grew out of early studies on the tissue distribution of normal cellular protein-tyrosine kinases. Work in our laboratory and others using synthetic, tyrosine-containing peptides as substrates indicated that the highest levels of peptide-phosphorylating activity were present in lymphocytes and organs of the immune system (1-3). This observation prompted several investigators to initiate studies aimed at the isolation and characterization of the enzymes responsible for this activity. Our own studies used a tyrosine-containing peptide corresponding in sequence to angiotensin I as a substrate. Angiotensin I was phosphorylated by an activity present in extracts of bovine thymus that was stimulated several-fold by high concentrations of NaCl (4). Fractionation of this activity by a sequence of column chromatographic steps led to the isolation of a 40 kDa enzyme, which was named p40 (5). A 40 kDa enzyme (CPTK40) with properties similar to p40 was subsequently isolated by Yamamura and coworkers from porcine spleen (6,7).

Initial studies on the effects of protease inhibitors on the recovery of the 40 kDa kinases from both thymus (5) and spleen (7) suggested that 40 kDa was, in fact, the native size of each enzyme. However, once polyclonal antibodies were raised against p40, it became apparent that this enzyme had actually arisen from the proteolytic cleavage of a higher molecular weight precursor. Anti-p40 antibodies recognized by immunoprecipitation and on Western blots a protein of 72 kDa that also had intrinsic protein-tyrosine kinase activity (8). This enzyme, originally named PTK72, was exquisitely sensitive to proteolysis, which generated a 40 kDa fragment with enhanced intrinsic phosphotransferase activity (8). A similar fate awaited the CPTK40 enzyme from porcine spleen. Oligonucleotide probes based on a partial amino acid sequence of the 40 kDa enzyme were used by Yamamura and coworkers (9) to clone a cDNA that encoded a protein with a predicted molecular weight of 72,000. This enzyme was named p72*syk*. Antibodies prepared against a peptide corresponding in sequence to a region located near the N-terminus of CPTK40 were used to verify that the 40 kDa spleen enzyme had also arisen by proteolysis of the larger molecular weight precursor. As of yet, no evidence exists for a physiological role for the proteolytic activation of Syk.

The original porcine Syk anti-peptide antibodies did not recognize an analogous 72 kDa protein in extracts of murine spleen (9), making it difficult to compare Syk with PTK72. However, these antibodies were prepared against a region of the porcine Syk sequence that we now know is not conserved in the murine enzyme. Antibodies prepared against a peptide corresponding to the C-terminal 28 amino acids of the porcine Syk sequence, which differs from the murine sequence by only a single amino acid, do recognize a 72 kDa tyrosine kinase from murine spleen. These antibodies recognize the same kinase that was identified by the original anti-p40 antibodies, indicating that Syk and PTK72 represent the same enzyme.

### Tissue Distribution of Syk

Examinations of the tissue distribution of Syk by either Western (8) or Northern blot analyses (9) indicate that the kinase is expressed at highest levels in spleen, and at lower levels in thymus, with little or no expression detected in most other organs. Syk is, however, widely distributed among hematopoietic cells. Among lymphocytes, Syk is found predominantly in B cells and is present in only very low levels in mature T cells (10). The expression of Syk has also been reported in basophils (11,12), monocytes (13,14), neutrophils (15), platelets (16) and red blood cells (17).

### Substrate Specificity

The amino acid sequences surrounding the sites of tyrosine phosphorylation on proteins that serve as substrates for Syk have been determined for a limited number of proteins. From our determinations, the best protein substrates for Syk contain tyrosines surrounded by an extensive number of acidic amino acids (18,19). The best substrate that we have identified is the cytoplasmic domain of the human erythrocyte anion transport channel, band 3 (18), which is phosphorylated near the N-terminus at a tyrosine residue contained within the following sequence:

Met-<u>Glu</u>-<u>Glu</u>-Leu-Asn-<u>Asp</u>-<u>Asp</u>-**Tyr**-<u>Glu</u>-<u>Asp</u>-<u>Asp</u>-Met-

An unrelated protein, glycogen synthase (19), is phosphorylated at an analogous site bearing the sequence:

Pro-<u>Glu</u>-<u>Glu</u>-<u>Asp</u>-Gly-<u>Glu</u>-Arg-**Tyr**-<u>Glu</u>-<u>Asp</u>-<u>Glu</u>-<u>Glu</u>-

Both of these peptides fit well into a proposed structure for the substrate binding site of Syk modeled on the published coordinates for the crystal structure of the cAMP-dependent protein kinase (PKA) bound to a 20-amino acid peptide inhibitor (20,21). An alignment of Syk and PKA sequences predict that a tight-binding substrate for Syk would have acidic amino acid residues at positions -2 and -6 (counting toward the N-terminus from the tyrosine) that would interact with $Arg^{502}$ and $Lys^{541}$ of the Syk catalytic domain. In PKA, the analogous amino acids $Glu^{170}$ and $Glu^{203}$ interact with Arg residues on the inhibitor peptide. Acidic amino acid residues distal to the tyrosine would also be expected to interact with a region of Syk containing the basic sequence $Gly^{536}$-Lys-Asn-Pro-Val-Lys-Trp-$Tyr^{543}$, which is present in place of the hydrophobic binding pocket of PKA that interacts with the hydrophobic amino acid at position +1 of the PKA inhibitor. Both of the substrates described above meet these predictions. However, as no crystal structure of a Syk/substrate complex is available, these alignments remain speculative. It is interesting to note, however, that recent evidence indicates that erythrocyte band 3 associates with and is phosphorylated by Syk in human red blood cells (17).

## A ROLE FOR SYK IN CELL SIGNALING

### Activation of Syk by the Engagement of Cell Surface Receptors

Clues to a possible physiological role for Syk came originally from studies on the mechanisms by which B cells are activated by the engagement of cell surface antigen receptors. The cross-linking of surface IgM by anti-IgM antibodies initiates a cascade of biochemical responses that include increases in the turnover of phosphatidylinositol lipids and in the intracellular concentration of free calcium (22). These responses are preceded by the activation of endogenous protein-tyrosine kinases and the phosphorylation on tyrosine of multiple cellular proteins. Among the prominent substrates identified in these early studies was a protein of 70-80 kDa (23,24) that Campbell and Sefton (24) predicted might represent a protein-tyrosine kinase. This observation prompted us to compare this endogenous B cell substrate with PTK72. Based on a combination of approaches involving immunoprecipitation, peptide mapping, affinity labeling and two-dimensional gel electrophoresis, we were able to confirm that PTK72 was, in fact, a protein that was rapidly tyrosine-phosphorylated in response to the engagement of antigen receptors (25). This tyrosine phosphorylation was accompanied by an increase in the intrinsic activity of the kinase (10).

The activation of Syk in response to receptor crosslinking is not restricted to B cells, a fact that might be predicted based on the widespread expression of Syk among different hematopoietic cells. The activation of a variety of immune cells occurs through the engagement of "multichain immune recognition receptors", which bear a number of structural and functional similarities (26). A common, early response to the aggregation of these receptors at the cell surface is an increase in the phosphorylation on tyrosine of proteins with molecular masses in the 70-80 kDa range. In many of these cells, this protein has now been identified as the Syk tyrosine kinase. This was first shown for the high affinity IgE receptor (FcεRI) on the rat mast cell line RBL-2H3, where receptor engagement leads to the phosphorylation and activation of Syk (11,12). Similarly, engagement of either high affinity (FcγRI) or low affinity (FcγRII) Fcγ receptors on monocytic cells leads to the stimulation of Syk (13,14). Activation of Syk has also been reported to occur in response to the treatment of platelets with thrombin (16) and of polymorphonuclear neutrophils with concanavalin A (15). In T cells, aggregation of the T cell antigen receptor leads to the rapid phosphorylation and activation of a 70 kDa kinase known as ZAP-70, which is distinct from, but structurally related to, Syk (27,28). Thus, the phosphorylation and activation of Syk-family enzymes appears to be a common mechanism in cells of the immune system for the transduction of signals initiated by the engagement of a variety of cell surface receptors.

### Association of Syk with Antigen Receptors

The mechanisms by which the activity of Syk is elevated following receptor aggregation are not yet well understood, but it is likely that phosphorylation plays a role since an increase in the state of tyrosine-phosphorylation of Syk in activated cells is invariably accompanied by an increase in its activity. Exogenous agents added to cells that alter the state of phosphorylation of Syk also result in an increase in Syk activity. For example, the treatment of cells with oxidizing agents such as $H_2O_2$ (29) or with pervanadate (formed from sodium orthovanadate and $H_2O_2$) (17), both of which inhibit the activities of intracellular phosphotyrosine phosphatases, leads to increases in both the tyrosine phosphorylation and activation of Syk. Chimeric proteins constructed with a CD16 extracellular domain and a Syk intracellular domain signal when crosslinked on the surface of transfected cytolytic T cells, suggesting that simple aggregation might be a sufficient signal to activate the Syk kinase (30). Such an activation might occur through the intermolecular transphosphorylation of aggregated Syk molecules or from aggregation-induced conformational changes in the enzyme.

For the aggregation of Syk to occur during the crosslinking of cell surface receptors, the enzyme must either be present as a component of the receptor complex or be recruited to the receptor following its engagement. In murine spleen B lymphocytes or in any one of several B cell lines that express sIgM, a small amount of Syk can be co-immunoprecipitated with the IgM complex prior to receptor crosslinking providing that mild detergents are used to lyse the cells (10). An example of this is shown in Fig. 1 where an anti-IgM immune-complex was isolated from L10.A B cells, which were lysed in the presence of either digitonin or Brij 96. In this experiment, the enzyme was detected in anti-IgM immune-complexes by its autophosphorylation in the presence of [$\gamma$-$^{32}$P]ATP. Phosphoproteins were separated by SDS-polyacrylamide gel electrophoresis, and detected by autoradiography. Several prominent phosphoproteins can be detected in these complexes that include the 72 kDa Syk kinase, a 53-56 kDa doublet of proteins that corresponds to the two alternatively spliced variants of the Src-family kinase Lyn, and the IgM-associated proteins B29 and MB-1. Similarly, Syk can be demonstrated to associate with the IgM receptor complex from human B cells (31) and the Fc$\varepsilon$RI receptor in mast cells (11). In addition to Syk and Lyn (32), multiple members of the Src-family of kinases have been reported to associate with the antigen receptor complex prior to activation (33).

The mechanism by which Syk associates with components of the antigen receptor complex are currently under investigation. Both Syk and ZAP-70 lack the SH3 domain, N-terminal myristoylation site and C-terminal negative regulatory site of tyrosine phosphorylation that distinguish the Src-family of kinases (34). However, Syk and ZAP-70 contain two tandem SH2 domains, which are protein motifs known to mediate protein-protein interactions that involve the recognition of sites containing phosphotyrosyl residues (35). For ZAP-70, these SH2 domains cooperate to mediate the interaction of ZAP-70 with the T cell receptor $\zeta$ and CD3 $\varepsilon$ chains, which become phosphorylated on tyrosine following receptor crosslinking (36). The MB-1 and B29 components of the B cell antigen receptor are also tyrosine-phosphorylated following the crosslinking of the receptor (37). While the enhanced binding of ZAP-70 to the T cell antigen receptor in activated cells has been clearly demonstrated (27,28), the analogous experiment has proven difficult to perform in B cells due to difficulties in the isolation of intact antigen receptor complexes from activated cells. For example, digitonin, which is a mild detergent used to solubilize intact receptor complexes from unactivated cells, does not readily solubilize such complexes from cells in which the receptors have been aggregated with anti-IgM antibodies (Fig. 1). Stronger detergents, such as NP-40, dissociate the components of the antigen receptor complex (38).

DeFranco and coworkers have circumvented this problem by expressing hybrid, transmembrane proteins bearing portions of the cytoplasmic domains of MB-1 (Ig-$\alpha$) and B29 (Ig-$\beta$) in a mIgM$^-$ B cell line (39). The crosslinking of these chimeric proteins resulted in their phosphorylation on tyrosine and enhanced their association with Fyn, Lyn and Syk. The stability of the Syk-receptor interaction is consistent with that expected of a SH2 domain-

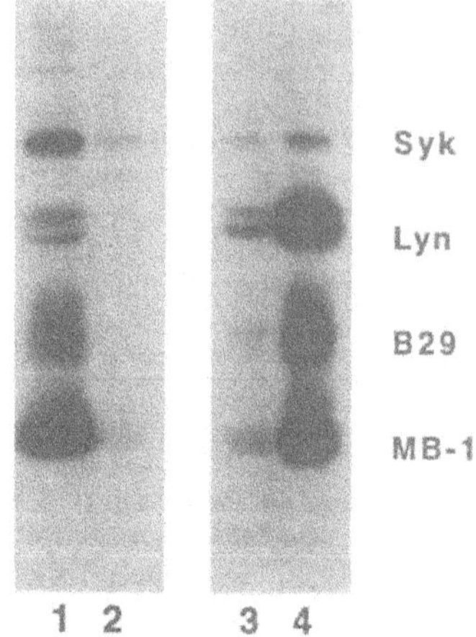


**Figure 1.** Effect of detergents on the immunoprecipitation of receptor-kinase complexes. Cells were lysed in the presence of buffers containing either 1% digitonin (*lanes 1 and 2*) or 1% Brij 96 (*lanes 3 and 4*) prior to (*lanes 1 and 3*) or following (*lanes 2 and 4*) the crosslinking of sIgM with anti-IgM antibodies. Proteins phosphorylated in anti-IgM immune-complexes with [$\gamma$-$^{32}$P]ATP were separated by SDS-PAGE and detected by autoradiography. Lanes 1 and 2 were exposed to X-ray film for a longer period of time than lanes 3 and 4; the actual intensities of bands in lanes 1 and 3 are comparable.

phosphotyrosyl peptide interaction. We have found that similar data can be obtained from cells bearing native antigen receptors if the detergent Brij 96 is used to solubilize IgM complexes from activated cells (Fig. 1). These complexes display enhanced levels of both Syk and Lyn activity. Kinetic analyses of the formation of these complexes indicate that Lyn is the first of the two kinases to appear. These data are consistent with a model in which receptor engagement results in the formation of a signaling complex to which is first recruited enzymes of the Src-family. Formation of this initial complex is then followed by the binding of Syk, perhaps to tyrosine-phosphorylated components of the receptor complex, and its subsequent phosphorylation and activation. It will be interesting in the future to determine if activated Syk then leaves the receptor complex to interact with and phosphorylate downstream mediators of the activation response.

## ACKNOWLEDGEMENTS

We thank Dr. Mark S. Cushman of Purdue University for the molecular modeling studies. This work was supported by NIH grant CA37372 to RLG. DLB was supported by a fellowship from the American Heart Association, Indiana Affiliate, Inc.

## REFERENCES

1. G. Swarup, J.D. Dasgupta, and D.L. Garbers, Tyrosine protein kinase activity of rat spleen and other tissues, *J. Biol. Chem.* 258:10341 (1983).
2. F.P.D. Tuy, J. Henry, C. Rosenfeld, and A. Kahn, High tyrosine kinase activity in normal nonproliferating cells, *Nature* 305:435 (1983).
3. M.L. Harrison, P.S. Low, and R.L. Geahlen, T and B lymphocytes express distinct tyrosine protein kinases, *J. Biol. Chem.* 259:9348 (1984).
4. D.L. Mason, M.L. Harrison, and R.L. Geahlen, Properties of a tyrosine protein kinase from calf thymus. Response to ionic strength and divalent cations, *Biochim. Biophys. Acta* 829:221 (1985).
5. T.F. Zioncheck, M.L. Harrison, and R.L. Geahlen, Purification and characterization of a protein-tyrosine kinase from bovine thymus, *J. Biol. Chem.* 261:15637 (1986).
6. K. Sakai, S.-i. Nakamura, K. Sada, T. Kobayashi, H. Uno, and H. Yamamura, Characterization of partially purified cytosolic protein-tyrosine kinase from porcine spleen, *Biochem. Biophys. Res. Commun.*.
7. T. Kobayashi, S.-i. Nakamura, T. Taniguchi, and H. Yamamura, Purification and characterization of a cytosolic protein-tyrosine kinase from porcine spleen, *Eur. J. Biochem.* 188:535 (1990).
8. T.F. Zioncheck, M.L. Harrison, C.C. Isaacson, and R.L. Geahlen, Generation of an active protein-tyrosine kinase from lymphocytes by proteolysis, *J. Biol. Chem.* 263:19195 (1988).
9. T. Taniguchi, T. Kobayashi, J. Kondo, K. Takahashi, H. Nakamura, J. Suzuki, K. Nagai, T. Yamada, S. Nakamura and H. Yamamura, Molecular cloning of a porcine gene Syk that encodes a 72-kDa protein-tyrosine kinase showing high susceptibility to proteolysis, *J. Biol. Chem.* 266:15790 (1991).
10. J.E. Hutchcroft, M.L. Harrison, and R.L. Geahlen, Association of the 72-kDa protein-tyrosine kinase PTK72 with the B cell antigen receptor, *J. Biol. Chem.* 267:8613 (1992).
11. J.E. Hutchcroft, R.L. Geahlen, G.G. Deanin, and J.M. Oliver, FcεRI-mediated tyrosine phosphorylation and activation of the 72-kDa protein-tyrosine kinase, PTK72, in RBL-2H3 rat tumor mast cells, *Proc. Natl. Acad. Sci. U.S.A.* 89:9107 (1992).
12. M. Benhamou, N.J.P. Ryba, H. Kihara, H. Nishikata, and R.P. Siraganian, Protein-tyrosine kinase $p72^{syk}$ in high affinity IgE receptor signaling, *J. Biol. Chem.* 268:23318 (1993).
13. A. Agarwal, P. Salem, and K.C. Robbins, Involvement of $p72^{syk}$, a protein-tyrosine kinase, in Fcγ receptor signaling, *J. Biol. Chem.* 268:15900 (1993).

14. P.A. Kiener, B.M. Rankin, A.L. Burkhardt, G.L. Shieven, L.K. Gilliland, R.G. Rowley, J.B. Bolen, and J.A. Ledbetter, Cross-linking of Fcγ receptor I (FcγRI) and receptor II (FcγRII) on monocytic cells activates a signal transduction pathway common to both Fc receptors that involves the stimulation of p72 Syk protein tyrosine kinase, *J. Biol. Chem.* 268:24442 (1993).
15. M. Asahi, T. Taniguchi, E. Hashimoto, T. Inazu, H. Maeda, and H. Yamamura, Activation of protein-tyrosine kinase $p72^{syk}$ with concanavalin A in polymorphonuclear neutrophils, *J. Biol. Chem.* 268:23334 (1993).
16. T. Taniguchi, H. Kitagawa, S. Yasue, S. Yanagi, K. Sakai, M. Asahi, S. Ohta, F. Takeuchi, S.-i. Makamura, and H. Yamamura, Protein-tyrosine kinase $p72^{syk}$ is activated by thrombin and is negatively regulated through $Ca^{2+}$ mobilization in platelets, *J. Biol. Chem.* 268:2277 (1993).
17. M.L. Harrison, C.C. Isaacson, D.L. Burg, R.L. Geahlen, and P.S. Low, Phosphorylation of human erythrocyte band 3 by endogenous $p72^{syk}$, *J. Biol. Chem.* (1993) In Press.
18. P.S. Low, D.P. Allen, T.F. Zioncheck, P. Chari, B.M. Willardson, R.L. Geahlen, and M.L. Harrison, Tyrosine phosphorylation of band 3 inhibits peripheral protein binding, *J. Biol. Chem.* 262:4592 (1987).
19. A.M. Mahrenholz, P. Votaw, P.J. Roach, A.A. DePaoli-Roach, T.F. Zioncheck, M.L. Harrison, and R.L. Geahlen, Phosphorylation of glycogen synthase by a bovine protein-tyrosine kinase, p40," *Biochem. Biophys. Res. Commun.* 155:52 (1988).
20. D.R. Knighton, J. Zheng, L.F. Ten Eyck, V.A. Ashford, N.-h. Xuong, S.S. Taylor, and J.M. Sowadski, Crystal structure of the catalytic subunit of cyclic adenosine monophosphate-dependent protein kinase, *Science* 253:407 (1991).
21. D.R. Knighton, J. Zheng, L.F. Ten Eyck, N.-h. Xuong, S.S. Taylor, and J.M. Sowadski, Structure of a peptide inhibitor bound to the catalytic subunit of cyclic adenosine monophosphate-dependent protein kinase, *Science* 253:414 (1991).
22. J. C. Cambier, and J.T. Ransom, Molecular mechanisms of transmembrane signaling in B lymphocytes, *Ann. Rev. Immunol.* 5:175 (1987).
23. M.R. Gold, D.A. Law, and A.L. DeFranco, Stimulation of protein tyrosine phosphorylation by the B-lymphocyte antigen receptor, *Nature* 345:810 (1990).
24 M.-A. Campbell, and B.M. Sefton, Protein tyrosine phosphorylation is induced in murine B lymphocytes in response to stimulation with anti-immunoglobulin, *EMBO J.* 9:2125 (1990).
25. J. E. Hutchcroft, M.L. Harrison, and R.L. Geahlen, B lymphocyte activation is accompanied by phosphorylation of a 72-kDa protein-tyrosine kinase, *J. Biol. Chem.* 266:14846 (1991).
26. A.D. Keegan, and W.E. Paul, Multichain immune recognition receptors: similarities in structure and signaling pathways, *Immunol. Today* 13:63 (1992).
27. A.C. Chan, B. Irving, J.D. Fraser, and A. Weiss, The TCR ζ chain associates with a tyrosine kinase and upon TCR stimulation associates with ZAP-70, a 70K Mr tyrosine phosphoprotein, *Proc. Natl. Acad. Sci. U.S.A.* 88:9166 (1991).
28. A.C . Chan, M. Iwashima, C. Turck, and A. Weiss, ZAP-70: a 70 kD protein tyrosine kinase that associates with the TCR ζ-chain, *Cell* 71:649 (1992).
29. G.L. Schieven, J.M. Kirihara, D.L. Burg, R.L. Geahlen, and J.A. Ledbetter, $p72^{syk}$ tyrosine kinase is activated by oxidizing conditions that induce lymphocyte tyrosine phosphorylation and $Ca^{2+}$ signals, *J. Biol. Chem.* 268:16688 (1993).
30. W. Kolanus, C. Romeo, and B. Seed, T cell activation by clustered tyrosine kinases, *Cell* 74:171-183.
31. C. Leprince, K.E. Draves, R.L. Geahlen, J.A. Ledbetter, and E.A. Clark, CD22 associates with the human sIgM/B cell antigen receptor complex," *Proc. Natl. Acad. Sci. USA* 90:3236 (1993).
32. Y. Yamanashi, T. Kakiuchi, J. Mizuguchi, Y. Tadashi, and K. Toyoshima, Association of B cell antigen receptor with protein tyrosine kinase Lyn, *Science* 251:192 (1991).
33. A.L. Burkhardt, M. Brunswick, J.B. Bolen, and J.J. Mond, Anti-immunoglobulin stimulation of B lymphocytes activates src-related protein-tyrosine kinases, *Proc. Natl. Acad. Sci. U.S.A.* 88:7410 (1991).
34. J.B. Bolen, P.A. Thompson, E. Eiseman, and I.D. Horak, Expression and interactions of the Src family of tyrosine protein kinases in T lymphocytes, *Adv. Cancer Res.* 57:103 (1991).

35. C.A. Koch, D. Anderson, M.F. Moran, C. Ellis, and T. Pawson, SH2 and SH3 domains: elements that control interactions of cytoplasmic signaling proteins, *Science* 252:668 (1991).
36. R.L. Wange, S.N. Malek, S. Desiderio, and L.E. Samelson, Tandem SH2 domains of ZAP-70 bind to T cell antigen receptor ζ and CD3ε from activated Jurkat T cells, *J. Biol. Chem.* 268:19797 (1993).
37. M.R. Gold, L. Matsuuchi, R.B. Kelly, and A.L. DeFranco, Tyrosine phosphorylation of components of the B-cell antigen receptors following receptor crosslinking, *Proc. Natl. Acad. Sci. U.S.A.* 88:3436 (1991).
38. J. Lin, and L.B. Justement, The MB-1/B29 heterodimer couples the B cell antigen receptor to multiple src family protein tyrosine kinases, *J. Immunol.* 149:1548 (1992).
39. D.A. Law, V. W.-F. Chan, S.K. Datta, and A.L. DeFranco, B-cell antigen receptor motifs have redundant signalling capabilities and bind the tyrosine kinases PTK72, Lyn and Fyn, *Curr. Biol.* 3:645 (1993).

# NONRECEPTOR TYROSINE KINASES IN AGGREGATION-MEDIATED CELL ACTIVATION

Brian Seed, Waldemar Kolanus, Charles Romeo, Ramnik Xavier

Department of Molecular Biology
Massachusetts General Hospital
Boston, MA 02114

## INTRODUCTION

Many of the cellular recognition events in the immune system which are mediated by cell surface receptors are initiated by cell-cell contacts which result in receptor aggregation. Almost universally, the relevant receptors lack intrinsic enzymatic activity. However a growing body of evidence suggests that tyrosine phosphorylation is an early concomitant of cellular activation initiated by aggregation of cell surface receptors, and in some cases a causal role for tyrosine kinases in the activation process can be identified.

PTK inhibitors block the calcium mobilization that is one of the earliest consequences of aggregation of B and T cell antigen receptors. Subsequent events such as cytokine release and cellular proliferation are also compromised by treatment of cells with PTK inhibitors[1-4]. Although the cellular programs initiated by receptor activation differ according to cell type, the early events are remarkably similar among cells from disparate hematopoietic lineages. Increases in PTK activity can be seen after crosslinking of antigen receptors[5-7] and certain forms of Ig Fc receptor[8,9], and one early consequence of the PTK activity appears to be phosphorylation of the $\gamma 1$ isoform of phosphatidylinositol-specific phospholipase C[9-14], which is known to be activated by tyrosine phosphorylation[15].

Among lymphocytes two classes of tyrosine kinase have been found to associate with cell surface receptors which recognize aggregated ligands, members of the large family of nonreceptor tyrosine kinases homologous to the prototypic oncogene Src; and members of the small family of nonreceptor kinases homologous to the prototypic enzyme syk. The kinases of the former class are constitutively associated with the cell surface by dint of their acylation at the amino terminus with tetradecanoic acid, also known as myristic acid, while the kinases of the latter class appear to associate with the intracellular domains of antigen and related receptors only following an initial

*Mechanisms of Lymphocyte Activation and Immune Regulation V*
Edited by S. Gupta *et al.*, Plenum Press, New York, 1994

phosphorylation event which creates an appropriate site for their binding.

Engagement of cellular effector programs mediated by the T cell and B cell antigen receptors, and various forms of Fc receptor, can be mimicked by crosslinking of chimeric proteins bearing the intracellular domains of individual chains of the receptor complexes[16-21]. The minimal effective trigger element appears to require a phylogenetically conserved[22] peptide sequence containing two tyrosine residues separated by 10 or 11 residues and embedded in a hydrophilic, typically acidic context[23,24].

The finding that a very short intracellular domain is the active element for initiation of signal transduction suggests that its principal role may be to provide a site for interaction with a larger protein which has enzymatic activity. The available evidence suggests that protein might be a tyrosine kinase, so it is reasonable to inquire which if any of the candidate tyrosine kinases can be activated by aggregation. We accordingly explored the possibility that nonreceptor kinases might be activated by clustering[25]. We created model receptor-like kinases by adding an extracellular domain and membrane spanning segment to nonreceptor kinases which consisted of the complete src or syk family kinase sequences. We then followed the consequences of aggregation of these synthetic receptors by antibodies directed against the extracellular domain. A clear distinction was observed between the syk and src family kinases, for aggregation of the src-type kinases did not lead to significant cellular activation, while the syk-type kinases elicited an increase in free intracellular calcium ion and, in the case of syk, of redirected cytolysis. The zap-70 chimera could not induce cytolysis by itself directly, but when paired with either a fyn or lck kinase chimera, could mediate an activity equal in potency to that of syk itself.

Two formal alternatives for enzymatic activation following receptor crosslinking can be advanced: that receptor aggregation leads to a conformational change or enzyme/substrate localization which results in increased enzymatic activity indirectly; or that the relevant enzymatic activities are themselves capable of being mobilized by aggregation. To test this hypothesis we created chimeric transmembrane forms of the known receptor-associated tyrosine kinases and artificially aggregated them by crosslinking the extracellular domains of the chimera with either antibodies or cellular targets recognized by the extracellular domains. If antigen receptor activation of tyrosine kinases proceeds through the indirect route, the relevant kinases will be refractory to activation by aggregation, while direct activation by aggregation should be demonstrable as a recapitulation of the cellular activation events following receptor aggregation.

## RESULTS

The chimeric transmembrane kinases were created by genetic fusion of a DNA fragment encoding the extracellular domain of CD16B, the phosphatidyl inositol-linked variant of a low affinity IgG receptor found on natural killer cells, neutrophils and macrophages. For the purposes of these studies, CD16 acts as a generic extracellular domain to which monoclonal antibodies and a suitable hybridoma can be conveniently found. The extracellular domain of CD16 was joined to a segment from the CD7 antigen which extends from the end of the single Ig-like domain of CD7 through the membrane and terminates just after the basic stop transfer sequences which are thought to anchor the transmembrane domain in the cytoplasm. Between the end of the Ig-like domain and the beginning of the transmembrane domain lies a proline rich sequence which we have suggested, without formal demonstration, might

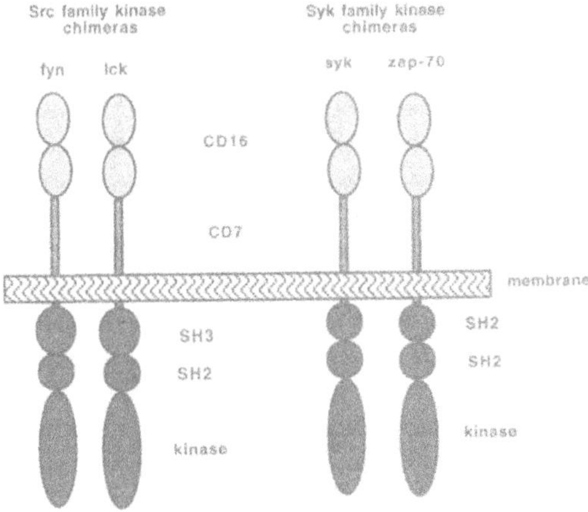


**Figure 1.** Schematic diagram of tripartite kinase chimeras designed to permit aggregation by external crosslinking with antibodies.

encode a stalk-like structure which projects the Ig-like domain away from the cell. The intracellular domain was joined in turn to the complete coding sequences of the human lck[26], murine fyn(T)[27], porcine[28] or human[25] syk, and human zap-70[29] tyrosine kinases (Fig. 1).

Genes encoding the CD16/CD7/kinase tripartite chimeras were introduced into recombinant vaccinia viruses by homologous recombination at the thymidine kinase locus with selection for coexpression of the *E. coli gpt* gene product. Cells infected with the recombinants showed efficient cell surface expression of all four kinase chimeras, and immunoprecipitation of the chimeras with anti- CD16 antibodies gave proteins of the expected molecular masses which were active in an in vitro phosphorylation assay[25].

Three types of assay were then applied to determine the ability of the kinase chimeras to behave in a manner reminiscent of the T cell receptor. The first was mobilization of free intracellular calcium ion, as measured by the change in either quantum yield or emission wavelength of calcium-chelating fluorescent dyes. Both spectrofluorimetric (bulk population) and flow cytometric (single cell) measurements were performed, with cells loaded with the dyes Fluo-3 and Indo- 1. Following infection with the vaccinia recombinants and loading with the acetomethoxy esters of the fluorophores, flow cytometric analyses were performed with gating for cells expressing CD16, using a phycoerythrin-conjugated second antibody to detect the CD16 expression. In this protocol the time course of crosslinking can be determined by following the increase in phycoerythrin fluorescence as a function of time, and saturation was routinely observed within ten seconds after addition of the fluorochrome conjugated second antibody. (This is well before the earliest indication of calcium mobilization). In Jurkat cells (a human T cell leukemia cell line) neither lck nor fyn chimeras could initiate an increase in calcium autonomously. Aggregation of fusion proteins based on zap-70 was highly effective in promoting the appearance of free cytoplasmic calcium ion, roughly as effective as aggregation of a similar chimera bearing the intracellular domain of the T cell receptor zeta chain. Cells expressing the syk chimera showed a high resting calcium ion concentration, as if the cells were constitutively activated. Above this high basal expression a minor increase in intracellular free calcium ion could be demonstrated. The calcium response to both zap-70 and syk kinase chimera crosslinking was delayed relative to the onset of calcium mobilization mediated by zeta chimera crosslinking.

The second measure of T cell receptor-mimetic action was the induction of redirected cytolysis against an anti-CD16 hybridoma cell line. The target cells were a subline of hybridoma cells which were selected for increased expression of cell surface IgG antibody[30]. A fixed number of hybridoma cells were labeled by incorporation of $^{51}$Cr-chromate and mixed with increasing numbers of effector cells. The effector cells were prepared by infection of a human allospecific cytotoxic T lymphocyte (CTL) line with vaccinia recombinants expressing the CD16/CD7/kinase fusion proteins. The relative efficiency of cytolysis in an assay of this sort is determined by recording the ratio of effector to target cells which is required to achieve a given proportion of release of the incorporated $^{51}$Cr-chromate. CTL expressing chimeric receptors bearing the lck or fyn kinases, or the syk family kinase zap-70, were incapable of redirecting cytolysis against the anti-CD16 hybridoma cells. However CTL expressing a syk kinase chimera were highly effective, approximately equipotent with CTL expressing a chimera composed of CD16 fused to the intracellular domain of the T cell receptor zeta chain. The cytolytic activity directed by the CD16/CD7/kinase chimeras was not due to nonspecific cytolysis (e.g., some form of bystander killing effect) because mixture of an irrelevant chromium- loaded cell with the CTL did not result in significant evidence of cytolysis.

The finding that syk has greater capacity to activate cytolysis than zap-70 is consistent with the high constitutive activity of syk in the calcium flux assay, but it was not expected that zap-70 would lack cytolytic potential entirely. In an effort to reconstitute cytolytic activity mediated by zap-70 we coinfected CTL with pairs of recombinant vaccinia viruses encoding chimeric zap-70 and lck, zap- 70 and fyn, or lck and fyn. CTL expressing both zap-70 and fyn, or zap-70 and lck were essentially equipotent with CTL expressing syk kinase chimera alone. CTL coexpressing lck and fyn chimeras did not show substantial cytolysis[25].

One possible explanation for the costimulatory activity of the src family kinases invokes their constitutive phosphorylation of either zap-70 or another intracellular

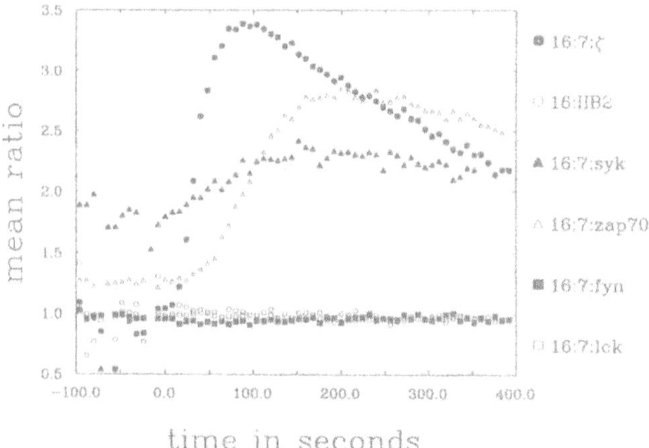


**Figure 2.** Calcium mobilization by cells expressing kinase chimeras after aggregation of extracellular domains with antibodies[25].

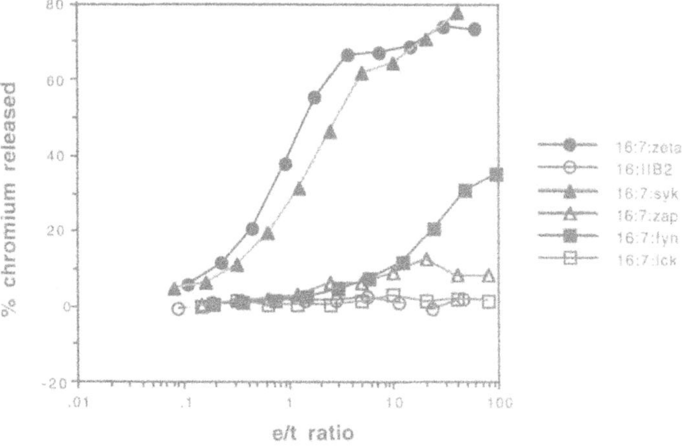


**Figure 3.** Redirected cytolysis against hybridoma cells expressing cell surface antibody to kinase chimeras[25].

target. Another invokes an aggregation-sensitive cascade. To distinguish between these alternatives we prepared a CD4/fyn chimera in which the extracellular and transmembrane domains of CD4 were joined to fyn in a similar fashion to that used for the CD16 chimeras. The CD4/fyn chimera was ten to twenty fold less active than the analogous CD16 chimera, suggesting that the second explanation, requiring coclustering of the zap-70 and src family kinases, better accounted for the costimulation mechanism[25]. However the cytolytic activity of cells expressing CD4/fyn and CD16/cd7/zap-70 was significantly greater than that of cells expressing chimeric zap-70 only. It may be that the increased activity in this case is due to nonspecific association of the CD4/fyn chimera with the CD16/CD7/zap-70 chimera.

Cellular activation following aggregation of chimeric kinases could be the result of association of the kinases with other signal transduction elements, such as the antigen receptors, rather than through the activation of the kinase domain itself. However in such a case it would be expected that the loss of kinase activity would be immaterial to the potency of the kinase chimeras for activation. However a point mutation which abolishes the ability of the kinase to transfer phosphate from ATP also abolishes the ability of the kinase to activate cells[25].

The third assay for T cell receptor-like action of kinase chimeras was the pattern of tyrosine phosphorylated proteins. T cell receptor negative cells infected with vaccinia recombinants bearing kinase chimeras were treated with crosslinking antibodies, and total cellular lysates of the treated cells were prepared by sodium dodecyl sulfate lysis and sonication. The cellular proteins were fractionated by electrophoresis, transfered to nitrocellulose membranes and probed with a monoclonal antibody specific for phosphotyrosine. Aggregation of chimeras bearing syk, zap-70, or fyn plus zap-70 resulted in the increased phosphorylation of several proteins which had apparent molecular masses similar to those of proteins which became phosphorylated on tyrosine after antigen receptor crosslinking[25]. The pattern of tyrosine phosphoproteins induced by aggregation of syk chimeras was particularly reminiscent of the pattern induced by anti-TCR antibody in TCR-positive cells (Fig. 4).

Because it is known that TCR crosslinking leads to phosphorylation of phospholipase C-$\gamma$1, we crosslinked chimeras, precipitated PLC-$\gamma$1 with a mixture of monoclonal antibodies, and analyzed nitrocellulose blots of the fractionated immunoprecipitates for phosphotyrosine. Clustering of syk gave an easily detectable increase in the phosphotyrosine content of PLC-γ1, and cocrosslinking of fyn plus zap-70 chimeras gave a smaller but readily detectable increase in PLC-γ1 phosphotyrosine[25]. Cells expressing CD16/CD7/fyn were found to exhibit a weak constitutive phosphorylation of PLC-γ1 which could not be induced by receptor aggregation[25].

These studies show that syk family kinase aggregation can lead to phosphorylation on cellular PLC-γ1, but do not establish whether the observed increases in tyrosine phosphorylation can be directly attributed to the syk kinases, or whether other kinases activated by syk aggregation act in turn to phosphorylate the species observed. To identify which targets of syk activation were substrates for syk directly we prepared bacterial fusions between glutathione-S- transferase and short peptide segments corresponding to the intracellular domains of cell surface molecules of interest, and to the relevant portion of PLC-γ1 known to be phosphorylated as a result of lymphocyte activation. The ability of the various fusions to act as substrates was evaluated by incubating purified soluble substrate proteins with immunoprecipitated kinase chimeras. In these assays many of the known substrates for src family kinases are not recognized, or very inefficiently utilized, by syk family kinases. However the intracellular domain of the low affinity IgG Fc receptor subtype IIA is a good substrate for both syk and src-type kinases, as is a peptide bearing the central tyrosine of PLC-γ1 thought to mediate phospholipase activation. Both TCR zeta chain and the intracellular domain of CD28 can act as substrates for syk and src-type kinases as well.

## DISCUSSION

One explanation for activation of immune system cells by nonreceptor tyrosine kinases requires that a receptor-associated kinase become activated either by prox-

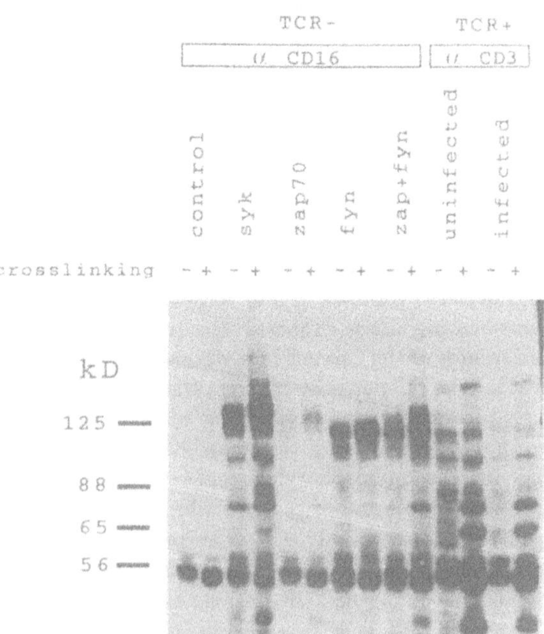


**Figure 4.** Change in pattern of tyrosine phosphorylation in cells expressing kinase chimeras after aggregation of extracellular domains with antibodies[25].

imity (e.g. by forming active kinase dimers) or by mutual enzymatic action (e.g., crossphosphorylation); the activated enzyme then acts on the intracellular substrates required for calcium mobilization and inositol phosphate synthesis. Support for the mutual enzymatic action mechanism can be found in studies reporting increases in receptor-associated kinase activity following receptor crosslinking[8,31–35]. However the reported changes in kinase activity are in most cases modest, and contrast with the dramatic changes in phosphotyrosine proteins seen in vivo. Because it is difficult to unambiguously rule out activation by weak allosteric interactions using in vitro kinase assays as a tool, we cannot at this point make a definitive statement about the relative importance of kinase activation in the initiation of signal transduction. But the data presented here suggest that aggregation- induced repartitioning of enzyme and substrate may be an important factor in directing an existing activity toward the appropriate physiological target.

Although in B cells the requirement for a src family kinase in activation has not been definitively established, in T cells two kinases, lck and fyn(T) have been shown by somatic or organismic genetics to play important roles[36–69]. At present we cannot conclusively establish whether the action of these kinases normally precedes or follows the action of the syk family kinases. One hypothesis accounting for the action of zap-70 in T cell activation invokes a receptor-associated src family kinase whose aggregation permits a transitory phosphorylation of receptor chains in turn leading to association of zap-70 and subsequent cellular activation.

The results discussed here suggest that syk family kinases act more directly on the effector apparatus of T cells than src family kinases. Src family kinases have been reported to associate with a number of cell surface molecules which do not belong to the antigen/Fc receptor family, including CD2[40], CD23[41], CD36[42] (and J. Papkoff, pers. comm.), IL-2 receptor beta chain[43] and various phosphatidylinositol anchored proteins[44,45], some of which are known to require the additional presence of at least one antigen receptor chain to induce activation in T cells. The most straightforward explanation for the antigen receptor component is that the motifs on the antigen receptor chains are phosphorylated by src family kinases activated or repartitioned by clustering of their affiliated coreceptors. Subsequent docking of the syk family kinases, followed perhaps by some modifying event such as phosphorylation, then facilitates activation. However the studies above suggest that one of the most important requirements for activation may be aggregation of the syk family kinases themselves.

## REFERENCES


1. June, C.H., M.C. Fletcher, J.A. Ledbetter, G.L. Schieven, J.N. Siegel, A.F. Phillips, and L.E. Samelson, Inhibition of tyrosine phosphorylation prevents T-cell receptor-mediated signal transduction, *Proc. Natl. Acad. Sci. USA* 87:7722 (1990).
2. Lane, P.J., J.A. Ledbetter, F.M. McConnell, K. Draves, J. Deans, G.L. Schieven, and E.A. Clark, The role of tyrosine phosphorylation in signal transduction through surface Ig in human B cells. Inhibition of tyrosine phosphorylation prevents intracellular calcium release, *J. Immunol.* 146:715 (1991).
3. Mustelin, T., K.M. Coggeshall, N. Isakov, and A. Altman, T cell antigen receptor-mediated activation of phospholipase C requires tyrosine phosphorylation, *Science* 247:1584 (1990).
4. Stanley, J.B., R. Gorczynski, C.K. Huang, J. Love, and G.B. Mills, Tyrosine phosphorylation is an obligatory event in IL-2 secretion, *J. Immunol.* 145:2189 (1990).
5. Gold, M.R., D.A. Law, and A.L. DeFranco, Stimulation of protein tyrosine phosphorylation by the B-lymphocyte antigen receptor, *Nature* 345:810 (1990).
6. Campbell, M.A., and B.M. Sefton, Protein tyrosine phosphorylation is induced in murine B lymphocytes in response to stimulation with anti-immunoglobulin, *EMBO J.* 9:2125 (1990).

7. June, C.H., M.C. Fletcher, J.A. Ledbetter, and L.E. Samelson, Increases in tyrosine phosphorylation are detectable before phospholipase C activation after T cell receptor stimulation, *J. Immunol.* 144:1591 (1990).
8. Eiseman, E., and J.B. Bolen, Engagement of the high-affinity IgE receptor activates src protein-related tyrosine kinases, *Nature* 355:78 (1992).
9. Li, W., G.G. Deanin, B. Margolis, J. Schlessinger, and J.M. Oliver, Fc epsilon R1-mediated tyrosine phosphorylation of multiple proteins, including phospholipase C gamma 1 and the receptor beta gamma 2 complex, in RBL-2H3 rat basophilic leukemia cells, *Mol. Cell Biol.* 12:3176 (1992).
10. Carter, R.H., D.J. Park, S.G. Rhee, and D.T. Fearon, Tyrosine phosphorylation of phospholipase C induced by membrane immunoglobulin in B lymphocytes, *Proc. Natl. Acad. Sci. USA* 88:2745 (1991).
11. Park, D.J., H.W. Rho, and S.G. Rhee, CD3 stimulation causes phosphorylation of phospholipase C-gamma 1 on serine and tyrosine residues in a human T-cell line, *Proc. Natl. Acad. Sci. USA* 88:5453 (1991).
12. Park, D.J., H.K. Min, and S.G. Rhee, IgE-induced tyrosine phosphorylation of phospholipase C-gamma 1 in rat basophilic leukemia cells, *J. Biol. Chem.* 266:24237 (1991).
13. Secrist, J.P., L.A. Burns, L. Karnitz, G.A. Koretzky, and R.T. Abraham, Stimulatory effects of the protein tyrosine phosphatase inhibitor, pervanadate, on T-cell activation events, *J. Biol. Chem.* 268:5886 (1993).
14. Weiss, A., G. Koretzky, R.C. Schatzman, and T. Kadlecek, Functional activation of the T-cell antigen receptor induces tyrosine phosphorylation of phospholipase C-gamma 1, *Proc. Natl. Acad. Sci. USA* 88:5484 (1991).
15. Nishibe, S., M.I. Wahl, S.M. Hernandez-Sotomayor, N.K. Tonks, S.G. Rhee, and G. Carpenter, Increase of the catalytic activity of phospholipase C-gamma 1 by tyrosine phosphorylation, *Science* 250:1253 (1990).
16. Irving, B.A., and A. Weiss, The cytoplasmic domain of the T cell receptor zeta chain is sufficient to couple to receptor-associated signal transduction pathways, *Cell* 64:891 (1991).
17. Kolanus, W., C. Romeo, and B. Seed, Lineage-independent activation of immune system effector function by myeloid Fc receptors, *EMBO J.* 11:4861 (1992).
18. Letourneur, F. and R.D. Klausner, Activation of T cells by a tyrosine kinase activation domain in the cytoplasmic tail of CD3$\epsilon$, *Science* 255:79 (1992).
19. Letourneur, F., and R.D. Klausner, T-cell and basophil activation through the cytoplasmic tail of T-cell-receptor zeta family proteins, *Proc. Natl. Acad. Sci. USA* 88:8905 (1991).
20. Romeo, C. and B. Seed, Cellular immunity to HIV activated by CD4 fused to T cell or Fc receptor polypeptides, *Cell* 64:1037 (1991).
21. Wegener, A.-M.K., F. Letourneur, A. Hoeveler, T. Brocker, F. Luton, and B. Malissen, The T cell receptor/CD3 complex is composed of at least two autonomous transduction molecules, *Cell* 68:83 (1992).
22. Reth, M, Antigen receptor tail clue, *Nature* 338:383 (1989).
23. Romeo, C., M. Amiot, and B. Seed, Sequence requirements for induction of cytolysis by the T cell antigen/Fc receptor zeta chain, *Cell* 68:889 (1992).
24. Irving, B.A., A.C. Chan, and A. Weiss, Functional characterization of a signal transducing motif present in the T cell antigen receptor zeta chain, *J. Exp. Med.* 177:1093 (1993).
25. Kolanus, W., C. Romeo, and B. Seed, T cell activation by clustered tyrosine kinases, *Cell* 74:171 (1993).
26. Koga, Y., N. Caccia, B. Toyonaga, R. Spolski, Y. Yanagi, Y. Yoshikai, and T.W. Mak, A human T cell-specific cDNA clone (YT16) encodes a protein with extensive homology to a family of protein-tyrosine kinases, *Eur. J. Immunol.* 16:1643 (1986).
27. Cooke, M.P., and R.M. Perlmutter, Expression of a novel form of the fyn proto-oncogene in hematopoietic cells, *New Biol.* 1:66 (1989).
28. Taniguchi, T., T. Kobayashi, J. Kondo, K. Takahashi, H. Nakamura, J. Suzuki, K. Nagai, T. Yamada, S. Nakamura, and H. Yamamura, Molecular cloning of a porcine gene syk that encodes a 72-kDa protein-tyrosine kinase showing high susceptibility to proteolysis, *J. Biol. Chem.* 266:15790 (1991).
29. Chan, A.C., M. Iwashima, C.W. Turck, and A. Weiss, ZAP-70: a 70 kd protein-tyrosine kinase that associates with the TCR zeta chain, *Cell* 71:649 (1992).
30. Shen, L., R.F. Graziano, and M.W. Fanger, The functional properties of Fc gamma RI, II and III on myeloid cells: a comparative study of killing of erythrocytes and tumor cells mediated through the different Fc receptors, *Mol. Immunol.* 26:959 (1989).
31. Bolen, J.B., P.A. Thompson, E. Eiseman, and I.D. Horak, Expression and interactions of the Src family of tyrosine protein kinases in T lymphocytes, *Adv Cancer Res.* 57:103 (1991).

32. Burkhardt, A.L., M. Brunswick, J.B. Bolen, and J.J. Mond, Anti-immunoglobulin stimulation of B lymphocytes activates src-related protein-tyrosine kinases, *Proc. Natl. Acad. Sci. USA* 88:7410 (1991).
33. Hutchcroft, J.E., R.L. Geahlen, G.G. Deanin, and J.M. Oliver, Fc epsilon RI-mediated tyrosine phosphorylation and activation of the 72-kDa protein-tyrosine kinase, P-TK72, in RBL-2H3 rat tumor mast cells, *Proc. Natl. Acad. Sci. USA* 89:9107 (1992).
34. Tsygankov, A.Y., B.M. Broker, J. Fargnoli, J.A. Ledbetter, and J.B. Bolen, Activation of tyrosine kinase p60fyn following T cell antigen receptor cross-linking, *J. Biol. Chem.* 267:18259 (1992).
35. Wong, S., A.B. Reynolds, and J. Papkoff, Platelet activation leads to increased c-src kinase activity and association of c-src with an 85-kDa tyrosine phosphoprotein, *Oncogene* 7:2407 (1992).
36. Appleby, M.W., J.A. Gross, M.P. Cooke, S.D. Levin, X. Qian, and R.M. Perlmutter, Defective T cell receptor signaling in mice lacking the thymic isoform of p59fyn, *Cell* 70:751 (1992).
37. Karnitz, L., S.L. Sutor, T. Torigoe, J.C. Reed, M.P. Bell, D.J. McKean, P.J. Leibson, and R.T. Abraham, Effects of p56lck deficiency on the growth and cytolytic effector function of an interleukin-2-dependent cytotoxic T-cell line, *Mol. Cell Biol.* 12:4521 (1992).
38. Stein, P.L., H.M. Lee, S. Rich, and P. Soriano, pp59fyn mutant mice display differential signaling in thymocytes and peripheral T cells, *Cell* 70:741 (1992).
39. Straus, D.B., and A. Weiss, Genetic evidence for the involvement of the ick tyrosine kinase in signal transduction through the T cell antigen receptor, *Cell* 70:585 (1992).
40. Bell, G.M., J.B. Bolen, and J.B. Imboden, Association of Src-like protein tyrosine kinases with the CD2 cell surface molecule in rat T lymphocytes and natural killer cells, *Mol. Cell Biol.* 12:5548 (1992).
41. Sugie, K., T. Kawakami, Y. Maeda, T. Kawabe, A. Uchida, and J. Yodoi, Fyn tyrosine kinase associated with Fc epsilon RII/CD23: possible multiple roles in lymphocyte activation, *Proc. Natl. Acad. Sci. USA* 88:9132 (1991).
42. Huang, M.M., Z. Indik, L.F. Brass, J.A. Hoxie, A.D. Schreiber, and J.S. Brugge, Activation of Fc gamma RII induces tyrosine phosphorylation of multiple proteins including Fc gamma RII, *J. Biol. Chem.* 267:5467 (1992).
43. Hatakeyama, M., T. Kono, N. Kobayashi, A. Kawahara, S.D. Levin, R.M. Perlmutter, and T. Taniguchi, Interaction of the IL-2 receptor with the src-family kinase p56lck: identification of novel intermolecular association, *Science* 252:1523 (1991).
44. Stefanova, I., V. Horejsi, I.J. Ansotegui, W. Knapp, and H. Stockinger, GPI-anchored cell-surface molecules complexed to protein tyrosine kinases, *Science* 254:1016 (1991).
45. Thomas, P.M., and L.E. Samelson, The glycophosphatidylinositol-anchored Thy-1 molecule interacts with the p60fyn protein tyrosine kinase in T cells, *J. Biol. Chem.* 267:12317 (1992).

# CONTROL OF LYMPHOPOIESIS BY NON-RECEPTOR PROTEIN TYROSINE KINASES

Roger M. Perlmutter and Steven J. Anderson

Howard Hughes Medical Institute and the Department
of Immunology, University of Washington
Seattle, WA 98195

## INTRODUCTION

Maintenance of a satisfactory immune system requires the daily generation of millions of lymphocytes from immature progenitor cells that reside (in adult mammals) in the bone marrow. Three fundamental processes underlie lymphopoiesis. First, a small population of hematopoietic stem cells (estimated to represent something less than 0.1% of bone marrow cells) gives rise continuously to mature daughter cells through successive, self-regenerating cell divisions (see ref. 1 for a review of hematopoietic stem cells). Second, cells committed to the lymphoid lineages must colonize specialized stromal cell environments wherein extrinsic cues are provided that direct maturation. The maturation of B cell precursors occurs in mammalian bone marrow and can be observed using in vitro culture systems containing well-characterized bone marrow-derived stromal cells (2). Similarly, T lymphocyte maturation, which for conventional T cells takes place in the thymus, can be modelled in vitro or in fetal thymic organ culture (3). Detailed analysis of T and B cell development has established that precursor cells proceed through a series of clearly defined maturation steps, giving rise to intermediate cell populations that display characteristic cell surface molecules. Thus the third fundamental process underlying lymphopoiesis is the regulated maturation of committed progenitors through discrete developmental checkpoints.

The existence of these discrete maturational stages in lymphocyte development suggests that it will be possible to define rate-limiting steps in the differentiative process. For example, the progress of lymphocyte maturation can be assessed by noting the status of gene rearrangement events required for satisfactory synthesis of the clonotypic antigen receptor. Recent evidence suggests that these gene rearrangements, by themselves, serve to orchestrate lymphopoiesis. Thus targeted disruption of the T cell receptor β chain locus yields cells blocked at an early point in T cell maturation (4), and mutations in the immunoglobulin μ heavy chain gene yield a similar defect in B cell development (5). How do gene rearrangement events, intrinsic to each precursor cell, trigger further maturation? Recent evidence suggests that expression of antigen receptors results in the engagement of a highly-specialized set of non-receptor protein tyrosine kinases. Indeed,

*Mechanisms of Lymphocyte Activation and Immune Regulation V*
Edited by S. Gupta *et al.*, Plenum Press, New York, 1994

at least some of the developmental checkpoints that subdivide the lymphocyte maturation sequence result from the ability of non-receptor protein tyrosine kinases to regulate lymphocyte development. This phenomenon is especially apparent in the case of T lymphocytes.

## T CELL MATURATION

Virtually all circulating T lymphocytes derive from immature progenitors that mature within the thymus, although an alternative maturation pathway that proceeds in gut-associated lymphoid tissue has also been defined (6). In the thymus, the process of T cell maturation can be readily studied by simply examining the expression of components of the T cell antigen receptor complex: the CD3 elements (CD3ε is typically used), and the CD4 and CD8 coreceptors. Figure 1 presents a simplified view of thymocyte maturation. Information regarding the sequence of thymocyte maturation has emerged from analyses of fetal thymocytes, and from the direct assessment of precursor activity achieved through injection of putative progenitors intrathymically (c.f. ref. 7, for a detailed review of this process). Most precursors that appear during early fetal thymic development, like immature cells found in the adult thymus, lack expression of CD3, CD4 and CD8 as detected using flow cytometric methods. Indeed, in these cells, gene rearrangement events required for satisfactory expression of the T cell antigen receptor complex have not yet taken place. These cells give rise to a transitional population of cells expressing low levels of CD8, which thereafter give rise to cells that simultaneously express both CD4 and CD8 (8). The thymus is a site of active mitogenesis, and in the adult animal most of the mitotic activity is detected in the $CD4^-8^-$ or $CD4^-8^{lo}$ cells (9). This expansion process results in the accumulation of $CD4^+8^+$ cells, such that about 85% of all thymocytes have this phenotype. Thymocyte mitogenesis is also accompanied by rearrangement of the T cell receptor loci; about half of the $CD4^+8^+$ cells also express the antigen receptor complex and hence are $CD3^+$ (successful rearrangement of the T cell receptor αβ or γδ genes is required for high-level expression of the CD3 elements).

The $CD3^+4^+8^+$ cells that emerge following successful assembly of T cell receptor genes are subject to two types of antigen receptor-mediated selection events: negative selection acts to eliminate self-reactive cells, while positive selection acts to permit maturation only of those cells bearing satisfactory antigen receptors capable of recognizing appropriate antigen presentation molecules. The products of these two types of selection events, which may differ primarily in terms of the strength of the T cell receptor signal that devolves following antigen encounter (10), emerge expressing CD4 or CD8 in a mutually exclusive fashion, along with high levels of the antigen receptor

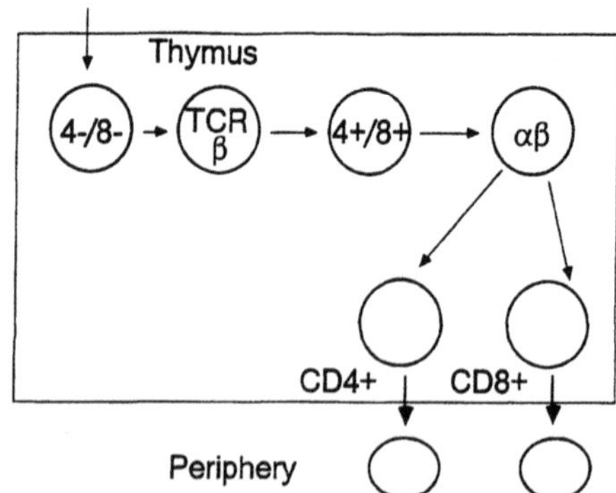


**Figure 1.** A simplified scheme for thymocyte maturation (see ref 7 for details).

complex. It is these cells that ultimately leave the thymus to populate the peripheral lymphoid organs.

## Gene Rearrangements During Thymocyte Maturation

Assembly of functional T cell receptor genes, like the analogous process that permits antibody gene assembly in B lymphocytes, occurs by site-directed juxtaposition of gene segments that are positioned discontinuously in germline DNA (11). This process is developmentally ordered in two ways. First, the stem cells that give rise to fetal thymocytes in the mouse support juxtaposition of V and J gene segments in the γ locus, and of Vδ, Dδ and Jδ gene segments. Hence the first T lymphocytes that emerge bear γδ antigen receptors, a receptor isoform that is rarely used in adult thymocytes. The process of T cell receptor gene rearrangement is also developmentally ordered within both the γδ lineage and the more productive αβ lineage. In the former, γ locus rearrangements proceed those in the δ locus. In cells destined to emerge as αβ T cells, juxtaposition of Dβ to Jβ, and thereafter of Vβ to DβJβ gene segments occurs first, with Vα to Jα joining representing the final rearrangement event. Although the mechanism underlying this developmentally ordered rearrangement phenomenon remains enigmatic, all evidence suggests that strict biochemical controls enforce this process. For example, expression of a functional T cell receptor β chain transgene in transgenic mice substantially blocks endogenous rearrangements at the β locus (12), even when the β transgene is engineered to encode a truncated protein lacking much of the V region (13).

The apparent control of β chain gene rearrangements by the β chain itself has been interpreted as a form of "feedback" regulation, and analogous experiments in B lymphocytes suggest that a similar process regulates antibody heavy chain gene rearrangements as well (14). Feedback regulation of this type provides an attractive explanation for the phenomenon of allelic exclusion, the fact that each individual T or B lymphocyte expresses a single type of antigen receptor heterodimer wherein each chain of the receptor is encoded by a single functional allele (15). Although this phenomenon could conceivably be explained by stochastic models which argue that the assembly of antigen receptors is sufficiently difficult so as to preclude successful rearrangement of two alleles simultaneously in the same cell, regulated models have recently gained considerable experimental support. Analyses of mice bearing targeted disruptions of T cell receptor genes, or of genes involved in antigen receptor gene rearrangement have proved especially revealing. For example, targeted disruption of either the RAG-1 or the RAG-2 genes, both of which are implicated in the enzymatic process that permits antigen receptor gene rearrangement, yields animals that lack both T and B lymphocytes. The thymuses in such mice contain less than 1% of the normal number of thymocytes, and the remaining cells fail to mature beyond an immature $CD3^{-}4^{-}8^{-}$ stage (16,17). This maturational defect reflects in part the failure to produce the T cell receptor β chain. Hence disruption of the T cell receptor β locus by itself yields animals with a nearly-identical thymic phenotype (4). Moreover, thymus cellularity, and the maturation of thymocytes to the $CD4^{+}8^{+}$ stage, can be reconstituted in $RAG^{null}$ mice simply by providing a functional T cell receptor β chain through introduction of a β chain transgene (18). These experiments suggest that production of a T cell receptor β chain polypeptide is somehow sensed intracellularly, leading to thymocyte maturation.

How, then, does synthesis of a T cell receptor β chain send a signal to the cell interior? Although the nature of this signaling process remains obscure, several observations implicate components of the signal transduction machinery that ordinarily supports T cell activation after antigen-mediated crosslinking of the T cell antigen receptor. Eichmann and colleagues have demonstrated that immature $CD4^{-}8^{-}$ thymocytes can be induced to differentiate into $CD4^{+}8^{+}$ cells following stimulation with anti-CD3ε antibodies (19). A similar effect was observed in thymocytes from T cell receptor β chain-deficient mice (20), which, of course, cannot assemble a functional antigen receptor. These observations indicate that although immature CD4⁻8⁻ cells bear insufficient CD3ε molecules on their surfaces to permit detection using conventional anti-CD3 staining protocols, signals can nevertheless be transmitted from the few CD3ε polypeptides that actually appear in the plasma membrane. Viewed in this way, it is reasonable to expect that mice engineered to lack other crucial components of the T cell antigen receptor complex, or the signaling mechanism with which this receptor complex interacts, would also exhibit defects in thymocyte maturation. This is in fact the case. Thus thymocytes from mice lacking the ζ chain gene of the antigen receptor complex fail to develop normally, exhibiting defects that resemble, in part, those seen in $RAG^{null}$ mice (21,22). Together these experiments focus attention on a signal transduction apparatus that could act to control thymocyte development by sensing the appearance of the T cell antigen receptor at the plasma membrane. Considerable evidence supports the view that the non-receptor protein tyrosine kinase $p56^{lck}$ plays a pivotal role in this signaling apparatus.

## STRUCTURE and FUNCTION of $p56^{lck}$

The *lck* gene was first identified by virtue of its activation through retroviral insertion in the mouse T cell lymphoma cell line LSTRA (23). It encodes a 509 amino acid, membrane-associated, *src*-family protein tyrosine kinase that is expressed only in lymphocytes, primarily T cells, and NK cells. Immunoprecipitation analyses have demonstrated that $p56^{lck}$ interacts with a variety of T cell surface proteins, and with a number of cytoplasmic enzymes that participate in growth factor-mediated signaling pathways (24). In particular, $p56^{lck}$ is physically and functionally associated with the CD4 and CD8 coreceptor molecules by virtue of a specialized motif in its amino-terminal region centering on cysteines 20 and 23 (see Fig. 2). A complementary cysteine-containing motif is found in the cytoplasmic tails of CD4 and CD8 (25), and in the 4-1BB protein that is a member of the tumor necrosis factor receptor family (26). In addition, $p56^{lck}$ interacts both physically and functionally with the β chain of the interleukin-2 receptor, and may assist in regulating normal interleukin--2 responses (27). Embedded within the amino-terminal half of $p56^{lck}$ are globular domains termed SH3 and SH2 which interact with proline-rich motifs (28) and with phosphotyrosine-containing peptides (29), respectively (Fig. 2). Hence there is reason to expect that $p56^{lck}$ will interact productively with a great many proteins, and indeed, some evidence suggests that the catalytic activity of $p56^{lck}$ may prove superfluous for some of its signaling functions (30).

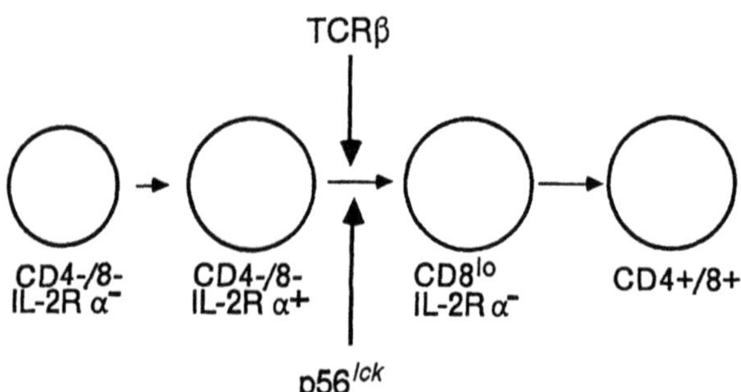


**Figure 2.** Structure of $p56^{lck}$ (adapted from Anderson et al., 1994).

Clear evidence for the importance of $p56^{lck}$ in T cell signaling has come from the study of mutant cell lines that fail to express this protein. In particular, the J.CaM-1 mutant, identified following mutagenesis of the human Jurkat T leukemic cell line, exhibits defective coupling of the T cell antigen receptor to intracellular signaling pathways. This defect can be rescued by transfection of functional *lck* expression constructs (31). A similar observation was made in a mouse CTLL T cell line which was noted serendipitously to lack *lck* expression (32). Moreover, experiments by two groups of investigators revealed that augmented expression of $p56^{lck}$ in T hybridoma cell lines could improve the sensitivity of these cells to low doses of antigen (33,34). Together these experiments support the view that $p56^{lck}$ acts as one component of the T cell receptor signaling apparatus, and hence could participate in controlling T cell development. In this context it is important to note that the *lck* gene is expressed in thymocytes from the very first time that hematopoietic elements first colonize the thymic rudiment during development, and in all T-lineage cells in the thymus (35).

## Developmental Defects in *lck* Transgenic Mice

Attempts to assess the importance of $p56^{lck}$ in lymphocyte development focussed first on the generation of *lck* transgenic mice. These studies made use of the fact that the *lck* gene contains two distinct promoter elements separated by about 35 kb of germline DNA (36). The 3' or proximal promoter is exclusively active in thymocytes and especially in immature cells. Hence by incorporating this 3' transcriptional regulatory element, it was possible to generate transgenic mice expressing very high levels of *lck* transcripts and of $p56^{lck}$ protein. Further augmentation of $p56^{lck}$ activity could be achieved by introducing a phenylalanine-for-tyrosine substitution at codon 505, thereby directing expression of a mutant protein lacking a negative regulatory phosphorylation site (Fig. 2).

Normal T cell development was severely disrupted in the *lck* transgenic mice, which exhibited a transgene dose-dependent reduction in the percentage of cells expressing the T cell antigen receptor complex (37). Further analysis revealed that the inability of these cells to express the T cell antigen receptor reflected an exquisitely specific developmental defect: the inability to rearrange Vβ gene segments to DβJβ elements and hence the failure to produce a T cell receptor β polypeptide (38). Indeed, satisfactory development could be completely rescued by simultaneous introduction of a functional β chain transgene.

Transgenic animals bearing a mutant *lck* construct encoding a catalytically inactive kinase (containing a lysine-for-arginine substitution at position 273) exhibited a more severe, and surprisingly complementary, defect. In this case, a transgene dose-dependent decrease in thymocyte cellularity was noted, such that when high levels of the transgene were expressed, thymocyte number was decreased by greater than 95% (39). The cells that remained in these thymuses were almost exclusively $CD3^-4^-8^-$ blast cells, arrested in the G1 phase of the cell cycle, that expressed a set of cell surface markers that positioned them within a discrete developmental stage defined by other investigators (Fig. 3; 8). Analysis of mice bearing a targeted disruption at the *lck* locus yielded similar results (40). Indeed, the thymuses in these mice resemble those that appear in mice bearing a disruption at the T cell receptor β locus (4).

Reviewing the phenotypes of the *lck* transgenic mice, it appears that $p56^{lck}$ acts to control thymocyte development at precisely the point where signals derived from expression of the T cell receptor β chain polypeptide are believed to hold sway. In particular, augmented expression of $p56^{lck}$ promoted the development of thymocytes to the $CD4^+8^+$ stage whilst simultaneously suppressing (apparently) Vβ to Dβ gene rearrangements. These events are precisely those that are ordinarily associated with β chain-derived signals (Fig. 3).

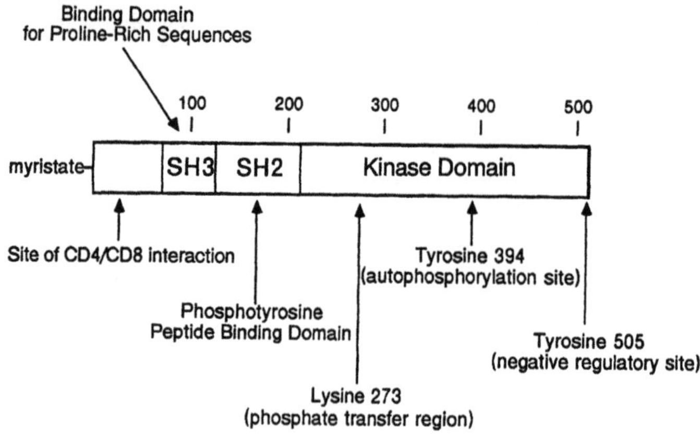


**Figure 3.** Positioning p56*lck* within the thymocyte developmental sequence.

The hypothesis that p56*lck* serves as the sensing mechanism for successful T cell receptor β chain gene rearrangement makes two strong predictions. The first of these is that allelic exclusion, which it has been argued results from a feedback mechanism elicited by satisfactory synthesis of a T cell receptor β chain polypeptide, should not occur if p56*lck* activity is compromised. Put another way, cells lacking p56*lck* function should synthesize functional T cell receptor β chains, but should not be able to perceive the fact that they have done so. This conjecture was tested directly by simultaneously introducing genes encoding catalytically inactive p56*lck* and a functional T cell receptor β chain into transgenic mice. As expected, the presence of the catalytically inactive p56*lck*R273 protein interfered with normal p56*lck* function, yielding mice with very few thymocytes, all of which expressed the β chain polypeptide. More importantly, endogenous T cell receptor β chain gene rearrangements were completely extinguished in mice expressing only the β chain transgene, while thymocytes from doubly-transgenic animals continued to support rearrangements at endogenous Vβ loci (41). These experiments provide extremely strong support for the view that p56*lck* ordinarily regulates allelic exclusion at the T cell receptor β chain locus.

A second strong prediction that follows from the hypothesis that p56*lck* serves as the sensing mechanism for T cell receptor β chain synthesis was already in part validated from the results described above. Specifically, our theory argues that artificial augmentation of p56*lck* activity should stimulate thymocyte maturation even in the absence of T cell receptor gene expression. This conjecture has now been confirmed directly. Introduction of an activated *lck*F505 transgene into RAG$^{null}$ mice yielded animals with normal (or even supra-normal) numbers of thymocytes, virtually all of which were CD4$^+$8$^+$. Thus the augmented p56*lck* activity functioned exactly as would a T cell receptor β chain polypeptide in contravening an early block in thymocyte maturation (P. Mombaerts, S. Anderson, R. Perlmutter, T. Mak, and S. Tonegawa, manuscript submitted for publication). Moreover, when the same *lck* transgene was introduced into mice bearing a targeted deletion of both Cβ genes, mice in which thymocyte number and development are usually profoundly attenuated, expression of activated p56*lck*F505 reconstituted both the number of thymocytes and the representation of CD4$^+$8$^+$ cells, while simultaneously extinguishing rearrangement of endogeneous Vβ gene segments and stimulating assembly of α chain genes (S. Anderson, P. Mombaerts, S. Tonegawa and R. Perlmutter, manuscript in preparation). By all available criteria, then, p56*lck* behaves as if it ordinarily serves as the signal transduction mechanism that promotes thymocyte maturation following assembly of a productive T cell receptor β chain gene.

### How Might p56$^{lck}$ Control Thymocyte Development?

Two fundamental questions emerge from the *lck* transgenic experiments outlined above. First, what is the nature of the receptor structure with which p56$^{lck}$ interacts in order to sense synthesis of a T cell receptor β chain, and second, what effector mechanisms permit p56$^{lck}$, once activated, to promote thymocyte development? Although neither question can be answered explicitly, it is possible to place boundary conditions on the sorts of molecules with which p56$^{lck}$ might interact to subserve its early functions.

Recent data indicate that the T cell receptor β chain appears on the surface of immature thymocytes in association with the CD3 complex and with another polypeptide which is believed to serve as a sort of surrogate α chain (42). This pre-T cell receptor complex can be viewed as an analogue of the pre-B cell receptor complex, which includes the antibody heavy chain and a "surrogate light chain", composed of VpreB and λ5 polypeptides, as well as signaling structures that are structural analogues of the CD3 proteins (43). Placement of the T cell receptor β chain within a receptor complex on the cell surface in $CD3^-4^-8^-$ thymocytes argues that this complex actually engages an extracellular ligand, suggesting that it may augment p56$^{lck}$ activity through a cross-linking process similar to that which is believed to occur following antigen recognition in mature T cells. This line of reasoning suggests that it may be possible to demonstrate a physical association of p56$^{lck}$ with the pre-T cell receptor complex. To date this has been difficult to confirm or refute because of the small number of immature thymocytes ordinarily available for study, and because of uncertainty surrounding the identity of the "surrogate α chain" (42).

As regards the mechanism whereby p56$^{lck}$ directs thymocyte maturation, it is apparent that the catalytic activity of p56$^{lck}$ is required for this effect. One proposed mechanism for signaling in mature T cells posits that initial phosphorylation of critical tyrosine residues in the CD3γ, CD3δ and CD3ε chains, and in the ζ chains, permits subsequent association of other signaling molecules via their SH2 domains. Hence it is possible that p56$^{lck}$ acts by directly phosphorylating these components of the T cell antigen receptor complex, thereby catalyzing the assembly of a more complicated signaling apparatus. This hypothesis, in its most restrictive form, clearly predicts that augmented p56$^{lck}$ activity will not correct the developmental abnormalities observed in mice lacking CD3 components. However if the CD3 components interact with p56$^{lck}$ in some fashion other than as substrates (which seems likely), this experiment may prove difficult to interpret.

Regardless of which substrates one views as most proximal in the p56$^{lck}$-initiated signaling cascade, two principal effects are observed in *lck* transgenic mice: alterations in gene expression, and enhanced mitogenesis. Indeed, activated p56$^{lck}$ is a potent transforming agent capable of conferring tumorigenicity on NIH 3T3 fibroblasts (44) and of stimulating malignant transformation in thymoblasts (45). Thus the mechanism whereby p56$^{lck}$ exerts its effects should be amenable to study by cataloguing transcriptional regulatory elements upon which it appears to act.

## FUTURE DIRECTIONS

Lastly we note that the control of thymocyte development by p56$^{lck}$ represents just one example of many wherein a non-receptor protein tyrosine kinase has been implicated in the regulation of hematopoietic cell differentiation. There are now more than 30 different non-receptor protein tyrosine kinases (46), and the majority of these are expressed preferentially if not exclusively in hematopoietic cells. Thus $p72^{btk}$, which is structurally related to $p56^{lck}$, appears to control an early step in B cell differentiation since defects in this gene either block mature B cell production (as in human X-linked agammaglobulinemia; 47) or alter it substantially (as in X-linked immunodeficiency disease in the mouse; 48). Studies in our laboratory have demonstrated that the *csk* gene is crucial for normal lymphopoiesis in both the T and B cell lineages (J. Gross, M. Appleby, S. Chien and R. Perlmutter, unpublished data), and $p60^{src}$ itself, the very first non-receptor protein tyrosine kinase ever identified, is required for proper differentiation of osteoclasts (49). In each of these cases, alterations in gene expression induced by activation of non-receptor protein tyrosine kinases produce irreversible alterations in the behavior of hematopoietic cells. The continued analysis of the $p56^{lck}$-mediated signaling pathways thus promises to illuminate broader aspects of the commitment process that regulates hematopoiesis.

## REFERENCES


1. Ikuta, K., Ichida, N., Friedman, J., and Weissman, I. L., 1992, Lymphocyte development from stem cells, *Ann Rev. Immunol.* 10:759.
2. Saffran, D.C., Faust, E.A., and Witte, O.N., 1992, Establishment of a reproducible culture technique for the selective growth of B cell progenitors, *Curr. Top. Microbiol. Immunol.* 182: 37.
3. Anderson, G., Jenkinson, E.J., Moore, N.C., and Owen, J.J., 1993, MHC class II-positive epithelium and mesenchyme cells are both required for T cell development in the thymus, *Nature* 362: 70.
4. Mombaerts, P., Clarke, A.R., Rudnicki, M.A., Iacomini, J., Itohara, S., Lafaille,J.J., Wang, L., Ichikawa, Y., Jaenisch, R., Hooper, M.L., and Tonegawa, S., 1992b, Mutations in T cell antigen receptor genes α and β block thymocyte development at different stages, *Nature* 360:225.
5. Kitamura, D., Roes, J., Kuhn, R., and Rajewsky, K., 1991, A B cell-deficient mouse by targeted disruption of the membrane exon of the immunoglobulin μ gene, *Nature* 350:423.
6. Rocha, B., Vassalli, P., and Guy-Grand, D., 1992, The extrathymic T cell development pathway, *Immunol. Today* 13:449.
7. Petrie, H.T., Hugo,P., Scollay, R., and Shortman, K., 1990, Linkage relationships and developmental kinetics of immature thymocytes: CD3, CD4 and CD8 acquisition in vivo and in vitro, *J. Exp. Med.* 172:1583.
8. Godfrey, D.I., and Zlotnik, A., 1993, Control points in early T cell development, *Immunol. Today* 14:547.
9. Egerton, M., Scollay, R., and Shortman, K., 1990, Kinetics of mature T cell development in the thymus, *Proc. Natl. Acad. Sci. USA* 87:2579.
10. Hogquist, K.A., Jameson, S.C., Heath, W.R., Howard, J.L., Bevan, M.J., and Carbone, F.R., 1994, T cell receptor antagonist peptides induce positive selection, *Cell* 76:17.
11. Kronenberg, M., Siu, G., and Hood, L.E., 1986, Organization and assembly of T cell receptor genes, *Ann. Rev. Immunol.* 4:529.
12. Uematsu, Y., Ryser, S., Dembic, Z., Borgulya, P., Krimpenfort, P., Berns, A., von Boehmer, H., and Steinmetz, M., 1988, In transgenic mice the introduced functional T cell receptor β gene prevents expression of endogenous β genes, *Cell* 52:831.
13. Krimpenfort, P., Ossendorp, F., Borst, J., Melief, C., and Berns, A., 1989, T cell depletion in transgenic mice carrying a mutant gene for TCRβ, *Nature* 341:742.
14. Nussenzweig, M.C., Shaw, A.C., Sinn, E., Danner, D.B., Holmes, K.L., Morse, H.C., III, and Leder, P., 1987, Allelic exclusion in transgenic mice expressing the membrane form of immunoglobulin μ, *Science* 236:816.

15. Malissen, M., Trucy, J., Jouvin-Marche, E., Cazanave, P.A., Scollay, R., and Malissen, B., 1992, Regulation of TCR α and β gene allelic exclusion during T cell development, *Immunol. Today* 13:315.
16. Mombaerts, P., Iacomini, J., Johnson, R.S., Herrup, K., Tonegawa, S., and Papaioannov, V.E., 1992a, RAG-1-deficient mice have no mature T and B lymphocytes, *Cell* 68: 869.
17. Shinkai, Y., Rathbun, G., Lam, K-P., Oltz, E.M., Stewart, V., Mendelsohn, M., Charron, J., Datta, M., Young, F., Stall, A.M., and Alt, F.W., 1992, RAG-2-deficient mice lack mature lymphocytes owing to inability to initiate V(D)J rearrrangement, *Cell* 68:855.
18. Shinkai, Y., Koyasu, S., Nakayama, K., Murphy, K.M., Loh, D.Y., Reinherz, E.L., and Alt, F.W., 1993, Restoration of T cell development in RAG-2-deficient mice by functional TCR transgenes, *Science* 259:822.
19. Levelt, C.N., Ehrfeld, A., and Eichmann, K., 1993a, Regulation of thymocyte development through CD3. I. Timepoint of ligation of CD3ε determines clonal deletion or induction of developmental program, *J. Exp. Med.* 177:707.
20. Levelt, C.N., Mombaerts, P., Iglesias, A., Tonegawa, S., and Eichmann, K., 1993b, Restoration of early thymocyte development in T cell receptor β-chain-deficient mutant mice by transmembrane signaling through CD3ε, *Proc. Natl. Acad. Sci. USA* 90:11401.
21. Malissen, M., Gillet, A., Rocha, B., Trucy, J., Vivier, E., Boyer, C., Kontgen, F., Brun, N., Mazza, G., Spanopoulou, E., Guy-Grand, D., and Malissen, B., 1993, T cell development in mice lacking the CD3-ζ/η gene, *EMBO J.* 12:4347.
22. Ohno, H., Aoe, T., Taki, S., Kitamura, D., Ishida, Y., Rajewsky, K., and Saito, T, 1993, Developmental and functional impairment of T cells in mice lacking CD3ζ chains, *EMBO J.* 12:4357.
23. Marth, J.D., Peet, R., Krebs, E.G. and Perlmutter, R.M., 1985, A lymphocyte-specific protein tyrosine kinase gene is rearranged and overexpressed in the murine T cell lymphoma LSTRA, *Cell* 43:393.
24. Anderson, S.J., Levin, S.D., and Perlmutter, R.M., 1994, Involvement of the protein tyrosine kinase $p56^{lck}$ in T cell signaling and thymocyte development, *Adv. Immunol.* 56:151.
25. Turner, J.M., Brodsky, M.H., Irving, B.A., Levin, S.D., Perlmutter, R.M., and Littman, D.R., 1990, Interaction of the unique N-terminal region of the tyrosine kinase $p56^{lck}$ with the cytoplasmic domains of CD4 and CD8 is mediated by cysteine motifs, *Cell* 60:755.
26. Kim, Y.J., Pollok, K.E., Zhou, Z., Shaw, A., Bolen, J.B., Fraser, M., and Kwon, B.S., 1993, Novel T cell antigen 4-1BB associates with the protein tyrosine kinase $p56^{lck}$, *J. Immunol.* 151:1255.
27. Hatakeyama, M., Kono, T., Kobayashi, N., Kawahara, A., Levin, S., Perlmutter, R.M. and Taniguchi, T., 1991, IL-2 receptor interacts with a *src*-family kinase, $p56^{lck}$; identification of novel intermolecular association, *Science* 252:1523-.
28. Booker, G.W., Gout, I., Downing, A.K., Driscoll, P.C., Boyd, J., Waterfield, M.D., and Campbell, I.D., 1993, Solution structure and ligand-binding site of the SH3 domain of the p85α subunit of phosphatidylinositol 3-kinase, *Cell* 73:813-822.
29. Songyang, Z., Shoelson, S.E., Chadhuri, M., Gish, G., Pawson, T., Haser, W.G., King, F., Roberts, T., Ratnofsky, S., Lechleider, R.J., Neel, B.G., Birge, R.B., Fajardo, J.E., Chou, M.M., Hanafusa, H., Schaffhausen, B., and Cantley, L.C., 1993, SH2 domains recognize specific phosphopeptide sequences, *Cell* 72:767.
30. Xu, H., and Littman, D.R., 1993, A kinase-independent function of lck in potentiating antigen-specific T cell activation, *Cell* 74:633.
31. Straus, D.B., and Weiss, A., 1992, Genetic evidence for the involvement of the lck tyrosine kinase in signal transduction through the T cell antigen receptor, *Cell* 70: 585.
32. Karnitz, L., Sutor, S.L., Torigoe, T., Reed, J.C., Bell, M.P., McKean, D.J., Leibson, P.J., and Abraham, R.T., 1992, Effects of $p56^{lck}$ deficiency on the growth and cytolytic effector function of an interleukin 2-dependent cytotoxic T-cell line, *Molec. Cell. Biol.* 12:4521.
33. Caron, L., Abraham, N., Pawson, T., and Veillette, A., 1992, Structural requirements for enhancement of T cell responsiveness by the lymphocyte-specific tyrosine protein kinase $p56^{lck}$, *Molec. Cell. Biol.* 12:2720.

34. Luo, K. and Sefton, B.M., 1992, Activated lck tyrosine protein kinase stimulates antigen-independent interleukin-2 production in T cells, *Molec. Cell. Biol.* 12:4724.
35. Perlmutter, R.M., Peet, R., Marth, J.D., Lewis, D.B., Ziegler, S.F., and Wilson, C.B., 1988 Conservation of function in the *src* gene family: the structure and expression of a human lymphocyte-specific protein tyrosine kinase (*lck*). *J. Cell. Biochemistry* 38:117.
36. Wildin, R.S., Garvin, A.M., Pawar, S., Lewis, D.B., Abraham, K.M., Forbush, K.A., Ziegler, S.F., Allen, J.M., Perlmutter, R.M., 1991, Developmental regulation of *lck* gene expression in T lymphocytes, *J. Exp. Med.* 173:383.
37. Abraham, K.M., Levin, S.D., Marth, J.D., Forbush, K.A., Perlmutter, R.M., 1991b, Delayed thymocyte development induced by augmented expression of $p56^{lck}$, *J. Exp. Med* 173:1421.
38. Anderson, S.J., Abraham, K.M., Nakayama T., Singer, A. and Perlmutter, R.M., 1992, Inhibition of T cell receptor β chain gene rearrangement by overexpression of the non-receptor protein tyrosine kinase $p56^{lck}$, *EMBO J.* 11: 4877-.
39. Levin, S.D., Anderson, S.J., Forbush, K.A., and Perlmutter, R.M., 1993, A dominant-negative transgene defines a role for $p56^{lck}$ in thymopoiesis, *EMBO J.* 12: 1671.
40. Molina, T.J., Kishihara, K., Siderovski, D.P., van Ewijk, W., Narendran, A., Timms, E., Wakeham, A., Paige, C.J., Hartmann, K.U., Veillette, A., Davidson,D., and Mak, T.W., 1992, Profound block in thymocyte development in mice lacking $p56^{lck}$, *Nature* 357:161.
41. Anderson, S.J., Levin, S.D., and Perlmutter, R.M., 1993, Protein tyrosine kinase $p56^{lck}$ controls allelic exclusion of the T cell receptor β chain genes, *Nature* 365: 552.
42. Groetrupp, M., Ungeweiss, K., Azogui, O., Palacios, R., Owen, M.J., Hayday, A.C., and von Boehmer, H., 1993, A novel disulfide-linked heterodimer on pre-T cells consists of the T cell receptor β chain and a 35 kd glycoprotein, *Cell* 75:283.
43. Rolink, A., and Melchers, F., 1993, B lymphopoiesis in the mouse, *Adv. Immunol.* 53:123.
44. Marth, J.D., Cooper, J.A., King, C.S., Ziegler, S.F., Tinker, D.A., Overell, R.W., Krebs, E.G. and Perlmutter, R.M., 1988 Neoplastic transformation induced by an activated lymphocyte-specific protein tyrosine kianse (pp56*lck*), *Mol. Cell. Biol.* 8: 540.
45. Abraham, K.M., Levin, S.D., Marth, J.D., Forbush, K.A., and Perlmutter, R.M, 1991a, Thymic tumorigenesis induced by overexpression of $p56^{lck}$, *Proc. Natl. Acad. Sci. USA* 88:3977-.
46. Bolen, J.B., 1993, Nonreceptor protein tyrosine kinases, *Oncogene* 8:2025.
47. Tsukada, S., Saffran, D.C., Rawlings, D.J., Parolini, O., Allen, R.C., Klisak, I., Sparkes, R.S., Kubagawa, H., Mohandas, T., Quan, S.,Belmont, J.W., Cooper, M.D., Conley, M.E., and Witte, O., 1993, Deficient expressionm of a B cell cytoplasmic tyrosine kinase in human X-linked agammaglobulinemia, *Cell* 72:279.
48. Thomas, J.D., Sidaras, P., Smith, C.I., Vorechovsky, I., Chapman, V., and Paul, W.E., 1993, Co-localization of X-linked agammaglobulinemia and X-linked immunodeficiency genes, *Science* 261: 355.
49. Boyce, B.F., Chen, H., Soriano, P., and Mundy, G.R., 1993, Histomorphometric and immunocytochemical studies of *src*-related osteopetrosis, *Bone* 14:335.

# INVOLVEMENT OF NONRECEPTOR PROTEIN TYROSINE KINASES IN MULTICHAIN IMMUNE RECOGNITION RECEPTOR SIGNAL TRANSDUCTION

Anne L. Burkhardt, Sandra J. Saouaf, Sandeep Mahajan, and Joseph B. Bolen

Department of Molecular Biology
Bristol-Myers Squibb Pharmaceutical Research Institute
Princeton, New Jersey 08543

## NONRECEPTOR PROTEIN TYROSINE KINASES

The nonreceptor protein tyrosine kinases (PTKs) represent cellular enzymes grouped together based upon their lack of defined extracellular sequences. The currently identified nonreceptor PTKs can be divided into nine different enzyme groups based upon predicted structural features. With the exception of the Focal Adhesion Kinase (Fak) and the Activated Cdc42Hs-associated Kinase (Ack) which are the only known members of these two individual PTK groups, the remaining PTKs appear to be members of distinct PTK families. They range in size from about 50 kDa for the C-src Kinase (Csk) family to approximately 150 kDa for the Abl kinase family. All of these PTKs are likely to be involved in one or more signaling pathways that modulate growth, differentiation, and mature cell function.

Figure 1 presents a comparison of the recognized major common domains of the nonreceptor PTK families. This comparison is significantly simplified and does not detail regions of the enzymes that might be unique for a given PTK type or PTK family. These

*Mechanisms of Lymphocyte Activation and Immune Regulation V*
Edited by S. Gupta *et al.*, Plenum Press, New York, 1994

unique sequences, which in some cases represent the bulk of the enzyme, are clearly important components of these kinases since they likely play a central role in mediating the specific interactions of the individual PTKs.

The major area of sequence homology between the nonreceptor PTKs and presumably the region of greatest shared structural identity is their catalytic domains[1]. The catalytic domains for the nonreceptor PTKs are the SH1 (src homology 1) domains since this is the region between the kinases that shares the greatest sequence similarity with the C-src catalytic domain. The Janus kinases (Jak1, Jak2, and Tyk2) are interesting in that they have two catalytic-like domains[2]. However, close scrutiny of the sequences in each of these domains indicates that the catalytic activity of the JAKs is contained within the most carboxy-terminal catalytic-like domain[2]. The role of the second catalytic-like domains has not been determined. A site for enzyme tyrosine autophosphorylation is also common for most of the nonreceptor PTKs corresponding to the tyrosine 416 residue of C-src. The Csk class of kinases does not possess a comparable autophosphorylation site[3] even though Csk and the other member of this family, Ctk, appear to be capable of tyrosine autophosphorylation when expressed as recombinant proteins[4]. The

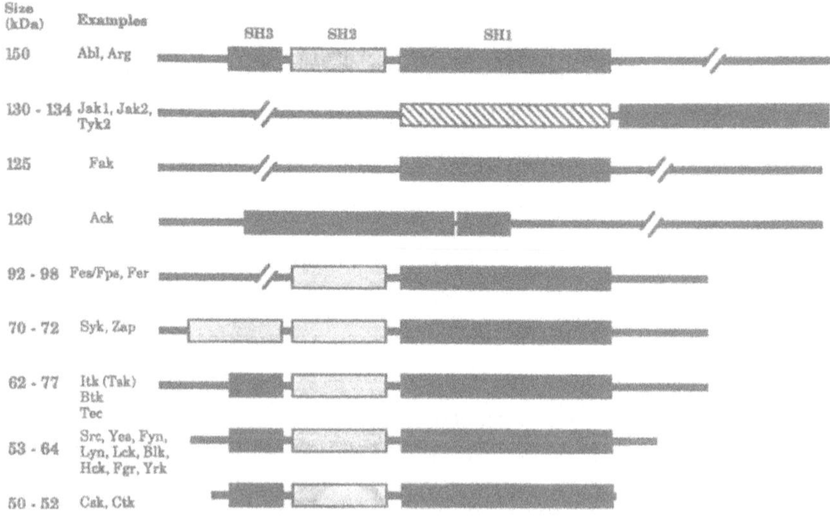


**Figure 1.** Common domain relationships for non-receptor protein tyrosine kinases.

Syk/Zap class of kinases possess tandemly arranged tyrosines at the predicted autophosphorylation site[5].

With the exception of the Jak and Fak groups, all other nonreceptor PTKs possess one or more SH2 domains and/or an SH3 domain[6,7]. In most cases, the SH2 and SH3 domains are positioned towards the aminoterminal end of the enzyme with respect to the catalytic domain. The exception is the Ack PTK in which the SH3 domain of this enzyme lies carboxyterminal to the catalytic domain[8]. In all cases where PTKs possess both SH3 and SH2 domains, the SH3 domain is positioned aminoterminal to the SH2 domain. The actual spacing between the SH2 and catalytic domains varies as does the spacing between SH3 and SH2. The significance of these spacing differences is not known. SH2 as well as SH3 domains exist together and independently in a variety of cellular proteins, some of which possess enzymatic activity, while others are implicated in enhancing specific protein-protein interactions thought to be important in signaling regulation. It is well established that individual SH2 domains are capable of specific binding to selected phosphotyrosine containing proteins[9] and are thought to play critical roles in the interactions between signaling components in PTK dependent pathways[10]. The SH3 domains are thought to recognize proline-rich peptide ligands which in some cases appear to be present in guanine nucleotide exchange factors and GTPase-activating proteins[11]. Both of these SH3 binding protein types are acknowledged in the regulation of the activity of small G proteins such as Ras/Rac/Rho. Recently the structures of several SH2 and SH3 domains with and without complexed peptides or phosphopeptides have been determined in solution by NMR and by X-ray crystallography[12-18]. These structures document the modular nature of the SH2 and SH3 domains and have defined important features of the specific interactions between these domains and their ligands.

Only the members of the src family, along with a single isoform of c-Abl/Arg (type IV), have an inherent mechanism for promoting their association with cell membranes. This feature of the src family and c-Abl type IV results from the posttranslational myristylation of a common glycine residue at position two[19-23]. Several members of the src family also appear to be palmitylated in addition to being myristylated[24]. Palmitylation of these enzymes occurs at one or more cysteines located at residue three and either residue five or six.

Current evidence suggests that palmitylated src PTKs are located in regions of plasma membranes that enahances their capacity to associate with glycosyl-phosphatidylinositiol (GPI)-anchored proteins[25-27]. Thus, these aminoterminal fatty acid modifications of the src PTKs distinguishes them (and c-Abl/Arg type IV) from the remaining nonreceptor PTKs and promotes the likelihood that these kinases will be located at cellular sites where a significant portion of PTK-mediated signaling reactions take place. It reasonable to assume that these are also the same sites where important physiologic substrates for the src enzymes exist.

## MULTICHAIN IMMUNE RECOGNITION RECEPTORS

The multichain immune recognition receptors (MIRR) include the antigen receptors for T cells and B cells as well as Fc receptors, such as the high affinity IgE receptors expressed on mast cells and basophils. Examples of multichain immune recognition receptors are shown in figure 2. This class of receptor typically includes antigen- or Fc- binding chain(s) expressed on the cell surface. The cytoplasmic content of the antigen/ligand binding proteins is usually limited, necessitating the association with other proteins that function from a signaling standpoint to couple the antigen/ligand binding proteins to cellular signal transduction enzymes. While the amino acid sequences of these signal transduction coupling proteins are diverse, common elements can be found in their cytoplasmic domains which include at least two tyrosine residues separated by approximately 10 amino acids (figure 2)[28]. Engagement of multichain immune recognition receptors results in rapid phosphorylation of these two tyrosines. These tyrosine residues represent phosphorylation sites for PTKs as well as binding sites for SH2 containing proteins and have been labeled as tyrosine activation motifs (TAMs)[29], antigen recognition activation motifs (ARAMs)[30], or antigen receptor homology 1domain (ARH1)[31].

The T cell antigen receptor (TCR) is comprised of six different gene products in the probable stoichiometry of $\alpha\beta\gamma\delta\varepsilon_2\zeta_2$, with the $\alpha$ and $\beta$ clonotypic chains forming the extracellular, antigen-recognition domain. The CD3 complex of the TCR includes the $\gamma$, $\delta$, and $\varepsilon$ proteins each containing an extracellular Ig-like domain, a transmembrane domain, and a cytoplasmic portion. A disulfide-linked homodimer of $\zeta$ chains or a heterodimer of $\zeta$-$\eta$ chains, with very short extracellular

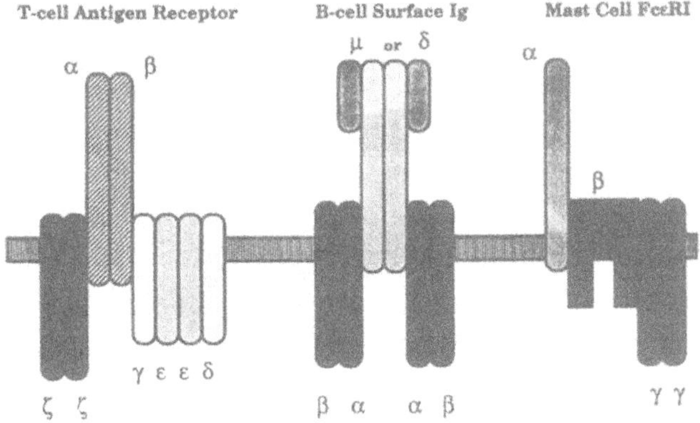


Alignment of Cytoplasmic Sequences

| | |
|---|---|
| *zeta* | ADAYSDIGTKGERRRGKGHDGLYQGLSTATKDTYDALHMQTLAPR |
| *ε gamma* | RLKIRVDKADIASREKSDAVYTGLNTDNQETYETLKHEKPPQ |
| *mb-1* (Ig α) | RKRWQNEKFGVDMPDDYEDENDYEGLNLDDCSMYEDISRGLQGTYQDVGNLHIGD |
| *b-29* (Ig β) | DKDDGKAGMEEDHTYEGLNIDQTATYEDIVTLRTGEVKWSVGEHPGQE |
| ε beta | YRIGQELE-SKKVPDDRLYEEL-NVYSPIY-SELEDKGETSSPVDS |
| *CD3 gamma* | GQDGVRQSRASDKQTLLPNDQLYQPLKDREDDQY-SHLQGNQLRRN |
| *CD3 delta* | GHETGRLRGAADTQALLRNDQVYQPLRDRDDAQY-SHLGGNWARNR |

| | |
|---|---|
| Consensus | DXXXXXXXDXXYXXLXXXXXXXYDXL |
| | E E E I |

**Figure 2.** Structural features of multichain immune recognition receptors and alignment of the tyrosine activation motifs (TAMs) within the cytoplasmic domains of the TCR (zeta, CD3 gamma, CD3 delta), the BCR (Igα, Igβ) and the FcεRI (ε gamma, ε beta).

domains, single transmembrane domains, and large cytoplasmic tails, completes the TCR. The tyrosine activation motif, appearing as a single copy in each of the CD3 chains and in triplicate in ζ, is rapidly phosphorylated following TCR stimulation.

The B cell antigen receptor (BCR) utilizes a membrane-bound immunoglobulin molecule (mIg) for its antigen-binding moiety. The mIg is found in association with disulfide-linked heterodimers of Igα and Igβ, molecules containing short, extracellular Ig-like domains, transmembrane domains, and cytoplamsmic tails. A single copy of the TAM is present in the cytoplasmic domains of both Igα and Igβ and is rapidly phosphorylated in response to mIg crosslinking.

The FcεRI complex includes the ligand-binding α chain, which contains two Ig-like extracellular domains responsible for high affinity binding of IgE, as well as a single transmembrane domain and a short cytoplasmic tail. The α chain associates with a β chain with four transmembrane regions and a disulfide-linked homodimer of γ chains. The TAM appears in the cytoplasmic domains of the β and γ chains and is rapidly phosphorylated upon crosslinking FcεRI.

The recurrence of the tyrosine activation motif in the cytoplasmic domains of multichain immune recognition receptors points towards a commonality of coupling to signal transduction pathways in the different immune effector cell types. The signaling functions of the TAM-containing receptor tails have been confirmed through experiments with chimeric molecules of an MIRR cytoplasmic tail directly linked to a ligand-binding extracellular moiety. A chimeric molecule expressing the IL2 receptor α subunit external domain fused to the CD3ε cytoplasmic tail was capable of activating a T cell hybridoma[32]. Similarly, CD8 chimeras expressing the cytoplasmic domains of the BCR Igα or Igβ[33] or chimeras constructed with sIg external domains directly linked to the cytoplasmic domains of Igα or Igβ[34] were able to initiate signal transduction in transfected B cells. Interestingly, comparison of the signaling efficiency of chimeras expressing single copies of the ζ TAM with chimeras expressing the intact ζ chain with its three TAMs revealed that multimerization of the TAM resulted in signal amplification[35]. The similarity of the signaling capacity of TAM-containing receptor tails is further underscored by the observation that chimeras expressing the MIRR cytoplasmic tail of one cell type are capable of utilizing signal transduction machinery of a different immune effector cell. The cytoplasmic domain of TCR ζ can substitute for the FcεRI γ chain in RBL-2H3 basophilic leukemia cells, eliciting degranulation upon FcεRI crosslinking[36]. In an analogous manner, the BCR-like mIg-Igβ chimera can be expressed in a T cell line and, upon crosslinking, elicits a signal that yields phosphorylation of proteins on tyrosine, activation of the the repertoire of PTKs that are present in T cells (which is different from that present in B cells), $Ca^{2+}$ flux, phosphoinositol turnover, and IL2 secretion[37].

Accumulated evidence indicates that the TAM is sufficient to couple receptors to events associated with immune cell activation. Mutational studies have revealed that the tyrosines are required for the functional activity of the TAM[32,34,35,37,38]. The sequences between the two tyrosines can be distinct, while the spacing between the tyrosines appears to be critical. It is not clear yet whether secondary or tertiary structure contributes to the function of this motif. Regardless, it appears likely that multiple SH2-containing proteins including PTKs, PTK substrates, protein tyrosine phosphatases, or other proteins co-localize at these sites of tyrosine phosphorylation.

## ACTIVATION OF NONRECEPTOR PROTEIN TYROSINE KINASES

Activation of nonreceptor PTKs appears to be necessary for multichain immune recognition receptor signaling. Several studies suggest that at least two families of nonreceptor PTKs are involved in this process. These include members of the src family as well as members of Syk family of PTKs.

As an initial step in determining the roles of different families of nonreceptor PTKs in the process of multichain immune recognition receptor signaling, we have investigated the temporal order of activation of the src PTKs and Syk PTKs following surface engagement of multichain immune recognition receptors and have compared the activation kinetics of these enzymes with the tyrosine phosphorylation of the receptor signaling subunits containing the TAMs. A general model summarizing the results for a generic multichain immune recognition receptor is presented in figure 3.

In this model, receptor engagement results in initial transient activation of one or more members of the membrane associated src family of PTKs followed by the subsequent activation of the Syk class of cytoplasmic PTKs. The mechanism for Src kinase activation is not known, but could include trans-phosphorylation at the enzyme's autophosphorylation site induced by clustering Src enzymes together or perhaps by juxtaposing one or more protein tyrosine phosphatases with Src enzymes leading to the initial dephosphorylation of their regulatory tyrosines. The location of the Src kinases and the early kinetics of their activation make this class of enzyme candidates for the PTKs responsible for initial phosphorylation of the TAMs present in the receptor.

From a variety of experimental approaches it is clear that Src kinases can promote the activation of the Syk class of PTK presumably through phosphorylation of Syk PTKs on as yet undefined tyrosine residues. Thus, it is thought that activation of Src kinases following receptor engagement leads directly to the

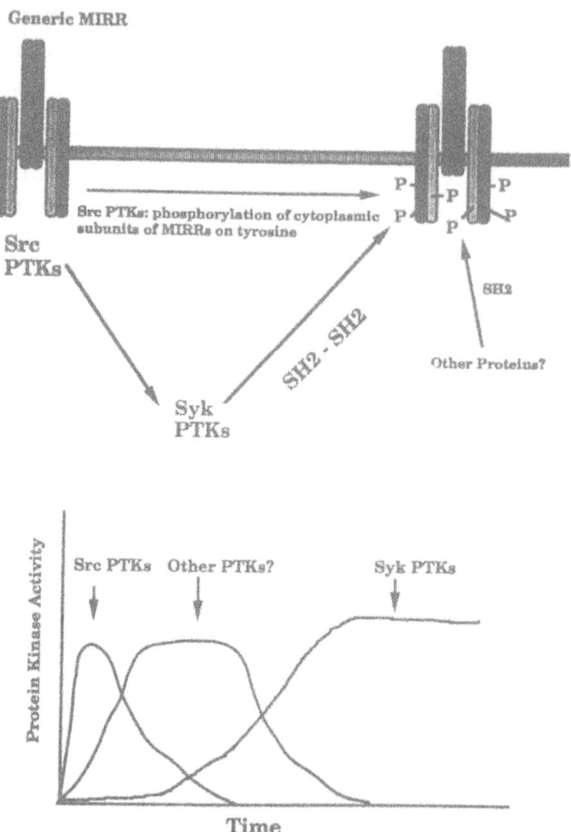


**Figure 3.** Model of signal transduction through multichain immune recognition receptors. The top portion illustrates activation of nonreceptor protein tyrosine kinases and tyrosine phosphorylation of the receptor cytoplasmic sequences and the graph depicts the temporal order of activation of the protein tyrosine kinases.

activation of the Syk PTKs. However, it is unclear how the membrane associated Src kinases interact with the normally cytoplasmic Syk-like PTKs and raise the possiblity that either intermediate kinases play some role in this communication process or that some population of Syk PTKs could be membrane associated. The Syk PTKs, possessing two tandem SH2 domains, are then translocated to the receptor TAMs where they bind thereby localizing the Syk class of enzyme to a membrane proximal position. It is possible that Syk association with TAMs can further stimulate the enzyme activity. The Syk PTKs would then be positioned to phosphorylate other substrates to further amplify and promote multichain immune recognition signaling.

## REFERENCES


1. S.K. Hanks, A.M. Quinn, and T. Hunter, The protein kinase family: conserved features and deduced phylogeny of the catalytic domains, *Science*, 241:42-52, (1988).
2. A.F. Wilks, A.G.. Harpur, R.R. Kurban, S.J. Ralph, G. Zurcher, and A. Ziemmeicki, Two novel protein-tyrosine kinases, each with a second phosphotransferase-related catalytic domain, define a new class of protein kinase, *Mol. Cell. Biol.* , 11:2057-2065, (1991).
3. S. Nada, M. Okada, A. MacAuley, J.A. Cooper, and H. Nakagawa, Cloning of a complementary DNA for a protein-tyrosine kinase that specifically phosphorylates a negative regulatory site of $p60^{c\text{-}src}$, *Nature*, 351:69-72, (1991).
4. S. Klages, D. Adam, K. Class, J. Fargnoli, J.B. Bolen, and R.C. Penhallow, Ctk: a novel protein tyrosine kinase related to Csk that defines a new enzyme family, *Proc. Natl. Acad. Sci. U.S.A.,* 91:2597-2601 (1994).
5. A.C. Chan, M. Iwashima, C.W. Turck, and A. Weiss., ZAP-70: a 70-kD protein tyrosine kinase that associates with the TCR ζchain, *Cell*, 71:649-662, (1992).
6. T. Pawson, and G.D. Gish, SH2 and SH3 domains: from structure to function, *Cell*, 71:359-362, (1992).
7. B.J. Mayer, and D. Baltimore, Signaling through SH2 and SH3 domains. *Trends in Cell Biol.*, 3:8-13, (1993).
8. E. Manser, T. Leung, H. Salihuddin, L. Tan, and L. Lim, A non-receptor tyrosine kinase that inhibits the GTPase activity of p21cdc42, Nature, 363:364-367 (1993).
9. Z. Songyang, S.E. Shoelson, M. Chaudhuri, G. Gish, T. Pawson, W.G. Haser, F. King, T. Roberts, S. Tarnofsky, R.J. Lechleider, B.G. Neel, R.B. Birge, J.E. Fajardo, M.M. Chou, H. Hanafusa, B. Schaffhausen, and L.C. Cantley, SH2 domains recognize specific phosphopeptide sequences, *Cell*, 72:767-778 (1993).
10. C.A. Koch, D. Anderson, M.F. Moran, C. Ellis, and T. Pawson, SH2 and SH3 domains: elements that control interactions of cytoplasmic signaling proteins, *Science*, 252:668-674, (1991).
11. R. Ren, B.J. Mayer, P. Cicchetti, and D. Baltimore. Identification of a ten amino acid proline rich SH3 binding site, *Science*, 259:1157-1161, (1993).
12. G. Waksman, D. Kominos, S.C. Robertson, N. Pant, D. Baltimore, R.B. Birge, D. Cowburn, H. Hanafusa, B.J. Mayer, M. Overduin, M.D. Resh,C.B. Rios, L. Silverman, and J. Kuriyan, Crystal structure of the phosphotyrosine recognition domain SH2 of v-src complexed with tyrosine-phosphorylated peptides, *Nature* 358:646-653, (1992).
13. M. Overduin, B. Mayer, C.B. Rios, D. Baltimore, and D. Cowburn, Secondary structure of src homology 2 domain of c-Abl by heteronuclear NMR spectroscopy in solution, *Proc. Natl. Acad. Sci. U.S.A.* 89:11673-11677, (1992).

14. G.W. Booker, A.L. Breeze, A.K. Downing, G. Panayatou, I. Gout, M.D.Waterfield, and I.D. Campbell, Structure of an SH2 domain of the p85α subunit of phosphatidylinositol-3-OH kinase, *Nature*, 358:684-687, (1992).
15. A. Musacchio, M. Noble, R. Paulpitt, R. Wierenga, and M. Saraste, Crystal structure of a Src-homology 3 (SH3) domain, *Nature*, 359:851-855, (1992).
16. H. Yu, M.K. Rosen, T.B. Shin, C. Seidel-Dugan, J.S. Brugge, and S.L. Schreiber, Solution structure of the SH3 domain of Src and identification of its ligand-binding site, *Science*, 258:1665-1668, (1992).
17. M.J. Eck, S.E. Shoelson, and S.C. Harrison, Recognition of a high-affinity phosphotyrosyl peptide by the Src homology-2 domain of $p56^{lck}$, *Nature*, 362:87-91, (1993).
18. G. Waksman, S.E. Shoelson, N. Pant, D. Cowburn, and J. Kuriyan, Binding of a high affinity phosphotyrosyl peptide to the Src SH2 domain: crystal structures of the complexed and peptide-free forms, *Cell*, 72:779-790, (1993).
19. J.E. Buss, M.P. Kamps, and B.M. Sefton, Myristic acid is attached to the transforming protein of Rous sarcoma virus during or immediately after synthesis and is present in both soluble and membrane-bound forms of the protein, *Mol. Cell. Biol.* 4:2697-2704, (1984).
20. M.P. Kamps, J.E. Buss, and B.M. Sefton, Mutation of NH2-terminal glycine of p60src prevents both myristylation and morphological transformation, *Proc. Natl. Acad. Sci. U.S.A.*, 82:4625-4628, (1985).
21. P. Jackson, and D. Baltimore, N-terminal mutations activate the leukemogenic potential of the myristolated form of c-Abl, *EMBO J.*, 8:449-456, (1989).
22. G.Q. Daley, R.A. Van Etten, P. Jackson, A. Bernards, and D. Baltimore, Nonmyristolated Abl proteins transform a factor-dependent hematopoietic cell line, *Mol. Cell. Biol.*12:1864-1871, (1992).
23. M.D. Resh, Membrane interactions of $pp60^{v\text{-}src}$: a model for myristylated tyrosine protein kinases, *Oncogene*, 5:1437-1444, (1990).
24. L.A. Paige, M.J.S. Nadley, M.L. Harrison, J.M. Cassady, and R.L. Geahlen, Reversible palmitylation of the protein tyrosine kinase $p56^{lck}$, J. Biol. Chem. 268: 8669-8674, (1993).
25. A.M. Shenoy-Scaria, J. Kwong, T. Fujita, M.W. Olszowy, A.S. Shaw, and D.M. Lublin, Signal transduction through decay-accelerating factor. Interaction of glycosyl-phosphatidylinositol anchor and protein tyrosine kinases $p56^{lck}$ and $p59^{fyn}$, *J. Immunol.* 149: 3535-3541 (1992).
26. P.M. Thomas, and L.E. Samelson, The glycophosphatidylinositol-anchored Thy-1 molecule interacts with the $p60^{fyn}$ protein tyrosine kinase. *J. Biol. Chem.* 267: 12317-12322, (1992).

27. A.M. Shenoy-Scaria, L.K. Timson Gauen, J. Kwong, A.S. Shaw, and D.M. Lublin, Palmitylation of an amino-terminal cysteine motif of protein tyrosine kinases $p56^{lck}$ and $p59^{fyn}$ mediates interaction with glycosyl-phosphatidylinositol-anchored proteins, *Mol. Cell. Biol.* 13: 6385-6392, (1993).
28. M. Reth, Antigen receptor tail clue, *Nature*, 338:383, (1989).
29. L.E. Samelson, and R.D. Klausner, Tyrosine kinases and tyrosine-based activation motifs - current research on activation via the T cell antigen receptor, *J. Biol. Chem.* 267:24913-24916, (1992).
30. A. Weiss, T cell antigen receptor signal transduction: a tale of tails and cytoplasmic protein-tyrosine kinases, *Cell*, 73:209-212, (1993).
31. J.C. Cambier, and K.S. Campbell, Membrane immunoglobulin and its accomplices: new lessons from an old receptor, *FASEB J.*, 6:3207-3217, (1992).
32. F. Letourner, and R.D. Klausner, Activation of T cells by a tyrosine kinase activation domain in the cytoplasmic tail of CD3 ε, *Science*, 255:79-82, (1992).
33. K.M. Kim, G. Alber, P. Weiser, and M.Reth, Differential signaling through the Igα and Igβ components of the B cell antigen receptor, *Eur. J. Immunol.*, 132:125-146, (1993).
34. M. Sanchez, Z. Misulovin, A.L. Burkhardt, S. Mahajan, T. Costa, R. Franke, J.B. Bolen, and M. Nussenzweig, Signal transduction by immunoglobulin is mediated through Igαand Igβ, *J. Exp. Med.*, 178:1049-1055, (1993).
35. B.A. Irving, A.C. Chan, and A. Weiss, Functional characterization of a signal transducing motif present in the T cell antigen receptor zeta chain, *J. Exp. Med.*, 177:1093-1103, (1993).
36. F. Letourner, and R.D. Klausner, T-cell and basophil activation through the cytoplasmic domain of T cell receptor ζfamily members, *Proc. Natl. Acad. Sci. U.S.A.*, 88:8905-8909, (1991).
37. A.L. Burkhardt, T. Costa, Z. Misulovin, B. Stealey, J.B. Bolen, and M.C. Nussenzweig, Igα and Igβ are functionally homologous to the signaling proteins of the T-cell receptor, *Mol. Cell. Biol.*, 14:1095-1103, (1994).
38. C. Romeo, M. Amiot, and B. Seed, Sequence requirements for the induction of cytolysis by the T cell antigen/Fc receptor ζchain, *Cell*, 68:889-897, (1992).

# PHORBOL MYRISTATE ACETATE-INDUCED CHANGES IN PROTEIN KINASE C ISOZYMES ($\alpha$, $\beta$, $\gamma$ and $\zeta$) IN HUMAN T CELL SUBSETS

Sudhir Gupta and William Harris

Division of Basic and Clinical Immunology
University of California
Irvine, CA 92717-4069

## INTRODUCTION

Protein kinase C (PKC), a serine/threonine kinase, plays an important role in downstream signaling pathway of T cell activation (1,2). Activation of PKC produces varied, and at times disparate, cell specific responses. In T cells, direct activation of PKC results in an induction of interleukin-2 receptor (IL-2R) expression and IL-2 production, initiation of DNA synthesis, and regulation of cytotoxic function (3-11). Although PKC was originally thought to be a single molecule, recent molecular and biochemical studies have revealed that PKC exists as a family of different isozymes that differ in their structure, dependence on $Ca^{++}$, tissue distribution and substrate specificity (12-15). They have been broadly categorized into three major subgroups; Group 1 including classical $Ca^{++}$-dependent PKCs ($\alpha$, $\beta$ and $\gamma$); Group 2 comprised of $Ca^{++}$-independent PKCs ($\delta$, $\epsilon$, $\eta$, $\theta$); and Group 3, consisting of atypical PKC ($\zeta$, $\lambda$). PKC$\zeta$ isozyme is considered to be resistant to the effect of phorbol esters (12). Ways et al (16), using a variety of non-lymphoid cell lines, have failed to demonstrate translocation of PKC$\zeta$ following stimulation with phorbol ester. To the best of our knowledge there is no study of PKC$\zeta$ in normal human peripheral blood T cells. Further there is a controversy regarding the role of PKC$\beta$ isozyme T cell functions (17-19) and regarding the presence of PKC$\gamma$ in T cells (22-26). In this study we examined the translocation of $Ca^{++}$-dependent (PKC$\alpha$, $\beta$, $\gamma$) and atypical (PKC$\zeta$) PKC isozymes in human peripheral blood T cells. This study demonstrates that PKC isozymes are differentially translocated among CD4+ and CD8+ T cells and PKC$\gamma$ isozyme is present in T cells and translocated following PMA activation.

## MATERIALS AND METHODS

### Materials

Rabbit polyclonal antibodies specific for PKC$\alpha$, $\beta$, $\gamma$ and $\zeta$ isozymes were purchased from BRL GIBCO, Gaithersberg, MD. PE-labeled anti-CD4 and anti-CD8 monoclonal antibodies and their isotypes were purchased from Becton Dickinson, San Jose, CA. Phorbol myristate acetate (PMA), normal rabbit serum and FITC-conjugated goat-antirabbit antibody were purchased from SIGMA chemicals, St. Louis, MO.

### Methods

Peripheral blood mononuclear cells (PBMC) were isolated on Ficoll-Hypaque density gradient. T cells were purified by passing the MNC over a nylon wool column. T cells were purified to >95% CD2+ cells. T cells were activated with 10nM of PMA for various lengths of time (15, 30, 45 and 60 minutes) and examined for PKC isozyme translocation in CD4 and CD8 T cells, using FACScan and PKC$\zeta$ isozyme in T cells by Western blot analysis.

**Flow Cytometry.** T cells with or without activation with PMA were first incubated with PE-CD4 or PE-CD8 monoclonal antibodies by incubating the cells with monoclonal antibodies and isotype controls for 45 minutes on ice. Cells were washed and then fixed with 0.25% Tween-20. The fixed cells were incubated with individual anti-PKC isozyme antibodies for 30 min on ice, washed x 3 with cold PBS and counterstained with FITC-conjugated goat-anti-rabbit igG. Normal rabbit serum was run as control. Five thousand cells were analyzed with dual color analysis, using FACScan. Data are expressed as mean-fluorescence channel numbers.

**Western Blot Analysis.** Stimulated and unstimulated T cells were washed and protein extracts from cytosol and membrane fractions were prepared. Electrophoresis was performed on a 12% SDS denaturing gel and transferred onto a nitrocellulose membrane. Non-specific binding was blocked with Tris buffer containing Tween 20 and 1% bovine serum albumin. Blots were incubated with PKC isozyme antibodies overnight. The membranes ware washed with biotin-labeled anti-rabbit Ig diluted in buffer containing 5% skim milk. The blots were washed and incubated with streptavidine alkaline phosphatase for an additional 1 hour. The blots were washed and immunoreactivity detected with the substrate 5-bromo-4-chloro-3-indolyl phosphate and 4 nitroblue tetrazolium (Boehringer Mannheim Biochem, Indianapolis, IN).

## RESULTS AND DISCUSSION

Several investigators have examined the distribution of PKC isozymes in human T cells and tumor T cell lines (17-26). In peripheral blood T cells, presence of PKC$\alpha$ and $\beta$ isozymes and lack of PKC$\gamma$ isozymes has been reported (17-22 ). In the present study, we detected all three PKC isozymes (PKC$\alpha$, $\beta$ and $\gamma$) in both CD4+ and CD8+ T cells. This is in agreement with the observation of Altman et al (23) and Harris and Gupta (24) who also detected PKC$\gamma$ in peripheral blood T cells and T cell subsets. Lucas et al (21) observed higher levels of PKC$\beta$ as compared to PKC$\alpha$ in mitogen-activated peripheral blood T cell blasts. In contrast, Jurkat cells displayed higher levels of PKC$\alpha$ compared to PKC$\beta$. In the present study, we also observed higher PKC$\beta$ as compared to PKC$\alpha$ in unstimulated CD4+ T cells; however, no difference was observed in these two PKC isozymes in unstimulated CD8+ T cells (Table 1).

We have observed that translocation of PKC isozymes from cytosol to plasma membrane following T cell activation is associated with increased in fluorescence intensity in flow cytometry. This correlates with translocation of PKC isozyme as demonstrated by confocal microscopy (25). Data on the effect of PMA on PKCα, β and γ in CD4+ and CD8+ T cell subsets are shown in Table 1. In CD4+ T cells PMA induced translocation of PKCα only, whereas in CD8+ T cells PMA induced translocation of PKCα, β and γ isozymes. Altman et al (23) also observed translocation of PKCα, β and γ isozymes following activation of T cells with TPA; however at the mRNA level only PKCβ mRNA

**Table 1.** Effect of PMA activation on PKC isoforms [1]

| PKC Isoforms on CD4+ T Cells | | | |
|---|---|---|---|
| Time (minutes) | α | β | γ |
| T0 | 188 | 225 | 197 |
| T15 | 175 | 201 | 199 |
| T30 | 217[2] | 206 | 171 |
| T60 | 165 | 151 | 169 |
| **PKC Isoforms in CD8+ T Cells** | | | |
| Time (minutes) | α | β | γ |
| T0 | 33 | 34 | 21 |
| T15 | 501 | 44 | 248 |
| T30 | 349 | 200 | 209 |
| T60 | 444 | 287 | 437 |

[1]Data are expressed as mean fluorescence channel numbers.
[2]Underlined numbers represent peak and signficant increased in fluoresecnce intensity.

was increased. This would suggest that PMA does not effect the transcription of PKCα and γ isozymes, but rather causes their translocation from cytosol to plasma membrane. In Jurkat T cell line, Tsutsumi et al (26) also observed translocation of PKCα and β following activation with PMA. Furthermore, the baseline fluorescence intensity of all PKC isozymes was significantly less in CD8+ T cells as compared to CD4+ T cells. To the best of our knowledge, this is the first report of PKC isozymes in human CD4+ and CD8+ T cells. Following stimulation with PMA, the peak levels of PKCα and PKCβ were comparable in CD4+ T cells, whereas peak PKCα levels in PMA-stimulated cells were significantly greater than those of PKCβ. The role of PKCβ in human T cell functions has been controversial and can be explained by the differences between normal peripheral T cells and tumor T cell lines used in these experiments. Koretzky et al (17), using a Jurkat T cell line, reported that PKCβ isozyme is not involved in PMA-induced IL-2 production. In contrast, Kelleher and Long (18) reported that a T cell line that was made PKCβ deficient, had a defect in IL-2 production. These investigators suggested that PKCβ

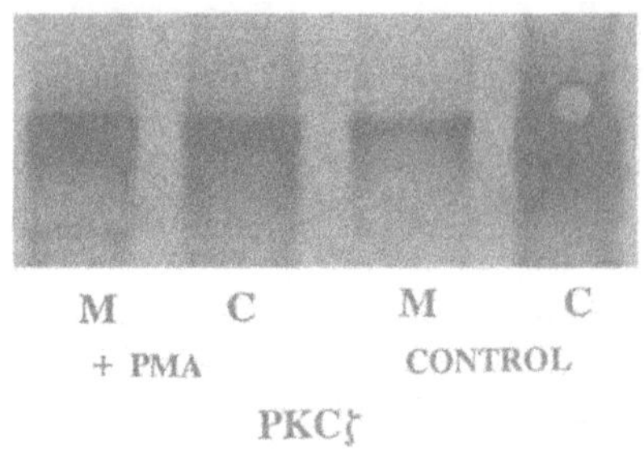


**Figure 1.** Effect of PMA stimulation on translocation of PKC$\zeta$ in T cells.

isozyme may be obligatory for IL-2 production. Szamel et al (27) demonstrated that the introduction of anti-PKC$\beta$ antibody in T cells by electroporation resulted in an inhibition of IL-2 production and lymphocyte proliferation. We have recently shown that 12-deoxyphorbol-13-O-phenylacetate 20 acetate (a PKC$\beta$ agonists) induces IL-2 production, IL-2R expression and DNA synthesis in peripheral blood T cells (19), further supporting a role of PKC$\beta$ in T cell functions.

PKC$\zeta$ lacks the C2 domain and is normally classified as a $Ca^{++}$-independent isozyme (12). In addition, it differs from other isozymes in containing only one set of cystine-rich zinc-finger motifs in the C1 domain that encodes the DAG-binding domain (12). Furthermore, PKC activity isolated from transfactants that overexpress PKC$\zeta$ is not regulated by DAG or PMA (16,28). Thus it was questionable whether PKC$\zeta$ could be induced to translocate by PMA. Ways et al (16) did not observe translocation or downregulation of PKC$\zeta$ in U937, HL-60, COS, or HeLa cells, following stimulation with TPA. Tsutsumi et al (26) observed translocation of 82-kd PKC$\zeta$ in Jurkat T cells and murine thymocytes, but failed to demonstrate any translocation of 75-kd band. Several studies have suggested that PKC$\zeta$ may be present in multiple forms (28,29). In fact, transfection studies with PKC$\zeta$ cDNA have yielded immunoreactive proteins of 64-kd and 80-kd (16,28). In the present study, we observed a significant translocation of 80-kd PKC$\zeta$ from cytosol (C) to membrane fraction (M) in peripheral blood T cells 15 minutes following stimulation with PMA (Figure 1). No translocation was observed at 5 minutes following stimulation with PMA (data not shown). Therefore, it appears that phorbol esters in cells of T-lineage induces translocation of at least 80-kd PKC$\zeta$ species. A role of PKC$\zeta$ in human T cell functions remains to be determined.

In summary, PKC$\alpha$ and $\beta$ are differentially expressed in CD4+ and CD8+ peripheral blood T cells. PKC$\gamma$ is also expressed in both CD4+ and CD8+ T cells. The extent of translocation of PKC$\alpha$, $\beta$ and $\gamma$ isozymes following PMA stimulation differs between CD4+ and CD8+ T cells; CD8+ T cells are more responsive than CD8+ T cells. Finally, PMA-induces translocation of 80-kd PKC$\zeta$ isozyme from cytosol to the plasma membrane.

## REFERENCES

1. A. Weiss, J.B. Imboden, K. Hardy, B. Manger, C. Terhorst, and J.D. Stobo. The role of T3/antigen receptor complex in T cell activation. *Ann. Rev. Immunol.* 4: 593 (1986).
2. S. Gupta, M. Shimizu, K. Ohira, and B. Vayuvegula. T cell activation via the T cell receptor: A comparison between WT31 (defining $\alpha/\beta$ TcR)-induced and anti-

CD3-induced activation of human T lymphocytes. *Cell. Immunol.* 132: 26 (1991).

3. M. Isokov and A. Altman. Human T cell activation by tumor promotors: role of protein kinase C. *J. Immunol.* 138: 3100 (1987).
4. A.M. Mastro and M.C. Smith. Calcium-dependent activation of lymphocytes by ionophore, A23187, and phorbol esters tumor promotor. *J. Cell Physiol.* 116: 51 (1993).
5. A.F. Truneh, F. Albert, P. Goldstein, and A-M. Schmitt-Verhulst. Early steps of lymphocyte activation by-passed by synergy between calcium ionophore and phorbol ester. *Nature (Lond)* 313: 318 (1985).
6. N. Berry, K. Ase, U. Kikkawa, A. Kishimoto and Y. Nishizuka. Human T cell activation by phorbol esters and diacylglycerol. *J. Immunol.* 143: 1417 (1989).
7. J.M. Depper, W.J. Leonard, M. Kronke, P.D. Noguchi, R.E. Cunningham, T.A. Waldmann, and W.C. Greene. Regulation of Interleukin 2 receptor expression: effects of phorbol diester, phospholipase C and reexposure to lectin or antigen. *J. Immunol.* 133: 3054 (1984).
8. T. Hirano, K. Fujimoto, T. Teranishi, N. Nishino, K. Onoue, S. Maeda and K. Shimada. Phorbol ester increases the level of interleukin 2 mRNA in mitogen-stimulated human lymphocytes. *J. Immunol.* 132: 2165 (1984).
9. H. Schrezenmeir, R. Kurrle, and B. Fleischer. Stimulus-dependent triggering or inhibition of cytotoxicity in human cytotoxic T-lymphocytes by activators of protein kinase C. *Immunology* 59: 359 (1986).
10. J.A. Ledbetter, L.E. Gentry, C.H. June, P.S. Rabinovitch and A.F. Purchio. Stimulation of T cells through the CD3/T-cell receptor complex: role of cytoplasmic calcium, protein kinase C translocation and phosphorylation of $pp^{60c\text{-}src}$ in the activation pathway. *Mol. Cell. Biol.* 7: 650 (1987).
11. J. Abb, G.J. Bayliss, and F. Deinhardt. Lymphocyte activation by the tumor promoting agent 12-O-tetradecanoyl-phorbol-13-acetate (TPA). *J. Immunol.* 122: 1639 (1979).
12. Y. Nishizuka. The molecular heterogenity of protein kinase C and its implication for cellular regulation. *Science* 334: 661 (1988).
13. P.M. Blumberg. Complexities of the protein kinase C pathway. *Mol. Carcinogenesis* 4: 339 (1991).
14. P.J. Parker, G. Kour, R. Marais, F. Mitchell, C. Pears, D. Schaap, S. Stabel and C. Webester. Protein kinase C-a family affair. *Mol. Cell. Endocr.* 65: 1 (1989).
15. Y. Nishizuka. Intracellular signalling by hydrolysis of phospholipids and activation of protein kinase C. *Science* 258: 607 (1992).
16. D.K. Ways, P.P. Cook, C. Webster, and P.J. Parker. Effect of phorbol esters on protein kinase C-$\zeta$. *J. Biol. Chem.* 267: 4799 (1992).
17. G.A. Koretzky, M. Wahi, M.E. Newton and A. Weiss. Heterogeneity of protein kinase C isozyme gene expression in human T cell lines. Protein kinase C-$\beta$ is not required for several T cell functions. *J. Immunol.* 143: 1692 (1989).
18. D. Kelleher D. and A. Long. Development and characterization of the protein kinase C$\beta$-isozyme-deficient T-cell line. *FEBS Lett.* 3: 310 (1992).
19. S. Aggarwal, S. Lee, A. Mathur, S. Gollapudi and S. Gupta. 12-deoxyphorbol-13-O-phenylacetate 20 acetate [an agonist of protein kinase C$\beta$1 (PKC$\alpha$1)] induces DNA synthesis, interleukin-2 (IL-2) production, IL-2 receptor $\alpha$-chain (CD25) and $\beta$-chain (CD122) expression, and translocation of PKC$\beta$ isozyme in human peripheral blood lymphocytes: Evidence for a role of PKC$\beta$1 in human T cell activation. *J. Clin. Immunol 14:* (1994) (in press).
20. M.S. Shearman, N. Berry, T. Oda, K. Ase, U. Kikkawa and Y. Nishizuka. Isolation of protein kinase C subspecies from a preparation of human T lymphocytes. *FEBS Lett.* 234: 387 (1988).

21. S. Lucas, R. Marais, J.D. Graves, D. Alexander, P. Parker and D.A. Cantrell Heterogeneity of protein kinase C expression and regulation in T lymphocytes. *FEBS Lett.* 260: 53 (1990).
22. A. Kvanta, M. Jondal and B.B. Freedholm. Translocation of the alpha-isoform and beta isoforms of protein kinase C following activation of human T lymphocytes. *FEBS Lett.* 283: 321 (1991).
23. A. Altman, M.I. Mally and N. Isakov. Phorbol ester synergizes with $Ca^{2+}$ ionophore in activation of protein kinase C (PKC) alpha and beta isozymes in human T cells and in induction of related cellular functions. *Immunology* 76: 465 (1992).
24. W.G. Harris and S. Gupta. Anti-CD3-mediated activation of PKC isoforms in human T cell subsets. *J. Allergy Immunol.* 91: 214 (1993).
25. W. Harris, S. Aggarwal, S. Gollapudi and S. Gupta. Anti-CD3 antibody-induced changes in PKC isozymes ($\alpha$, $\beta$, $\gamma$, $\delta$, $\epsilon$ and $\zeta$) in human T cell subsets. *(Submitted)*
26. A. Tsutsumi, M. Kubo, H. Fujii, J. Freire-Moar, C.W. Turck and J.T. Ransom. Regulation of protein kinase C isoform proteins in phorbol ester-stimulated Jurkat T lymphoma cells. *J. Immunol.* 150: 1746 (1993).
27. M. Szamel, F. Bartels, and K. Resch. Cyclosporin A inhibits T cell receptor-induced interleukin-2 synthesis in human T lymphocytes selectively preventing a transmembrane signal transduction pathway leading to sustained activation of a protein kinase C isozyme, protein kinase C-beta. *Eur. J. Immunol.* 23: 3072 (1993).
28. J. Terajima, A. Tsutsumi, J. Freire-Moar, H.M. Cherwinski and J.T. Ramson. Evidence of clonal heterogeneity of the expression of six protein kinase C isoforms in murine B and T lymphocytes. *Cell. Immunol.* 142: 197 (1992).
29. T. Ono, T. Fuji, K. Ogata, U. Kikkawa, K. Igarashi and Y. Nishizuka. The structure, expression, and properties of additional members of the protein kinase C family. *J. Biol. Chem.* 263: 6927 (1988).

# SPECIFIC CD45 ISOFORMS REGULATE T CELL ONTOGENY AND ARE FUNCTIONALLY DISTINCT IN MODIFYING IMMUNE ACTIVATION

Jamey D. Marth, Christopher J. Ong and Daniel Chui

The Biomedical Research Centre and the Department of Medical Genetics
2222 Health Sciences Mall, University of British Columbia
Vancouver, B.C. V6T 1Z3 Canada

## INTRODUCTION

The antigen receptor complex on the cell surface of T lymphocytes is one of the most modular signal transduction systems yet defined. This characteristic emanates from various mechanisms that promote intermolecular associations between the αβ T cell receptor and the enzymes that transduce the intracellular biological signal cascade (reviewed in 1). Among the multiple proteins that regulate cellular responses following T cell receptor (TCR) stimulation, the CD45 tyrosine phosphatase is a crucial effector. T lymphocytes that lack CD45 expression at the cell surface are unable to transmit immunologic activation signals that initiate from TCR interaction with antigen and major histocompatibility (MHC) molecules (2).

Understanding the role of CD45 in TCR signal transduction is complicated by the variation in CD45 isoform expression that occurs as a result of alternative RNA splicing among extracellular exons (reviewed in 3 and 4). Eight distinct isoforms may be expressed from the single gene locus encoding CD45 (Figure 1). Moreover, these may be variably present on the same T cell at a given time. Nevertheless, T cells are known to preferentially display certain CD45 isoforms at the cell surface and to modify expression of

*Mechanisms of Lymphocyte Activation and Immune Regulation V*
Edited by S. Gupta *et al.*, Plenum Press, New York, 1994

distinct CD45 isoforms in response to thymic differentiation and peripheral T cell activation (5-11). While exon-specific CD45 antibodies are available, there is not any conclusive evidence that CD45 isoform-specific antibodies presently exist; hence it is difficult to clearly identify alterations in CD45 isoform expression. However, variations in CD45 glycosylation may be resolved by certain glycoform-specific CD45 antibodies, such as the sialic acid-binding motif of UCHL-1 (12). Nevertheless, CD45 antibodies can assist in defining the possible repertoire of CD45 isoforms likely to be present on various T cell populations.

Although thymic differentiation events are required for the emergence of peripheral immunocompetent T cells, differentiation may continue in the periphery. Peripheral T cells are heterogeneous in subtypes that can be defined by functional consequences following TCR activation. For example, $T_H1$ and $T_H2$ $CD4^+$ T cell subpopulations differ in lymphokine production profiles (13). Additionally, T cells that have previously undergone antigen-mediated activation may become memory cells and acquire distinct activation requirements (14,15). Since allelic exclusion in the thymus produces a stable singular TCR heterodimeric structure on any one peripheral T cell, modifications within the TCR complex may result in functional disparity among otherwise identical T cell populations. Hence, the thymic TCR selection program is likely to be only one method by which organisms may generate different immunological responses to identical stimuli. Thus it is of importance to understand the biological role of enzymes that are variously altered, whether through changes in activity or expression, in association with the TCR complex.

Although no functional links have been previously established, alterations in T cell activation potential occur in association with changes in CD45 isoform expression. Levels of CD45 exon A, B and C expression correlate with differences in T cell activation responses, lymphokine production and peripheral life-span (16-20). These provocative observations are only correlative in nature since isoform-specific CD45 gene transfer studies have not previously established cause and effect relationships. However, this laboratory has recently succeeded in altering the repertoire of CD45 isoforms in a defined manner among genetically-comparable thymic T cell populations *in vivo*. Our results show that levels of CD45 expression can greatly influence the process of apoptosis and MHC-restricted negative selection (21). Furthermore, mature thymic T cells undergo enhanced immunologic activation in response to specific changes in CD45 isoform expression (22). Experiments described herein have therefore provided genetic evidence for a physiologic distinction among CD45 isoforms and imply a mechanism by which the TCR complex can be functionally modulated by alterations in CD45 isoform expression.

## RESULTS AND DISCUSSION

### Thymic CD45 transgene expression and endogenous levels of CD45 extracellular exon expression

Experiments to alter the repertoire of endogenous CD45 isoform expression can be accomplished either by gene-targeting approaches to delete certain CD45 exons or by

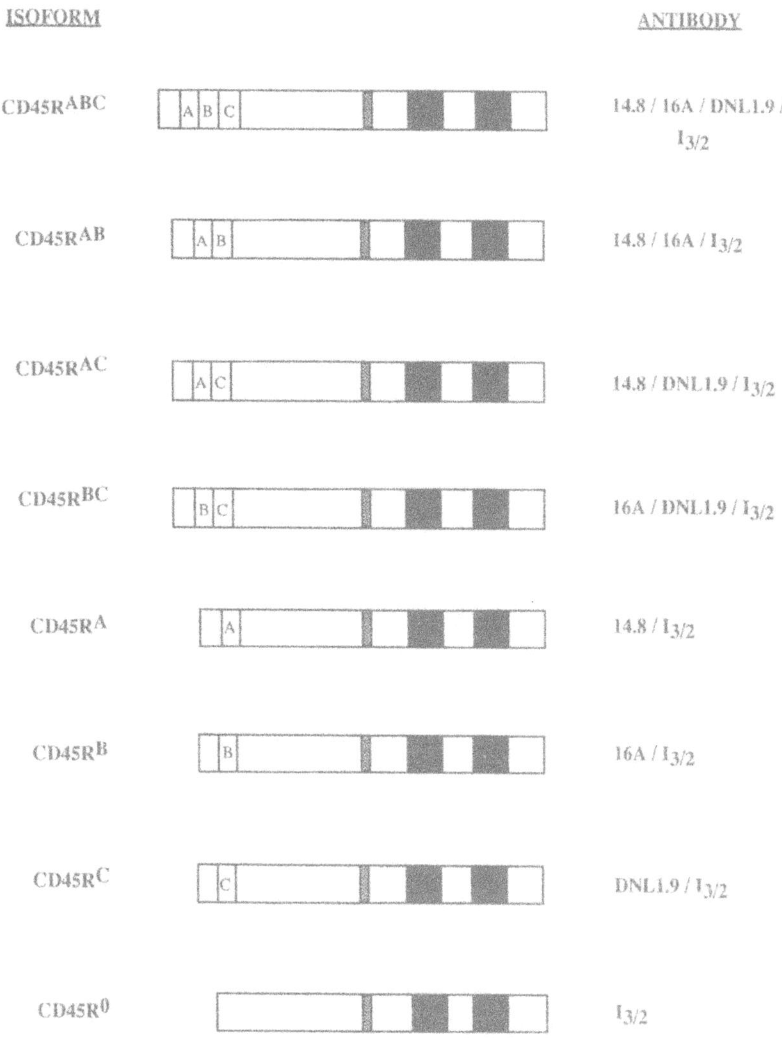


**Figure 1.** CD45 isoform variation occurs in the extracellular domain as a result of alternate RNA splicing. Exon-specific mouse CD45 antibodies are available, as are pan-specific anti-CD45 antibodies. Not all such antibodies are listed above. The vertical hatched region denotes the transmembrane domain. Solid black boxes represent the two intracellular phosphotyrosine phosphatase catalytic domains.

augmenting expression of specific CD45 isoforms. We chose to pursue the latter strategy for two reasons. First, models of CD45 isoform function via gene-targeting approaches are rather limited as they must rely upon exon-specific deletions and can not provide for any one isoform-specific deletion. Secondly, loss of all CD45 expression, by targeting an exon required for the production of all isoforms, will generate a model that will require subsequent CD45 isoform-specific reconstitution using mice that harbor and express isoform-specific CD45 transgenes. Hence we generated thymus-specific CD45 transgene expression vectors that would augment expression of two CD45 isoforms normally found on quiescent T cells - $CD45R^0$ and $CD45R^{ABC}$ (5-11 and see below). Augmenting expression of nonmutated transgenes in anatomic compartments in which they are normally expressed can result in phenotypes that reflect normal biological roles, as exemplified by other studies (23,24).

Mouse CD45 cDNAs encoding $CD45R^0$ and $CD45R^{ABC}$ were subcloned immediately 3' to the thymocyte-specific proximal *lck* promoter and 5' of the human growth hormone (hGH) gene (Figure 2). This particular transgene expression vector has successfully generated high level expression of various cDNA constructs in thymocytes with resulting phenotypes that are unique to the gene expressed (23, 25-27). Several approaches were available to monitor transgene expression in founders that were established. In RNA analyses, we were able to conclude that the transgenes were expressed by using hGH sequence as a probe of total thymocyte RNA (data not shown). Two transgene-expressing founder lines for each CD45 isoform were further characterized. These lines are denoted 116Q and 116I ($CD45R^{ABC}$) and 120 and 126 ($CD45R^0$). In $CD45R^{ABC}$ transgenic mice, increased expression of exons A, B and C could be clearly observed by using these sequences as a probe of RNA blots, while $CD45R^0$ transgene RNA expression could only be easily monitored by hGH sequence (data not shown).

Following cell surface iodination and quantitation of anti-CD45 immunoprecipitates, levels of CD45 protein were found to be increased at the cell surface on all transgenic thymocytes. In comparison to non-transgenic littermates, relatively small increases were observed and reflected between 10-30% of endogenous CD45 expression (21,22). CD45 transgene expression was highest in line 126 which expressed $CD45R^0$ levels that represented 20-30% to cell-surface CD45 on controls. Further evidence for CD45 transprotein expression was obtained following fluorescent-activated cell sorting (FACS) analyses in the Ly5.1 ($CD45^a$) allelic background (21). CD45 transgenic mice were mated for over five generations with S/JL ($CD45^a$) mice prior to expression analyses of the Ly5.2 or $CD45^b$ allelic form, as encoded by the mouse CD45 cDNAs used in the generation of transgenic mice (21). Additional experiments have shown that only $CD45R^{ABC}$ transgenic mice harbored increased expression of exons A, B and C and that intracellular levels of CD45 were not affected by CD45 transgene expression (22).

The observed linear FACS-peak increase in exon A, B and C expression in $CD45R^{ABC}$ transgenic mice (22) implied that these exon-specific CD45 antibodies could

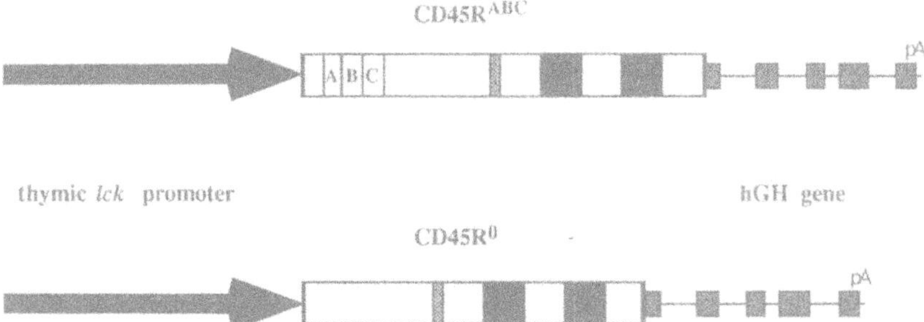


**Figure 2.** Structure of thymocyte-specific transgene expression vectors harboring either $CD45R^{ABC}$ or $CD45R^{0}$ mouse cDNAs. The construction of these vectors has been previously described (22).

**Table 1.** CD45 expression in murine thymocyte subpopulations[1]

| | Total CD45 | Exon A | Exon B | Exon C | $CD45R^{0}$ |
|---|---|---|---|---|---|
| $CD4^{-}/CD8^{-}$ | 550U | 20U (4%) | 150U (26%) | 30U (6%) | 64% |
| $CD4^{+}/CD8^{+}$ | 1000U | 20U (2%) | 200U (20%) | 45U (5%) | 73% |
| $CD4^{+}/CD8^{-}$ | 1050U | 20U (2%) | 250U (24%) | 60U (6%) | 68% |
| $CD8^{+}/CD4^{-}$ | $1050^{+}$U | 35U (3%) | 350U (33%) | 65U (6%) | 57% |

[1]Values indicated reflect peak fluorescent units (U) observed on various thymocyte subpopulations defined by CD4 and CD8 expression. Average percentages of CD45 molecules bearing exons A, B, and C are indicated. $CD45R^{0}$ levels are inferred by subtracting alternate exon abundance. Values assume stoichiometric antibody binding, as implied by linear increases obtained with isoform-specific CD45 transgene expression, and were partially confirmable by immunoprecipitation and immunoblotting experiments (22). Antibodies included 14.8 (exon A), 16A and 4B4 (exon B), DNL1.9 (exon C), and $I_{3/2}$ or 30F11 (pan-specific) as previously described (22).

provide quantitative measurements of thymic levels of alternate CD45 exon abundance in thymocyte subpopulations defined by CD4 and CD8 expression. This is important since RNA levels may not completely reflect CD45 protein expression at the cell surface as CD45 may be intracellularly sequestered and turnover rates can be slow (28,29). We therefore generated data that have likely provided an accurate assessment of the percentage of CD45 molecules that harbor alternate exons A, B and C. These data permit certain conclusions regarding the presence and abundance of specific CD45 isoforms (Table 1).

Highest levels of CD45 in thymocytes is a result of $CD45R^{0}$ expression, regardless of the thymocyte subpopulation (Table 1; refs. 6, 10, 22). Furthermore, relatively high level expression of exon B implies that the next most abundant CD45 isoform at the cell surface is $CD45R^{B}$. Exon C expression levels infer the presence of $CD45R^{BC}$ and/or $CD45R^{C}$, while exon A abundance is quite low and hence can reflect expression of $CD45R^{A}$, $CD45R^{AB}$, $CD45R^{AC}$ and $CD45R^{ABC}$ isoforms. It is difficult to conclude which isoforms are actually expressed when multiple variations are accounted for by FACS. However, immunoblotting experiments with exon-specific and pan-specific CD45 antibodies have provided further evidence for high-level expression of the 175-180 kilodalton (kD) $CD45R^{0}$ isoform followed by a 190 kD $CD45R^{B}$ isoform (5-11, 22 and data not shown). Only one exon C-specific mouse CD45 antibody exists at present (DNL1.9) and this reagent has not worked well in immunoblotting experiments thus far. Nevertheless, low level expression of 205 kD and 220 kD thymic CD45 isoforms in nontransgenic mice has been observed in immunoblotting experiments with the pan-specific CD45 antibody $I_{3/2}$, thus implying exon C may be present in these molecules (data not shown).

Together, these studies demonstrate that CD45 transgenic mice harbor increases in CD45 that are a result isoform-specific cDNA transgene expression. We did not find any evidence that endogenous CD45 alleles are aberrantly regulated in the presence of augmented expression of either $CD45R^{ABC}$ or $CD45R^{0}$. We have therefore analyzed the ability of thymocytes to differentiate and undergo activation through the TCR complex in the presence of these experimentally-defined changes in extracellular exon use as further controlled by comparable augmentation of the intracellular PTPase domains. As previous experiments have shown that distinct CD45 isoforms (including $CD45R^{0}$ and $CD45R^{ABC}$) all harbor identical PTPase activities (reviewed in 4). The results of our comparative studies conclusively demonstrate functional distinctions encoded for within alternate CD45 extracellular domains.

**Role of thymic CD45 expression in apoptosis and MHC-restricted negative selection**

A common result with increased CD45 expression, regardless of the isoform expressed, included a diminution in thymocyte number. Although this phenotype is not unique to CD45 as a thymic transgene, we observed an invariable and specific loss of

CD4$^+$CD8$^+$ thymocytes that harbored higher levels of TCR expression. This occurred while CD4$^-$CD8$^-$ thymocytes were unaffected in cell number or viability (21). Moreover, mature CD4$^+$ or CD8$^+$ thymic T cells emerged in comparably similar ratios and harbored an unaltered TCR $V_\beta$ repertoire ($V_\beta$2-14 and 17, ref. 22, and data not shown).

In response to anti-CD3 (2C11, ref. 30) stimulations, we noted that a larger proportion of CD4$^+$CD8$^+$ thymocytes from CD45 transgenic mice were engaged in apoptosis as evidenced by increased percentages of CD4$^+$CD8$^+$ thymocytes entering the CD4$^{dull}$CD8$^{dull}$ population and increased DNA fragmentation (21). Moreover, upon matings to produce double transgenic mice, augmented levels of CD45 expression significantly increased the efficacy of MHC-restricted negative selection during post-natal development in HY TCR transgenic mice (21).

The above experiments therefore identify a long sought-after enzymatic signaling pathway, emanating from the cell surface, that is capable of regulating thymocyte apoptosis and MHC-restricted negative selection. No other thymic transgene has been shown to encode these properties, including *bcl-2* (27). The MHC-restricted nature of our results is further evidence that these phenotypes likely reflect a normal physiologic role for thymic CD45. Although CD45 deficient thymocytes underwent negative selection for TCR superantigen specificity (31), it is interesting that a widespread block in thymocyte development occured at the CD4$^+$CD8$^+$ thymocyte stage. Additional experiments must be accomplished to determine whether CD4 or CD8 co-receptor-dependent TCR structures, such as the HY TCR, are effectively deleted in the absence of thymic CD45. Nevertheless, alterations in CD45 expression levels may influence the efficacy of the TCR selection process, regardless of a requirement for CD45 expression *per se*.

### Does p56$^{lck}$ activation play a role in CD45-mediated augmentation of MHC-restricted negative selection?

Since CD45 is known to regulate p56$^{lck}$ (32,33), we analyzed the activity of this tyrosine kinase in CD4$^+$CD8$^+$ thymocytes from CD45 transgenic mice, especially since HY TCR selection is CD8 co-receptor dependent and may require p56$^{lck}$ (34,35). We found that p56$^{lck}$ is activated by 2-4 fold in immunoprecipitates from CD4$^+$CD8$^+$ thymocytes bearing increased levels of CD45 (21). This is a unique and important result that provides a model for endogenous p56$^{lck}$ activation. Moreover, this occurred in the absence of global alterations in cellular phosphotyrosine, unlike results with expression of exogenously-derived and mutated *lck* molecules (23,26). While p56$^{lck}$ activation may be only one of many possible consequences of augmented CD45 expression, the potential role of p56$^{lck}$ activity in TCR selection led us to determine if p56$^{lck}$ activation occurs with MHC-restricted negative selection of the HY TCR, as has been reported during positive selection of the HY TCR (36).

As shown in Figure 3, p56$^{lck}$ is increased in activity in HY TCR transgenic CD4$^+$CD8$^+$ thymocytes that are undergoing negative selection in the selecting H2$^b$ MHC background. Since this result occurs with increasing levels of cell surface HY TCR expression (21), it is therefore unlikely that the p56$^{lck}$ activation previously noted in CD4$^+$CD8$^+$ thymocytes derived from CD45 transgenic mice was the result of a loss of the CD4$^+$CD8$^+$ TCR$^{hi}$ subpopulation which harbored relatively lower levels p56$^{lck}$ activity. Moreover, augmented CD45 expression significantly increased HY TCR expression in ontogeny and during negative selection in the H-2$^b$ MHC background (21). Hence p56$^{lck}$ activation appears to be a marker of HY TCR thymocytes engaged in MHC-restricted negative selection.

Although there remains controversy over a role for p56$^{lck}$ in apoptosis and MHC-restricted negative selection, a previous study prematurely concluded that p56$^{lck}$ does not participate in this process (37). That study noted a loss of p56$^{lck}$ activity at 20 hours post-stimulation in the presence of tyrosine kinase inhibitors, yet without a reduction in apoptosis in negative selection *in vitro* (37). However, those results did not provide controls for p56$^{lck}$ protein levels. Our experiments reveal that p56$^{lck}$ protein is in fact eliminated by 20 hours post-activation under identical cell culture conditions in HY TCR transgenic thymocytes (Figure 4A). Additionally, an apoptotic signal could be transmitted rapidly and transiently upon TCR stimulation, and prior to the 20 hour time point for DNA fragmentation analysis. In this regard, we observed the presence of p56$^{lck}$ activity and DNA fragmentation in cells assessed 2 hours post-treatment with either genistein, herbamycin A or DMSO alone (Figure 4A, B). Obviously, this experimental design does not add genistein or herbamycin A to cell extracts and immunoprecipitates, nor would such an addition to this protocol reproduce intact cell conditions. Therefore it is not possible to conclude that p56$^{lck}$ activity is either inhibited or not participating in apoptosis and negative selection by using such an experimental strategy. Thus a role for p56$^{lck}$ activity in MHC-restricted negative selection remains possible and cannot be discounted at present. Future studies with described methods for inducibly-regulating gene inactivation (38) will likely provide conclusive evidence either for or against an involvement of p56$^{lck}$ in thymic CD4- and CD8-dependent TCR selection.

**CD45 isoforms are functionally distinct in mature thymic T cell activation**

Remarkably, we found that mature CD4$^+$ and CD8$^+$ thymic T cells were functionally distinct depending upon the CD45 isoform expressed as a transgene. Mature thymic T cells are immunologically-responsive and can be monitored following activation protocols. We were thus able to define TCR signaling events that were specifically altered among otherwise genetically-comparable host T cell populations.

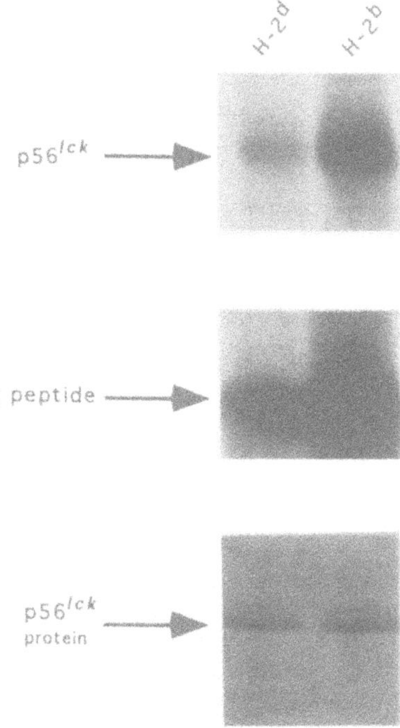


**Figure 3.** Activation of endogenous p56$^{lck}$ occurs during MHC-restricted negative selection in HY TCR transgenic mice. $CD4^+CD8^+$ thymocytes were FACS-isolated sorted from H-Y TCR transgenic mice bearing either H-$2^b$ or H-$2^d$ MHC backgrounds. p56$^{lck}$ immunoprecipitates from 6 x $10^6$ cells were divided equally with 2 x $10^6$ $CD4^+CD8^+$ thymocytes analyzed for p56$^{lck}$ autophosphorylation (top), ζ-peptide phosphorylation (middle) and p56$^{lck}$ protein abundance (bottom). A 24 hour autoradiography exposure is represented in the top and middle panels. p56$^{lck}$ protein was assessed by immunoblotting as previously described (21).

Expression of CD45R$^{ABC}$ significantly increased cellular proliferation of $CD4^+CD8^-$ thymic T cells in a mixed lymphocyte reaction (MLR) and following anti-CD3 (2C11) or anti-TCR V$_\beta$ (H57, ref. 39) antibody stimulation (Table 2 and ref. 22). In additional studies, we have found that $CD8^+CD4^-$ thymic T cells derived from CD45R$^{ABC}$ transgenic mice are also hyperesponsive in anti-TCR-mediated cellular proliferation, while other immunologic analyses await access to adequate cell numbers (data not shown). Since the TCR V$_\beta$ repertoire was unaltered, as were expression levels of cell surface molecules including interleukin-2 receptor, TCR and CD4 (22 and data not shown), these results implied that the T cell population was functionally enhanced by CD45R$^{ABC}$ expression.

In response to soluble anti-CD3 antibody on single cell suspensions of $CD4^+$ thymic T cells, $Ca^{2+}$ mobilization was greatly augmented by CD45R$^{ABC}$ expression, but not by CD45R$^0$. Although $Ca^{2+}$ mobilization was slightly, yet significantly, enhanced in

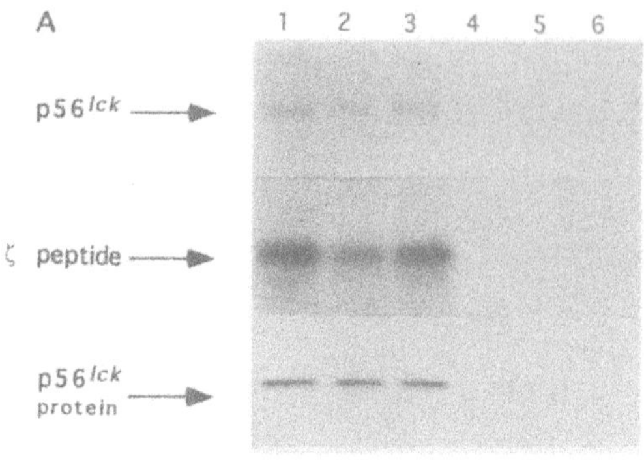


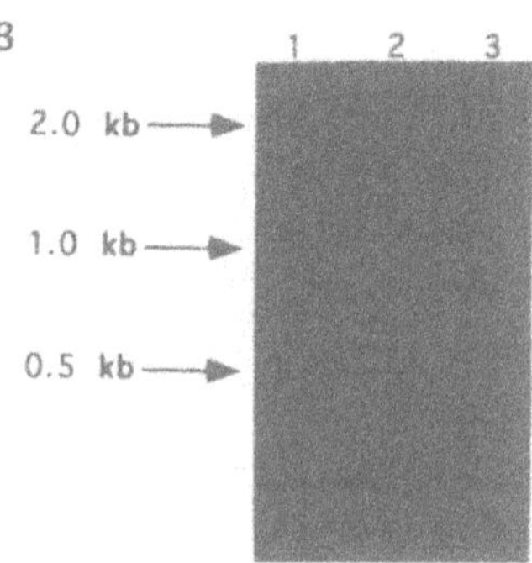


**Figure 4.** p56*lck* activity is not dissociable from thymocyte apoptosis. (A) HY TCR trangenic thymocytes bearing H-$2^d$ MHC (H-$2^b$ MHC-negative) were isolated from female mice and treated with DMSO (Lanes 1 and 4), 37 μM genistein (Lanes 2 and 5) or 400 nM herbamycin A (Lanes 3 and 6) for either two hours (lanes 1-3) or twenty hours (Lanes 4-6). Results identical to those shown in Lanes 1-3 were also obtained following incubations for five hours (data not shown). Following anti-*lck* immunoprecipitation from 6 x $10^6$ thymocytes, reactions were divided equally into thirds and p56*lck* activity was analyzed by autophosphorylation (top), ζ-peptide phosphorylation (middle) and for protein abundance by immunoblotting (bottom). (B) In parallel experiments, following incubations with DMSO (Lane 1), 37 μM genistein (Lane 2) or 400 nM herbamycin A (Lane 3) for two hours, male H-$2^d$ MHC-positive dendritic cells were added to thymocytes (1:1) for five hours co-incubation in vitro prior to analyzing DNA fragmentation. As shown, all samples displayed significant DNA fragmentation. Subsequent FACS analyses of these cultures revealed that $CD4^+CD8^+$ thymocytes were specifically depleted (completely by 20 hours) while DNA fragmentation in the absence of dendritic cells was significantly lower and more similar to that observed upon isolation of thymocytes from intact thymi (data not shown).

$CD4^+CD8^+$ thymocytes bearing augmented levels of $CD45R^0$ (21), $CD45R^{ABC}$ expression was much more efficient in this regard and increased $Ca^{2+}$ mobilization to a much greater extent (data not shown).

$CD45R^0$ and $CD45R^{ABC}$ isoforms differentially modified phosphotyrosine accumulation in $CD4^+$ T cells in the absence and presence of anti-CD3 stimulation.

Although the identities of these differentially-modified proteins are unclear at present, phosphorylation kinetics and SDS-PAGE mobility imply that ZAP-70 and the previously observed CD45-associating 30-32 kD molecule may be included (21,22, 40-44). While $p56^{lck}$ activity was enhanced in $CD4^+CD8^+$ thymocytes bearing augmented levels of CD45, we did not find reproducible alterations in $p56^{lck}$ activity that reflected variations in CD45 isoform-specific expression in $CD4^+$ thymic T cells. Moreover, augmented CD45 expression increased apoptosis in $CD4^+CD8^+$ thymocytes yet enhanced cellular proliferation in mature thymic T cells (21 and see below). It is therefore likely that T cell differentiation proceeds with expression of novel intracellular substrates that may provide for differences in the action of cell-surface signal transduction molecules that are expressed throughout multiple stages of ontogeny. Future studies can focus upon identifying phosphotyrosine-containing proteins that are differentially affected by $CD45R^{ABC}$ and $CD45R^0$ expression since these likely provide the molecular basis for CD45 isoform-specific function.

While $CD45R^{ABC}$ expression was capable of improving TCR signaling and T cell activation in response to antibody stimulation alone, increases in $CD45R^0$ expression were able to generate a similarly enhanced proliferative response in the presence of the phorbol ester PMA when added as a TCR co-stimulus (22 and Table 2). In comparison, co-activation of the CD28/B7 TCR pathway (reviewed in 45) did not differentially affect $CD4^+$ T cell proliferation, regardless of CD45 expression levels or CD45 isoform-specific alterations. These results reveal that $CD45R^0$ extracellular structure is functionally restricted by requiring a TCR co-stimulus to improve T cell activation. In the absence of the PMA co-stimulus, the extracellular structure of $CD45R^{ABC}$ is sufficient to improve T cell activation. While these results imply that $CD45R^{ABC}$ may integrate signals that are commonly provided by interactions with certain antigen-bearing accessory cells (46), we found that $CD45R^{ABC}$ may exert it's specific function in the absence of cell-cell interaction.

Enhanced $Ca^{2+}$ mobilization in single cell suspensions suggests that cell-cell interaction is not required for $CD45R^{ABC}$-specific function. Alternatively, modifications to the TCR complex that permit such enhanced signaling may be relatively stable events that occurred previously in the thymic milieu. Nevertheless, we found an additional $CD45R^{ABC}$-specific phenotype that further implied a cell-intrinsic modification to the TCR complex. Anti-CD4 antibodies (as well as CD45 antibodies) can act to inhibit T cell activation, likely through steric interference with the ability of CD4 (and CD45) to subsequently functionally associate with the TCR complex (47-49). We thus felt it was important to determine if FACS-derived, and positively-sorted, $CD4^+$ T cells were differentially affected by anti-CD4 antibodies in a CD45 isoform-specific manner.

$CD4^+CD8^-HSA^{low}$ thymic T cells were isolated by a negative-sorting strategy that resulted in a highly purified (>90%) subpopulation that did not harbor antibody complexes on the cell surface (22). Remarkably, $CD45R^{ABC}$ expression specifically suppressed the

**Table 2.** CD45R$^{ABC}$ and CD45R$^{0}$ differentially regulate CD4$^{+}$ thymic T cell activation[2]

| | Littermate controls | CD45R$^{ABC}$ transgenic line 116Q | line 116I | CD45R$^{0}$ transgenic line 126 | line 120 |
|---|---|---|---|---|---|
| Anti-CD3 proliferation | + | +++ | +++ | + | + |
| Anti-TCR Vβ proliferation | +/- | ++++ | +++ | +/- | +/- |
| MLR response | + | +++ | +++ | + | + |
| Anti-CD3 $Ca^{2+}$ mobilization | + | +++ | +++ | + | + |
| Anti-CD3/PMA proliferation | + | +++ | +++ | +++ | +++ |
| Anti-CD3/CD28 proliferation | + | + | + | + | + |
| Anti-CD3 phosphotyrosine accumulation | + | ++ | ++ | +/- | +/- |
| Anti-CD4 inhibition of anti-CD3-induced proliferation | + | - | - | + | + |

[2]CD4$^{+}$ thymic T cells were isolated by FACS or by a negative sorting strategy and analyzed for T cell activation responses described in the text and as published (22).

inhibitory effect of anti-CD4 antibodies in anti-TCR-mediated activation as judged by cellular proliferation (22 and Table 2). In understanding the potential mechanism for this specific effect of $CD45R^{ABC}$, we may gain clues from previous results that investigated the role of CD4 in modulating T cell activation.

Physiologic $CD4^{+}$ TCR interaction with antigen and class II MHC acts to proximate the CD4-$p56^{lck}$ molecule with the TCR complex (50). This proximation greatly improves T cell activation, as further obtained by antibody-mediated juxtapositioning *in vitro* (48, 51-53). Interestingly, in the absence of CD4 expression, or $p56^{lck}$ binding and sequestration to the cytoplasmic tail of CD4, T cell activation is also greatly improved (48). When these observations are considered with the inhibitory effects of anti-CD4 and anti-CD45 antibodies, it becomes plausible to hypothesize that extracellular domains may play crucial roles in restricting and promoting the association of intracellular enzymatic domains to particular complexes and substrates, in addition to a function in ligand-induced activation of enzymatic activity. In this regard, previous complementation of TCR signaling defects by CD45 molecules that either lacked extracellular sequence or harbored MHC class I extracellular domains (54,55) may thus be explained by hypothesizing that CD45 extracellular domains act to compartmentalize PTPase activity and, therefore, variously modify TCR activation responses. As would be predicted by this view, intracellular signaling events generated by chimeric receptor molecules, at least in one case, were found to be dependent upon the identity of the extracellular domain and not by virtue of a unique, yet related, intracellular enzymatic domain (56).

We have observed that $CD45R^{ABC}$ can specifically improve TCR signaling by enhancing phosphotyrosine accumulation, $Ca^{2+}$ mobilization, cellular proliferation and by suppressing anti-CD4-mediated inhibition. This spectrum of cellular responses is quite similar to those observed following proximation of CD4-$p56^{lck}$ with the TCR. Our results may be explained by hypothesizing that $CD45R^{ABC}$, but not $CD45R^{0}$, can improve T cell activation by promoting association of CD4-$p56^{lck}$ (and perhaps CD8-$p56^{lck}$) with the TCR complex. Once CD4-$p56^{lck}$-TCR complexes have formed, anti-CD4 antibodies may no longer efficaciously suppress TCR activation. Alternatively, $CD45R^{ABC}$ might act to uncouple $p56^{lck}$ from sequestration by CD4, an effect previously demonstrated to occur with PMA addition (57). As observed in our studies, either $CD45R^{ABC}$ or $CD45R^{0}$ may indeed hyperactivate T cells in the presence of the TCR co-stimulus PMA. Importantly, these models are easily tested. Still, the relatively minor changes in CD45 PTPase levels obtained in our studies may result in modifications to the activation state or localization of other substrates and enzymes.

Regardless of the exact mechanisms whereby CD45 isoforms function differentially, minimal structural requirements for isoform-specific CD45 function *in vivo* can now be investigated. Moreover, our results imply that the highly regualted changes in T cell CD45 RNA splicing and CD45 isoform variation may be an important physiological mechanism for modulating the immunologic responsiveness of the thymic-derived T cell repertoire.

## MATERIALS AND METHODS

All materials and methods contained within this manuscript have been previously published or are presently in the press (22, 38, 58-61).

## ACKNOWLEDGEMENTS

We thank Drs. Hung-Sia Teh, Pauline Johnson and John Schrader for their interest and assistance in these studies. This research is supported by awards to J.D.M. from the Medical Research Council of Canada, the National Cancer Institute of Canada and the National Centres of Excellence Genetics program. J.D.M. is a recipient of an M.R.C. scholarship.

## REFERENCES

1. C.A. Janeway. The T cell receptor as a multicomponent signaling machine: CD4/CD8 coreceptors and CD45 in T cell activation. *Ann. Rev. Immunol.* 10:645 (1992).
2. J.T. Pingel and M. L. Thomas. Evidence that the leukocyte-common antigen is required for antigen-induced T lymphocyte proliferation. *Cell* 58:1055 (1989).
3. M.L. Thomas and L. Lefrancois. Differential expression of the leukocyte common antigen family, *Immunol. Today* 9:320 (1988).
4. I.S. Trowbridge, H. Ostergaard, and P. Johnson. CD45: A leukocyte-specific member of the protein tyrosine phosphatase family. *Biochem. Biophys. Acta.* 1095:46 (1991).
5. L. Lefrancois and T. Goodman. Developmental sequence of T200 antigen modificationsin murine T cells. *J. Immunol.* 139:3718 (1987).
6. A.N. Akbar, L. Terry, A. Timms, P.C.L. Beverley, and G. Janossy. Loss of CD45R and gain of UCHL1 reactivity is a feature of primed T cells. *J. Immunol.* 140:2171 (1988).
7. H.M. Serra, J.F. Krowka, J.A. Ledbetter, and L.M. Pilarski. Loss of CD45R (Lp220) represents a post-thymic T cell differentiation event. *J. Immunol.* 140:1435 (1988).
8. M.L. Birkeland, P. Johnson, I.S. Trowbridge, and E. Puré. Changes in CD45 isoform expression accompany antigen-induced murine T-cell activation. *Proc. Natl. Acad. Sci. USA* 86:6734 (1989).
9. D.M. Rothstein, A. Yamada, S.F. Schlossman, and C. Morimoto. Cyclic regulation of CD45 isoform expression in a long term human $CD4^+CD45RA^+$ T cell line. *J. Immunol.* 146:1175 (1991).
10. K.S. Hathcock, G. Laszlo, H.B. Dickler, S.O. Sharrow, P. Johnson, I.S. Trowbridge, and R.J. Hodes. Expression of variable exon A-, B-, and C-specific CD45 determinants on peripheral and thymic T cell populations. *J. Immunol.* 148:19 (1992).
11. D.M. Rothstein, H. Saito, M. Streuli, S.F. Schlossman, and C. Morimoto. The alternative splicing of the CD45 tyrosine phosphatase is controlled by negative regulatory trans-acting splicing factors. *J. Biol. Chem.* 267:7139 (1992).

12. R. Pulido and F. Sanchez-Madrid. Biochemical nature and topographic localization of epitopes defining four distinct CD45 antigen specificities. *J. Immunol.* 143:1930 (1989).

13. T.R. Mosmann, H. Cherwinski, M.W. Bond, M.A. Giedlin, and R.L. Coffman. Two types of murine helper T cell clone: I. Definition according to profiles of lymphokine activities and secreted properties. *J. Immunol.* 136:2348 (1986).

14. J.A. Byrne, J.L. Butler, and M.D. Cooper. Differential activation requirements for virgin and memory T cells. *J. Immunol.* 141:3249 (1988).

15. S. Huet, L. Boumsell, J. Dausset, L. Degos, and A. Bernard. The required nteraction between monocytes and perpiheral blood T-lymphocytes (T-PBL) upon activation via CD2 or CD3. Role of HLA class I molecule from accessory cells and the differential response of T-PBL subsets. *Eur. J. Immunol.* 18:1187 (1988).

16. K, Bottomly, M. Luqman, L. Greenbaum, S. Carding, J. West, T. Pasqualini, and D.B. Murphy. A monoclonal antibody to murine CD45R distinguishes CD4 T cell populations that produce different cytokines. *Eur. J. Immunol.* 19:617 (1989).

17. U. Dianzani, M. Luqman, J. Rojo, J. Yagi, J.L. Baron, A. Woods, C.A. Janeway, and K. Bottomly. Molecular assocations on the T cell surface correlate with immunological memory. *Eur. J. Immunol.* 20:2249 (1990).

18. M. Luqman, P. Johnson, I. Trowbridge, and K. Bottomly. Differential expression of the alternatively spliced exons of murine CD45 in $T_h1$ and $T_h2$ cell clones. *Eur. J. Immunol.* 21:17 (1991).

19. M. Luqman and K. Bottomly. Activation requirements for $CD4^+$ T cells differing in CD45R expression. *J. Immunology* 149:2300 (1992).

20. C.A. Michie, A. McLean, C. Alcock, and P.C.L. Beverley. Lifespan of human lymphocyte subsets defined by CD45 isoforms. *Nature* 360:264 (1992).

21. C.J. Ong, D. Chui, H.-S. Teh, and J.D. Marth. Thymic CD45 tyrosine phosphatase regulates apoptosis and MHC-restricted negative selection. J. Immunol. 152, in press (1994).

22. D. Chui, C.J. Ong, P. Johnson, H.-S. Teh, and J.D. Marth. Specific CD45 isoforms differentially regulate T cell receptor signaling. *EMBO J.*, 13:798 (1994).

23. M.P. Cooke, K.M. Abraham, K.A. Forbush, and R.M. Perlmutter. Regulation of T cell receptor signaling by a *src* family protein-tyrosine kinase ($p59^{fyn}$). *Cell* 65:281 (1991).

24. M. Appleby, J.A. Gross, M.P. Cooke, S.D. Levin, X. Qian, and R.M. Perlmutter. Defective T cell receptor signaling in mice lacking the thymic form of $p59^{fyn}$. *Cell* 70:751 (1992).

25. K.E. Chaffin, C.R. Beals, K.A. Forbush, T.M. Wilkie, M.I. Simon, and R.M. Perlmutter. Dissection of thymocyte signaling pathways by *in vivo* expression of pertussis-toxin ADP ribosyltransferase. *EMBO J.* 9:3821 (1990).

26. K.M. Abraham, S.D. Levin, J.D. Marth, K.A. Forbush, and R.M. Perlmutter. Delayed thymocyte development induced by augmented expression of $p56^{lck}$. *J. Exp. Med.* 173:1421 (1991).

27. C.L. Sentman, J.R. Shutter, D. Hockenbery, O. Kanagawa, and S.J. Korsmeyer. *bcl-2* inhibits multiple forms of apoptosis but not negative selection in thymocytes. *Cell* 67:879 (1991).

28. J.P. Deans, A.W. Boyd, and L.M. Pilarski. Transitions from high to low molecular weight isoforms of CD45 (T200) involve rapid activation of alternate mRNA splicing and slow turnover of surface CD45R. *J. Immunol.* 143:1233 (1989).

29. Y. Minami, F.J. Stafford, J. Lippincott-Schwartz, L.C. Yuan, and R.D. Klausner. Novel redistribution of an intracellular pool of CD45 accompanies T cell activation. *J. Biol. Chem.* 266:9222 (1991).

30. O. Leo, M. Foo, D.H. Sachs, L.E. Samelson, and J.A. Bluestone. Identification of a monoclonal antibody specific for a murine T3 polypeptide. *Proc. Natl. Acad. Sci. USA* 84:1374 (1987).

31. K. Kishihara, J. Penninger, V.A. Wallace, T.M. Kundig, K. Kawai, A. Wakeham, E. Timms, K. Pfeffer, P.S. Ohashi, M.L. Thomas, C. Furlonger, C.J. Paige, and T.W. Mak. Normal B lymphocyte development but impaired T cell maturation in CD45-exon6 protein tyrosine phosphatase-deficient mice. *Cell* 74:143 (1993).

32. H.L. Ostergaard, D.A. Shackelford, T.R. Hurley, P. Johnson, R. Hyman, B.M. Sefton, and I.S. Trowbridge. Expression of CD45 alters phosphorylation of the *lck*-encoded tyrosine protein kinase in murine lymphoma cells. *Proc. Natl. Acad. Sci. USA* 86:8959 (1989).

33. E.D. Cahir McFarland, T.R. Hurley, J.T. Pingel, B.M. Sefton, A. Shaw, and M.L. Thomas. Correlation between Src family member regulation by the protein-tyrosine-phosphatase CD45 and transmembrane signaling through the T-cell receptor. *Proc. Natl. Acad. Sci. USA* 90:1402 (1993).

34. N. Killeen, A. Moriarty, H.-S. Teh, and D.R. Littman. Requirement for CD8-major histocompatibility complex class I interaction in positive and negative selection of developing T cells. *J. Exp. Med.* 176:89 (1992).

35. N.S.C. van Oers, A.M. Garvin, C.B. Davis, K.A. Forbush, D.A. Carlow, D.R. Littman, R.M. Perlmutter, and H.-S. Teh. Disruption of CD8-dependent negative and positive selection is correlated with a decrease in association between CD8 and the protein tyrosine kinase, $p56^{lck}$. *Eur. J. Immunol.* 22:735 (1992).

36. A.C. Carrera, C. Baker, T.M. Roberts, and D.M. Pardoll. Tyrosine kinase triggering in thymocytes undergoing positive selection. *Eur. J. Immunol.* 22:2289 (1992).

37. K. Nakayama and D.Y. Loh. No requirement for $p56^{lck}$ in the antigen-stimulated clonal deletion ot thymocytes. *Science* 257:94 (1992).

38. P.C. Orban, D. Chui, and J.D. Marth. Tissue- and site-specific DNA recombination in transgenic mice. *Proc. Natl. Acad. Sci. USA* 89:6861 (1992).

39. R.T. Kubo, W. Born, J.W. Kappler, P. Marrack, and M. Pigeon. Characterization of a monoclonal antibody which detects all murine αβ T cell receptors. *J. Immunol.* 142:2736 (1989).

40. A.C. Chan, B.A. Irving, J.D. Fraser, and A. Weiss. The ζ chain is associated with a tyrosine kinase and upon T-cell antigen receptor stimulation associates with ZAP-70, a 70-kDa tyrosine phosphoprotein. *Proc. Natl. Acad. Sci USA* 88:9166 (1991).

41. A.C. Chan, M. Iwashima, C.W. Turck, and A. Weiss. ZAP-70: A 70 kd protein-tyrosine kinase that associates with the TCR ζ chain. *Cell* 71:649 (1992).

42. B. Schraven, H. Kirchgessner, B. Gaber, Y. Samstag, and S. Meuer. A functional complex is formed in human T lymphocytes between the protein tyrosine phosphastase CD45, the protein tyrosine kinase p56$^{lck}$ and pp32, a possible common substrate. *Eur. J. Immunol.* 21:2469 (1991).

43. B. Schraven, A. Schirren, H. Kirchgessner, B. Siebert, and S.C. Meuer. Four CD45/P56$^{lck}$ associated phosphoproteins (pp29-pp32) undergo alterations in human T cell activation. *Eur. J. Immunol.* 22:1857 (1992).

44. A. Takeda, J.J. Wu, and A.L. Maizel. Evidence for monomeric and dimeric forms of CD45 associated with a 30-kDa phosphorylated protein. *J. Biol. Chem.* 267:16651 (1992).

45. P.S. Linsley and J.A. Ledbetter. The role of the CD28 receptor during T cell responses to antigen. *Ann. Rev. Immunol.* 11:191 (1993).

46. D.L. Rosenstreich and S.B. Mizel. Signal requirements for T lymphocyte activation. I. Replacement of macrophage function with phorbol myristic acetate. *J. Immunol.* 123:1749 (1979).

47. M.K. Newell, L.J. Haughn, C.R. Maroun, and M.H. Julius. Death of mature T cells by separate ligation of CD4 and the T cell receptor for antigen. *Nature* 347:286 (1990).

48. L. Haughn, S. Gratton, L. Caron, R.-P. Sekaly, A. Veillette, and M. Julius. Association of tyrosine kinase p56$^{lck}$ with CD4 inhibits the induction of growth through the αβ T-cell receptor. *Nature* 358:328 (1992).

49. E. Shivnan, M. Biffen, M. Shiroo, E. Pratt, M. Glennie, and D. Alexander. Does co-aggregation of the CD45 and CD3 antigens inhibit T cell antigen receptor complex-mediated activation of phospholipase C and protein kinase C? *Eur. J. Immunol.* 22:1055 (1992).

50. N. Glaichenhaus, N. Shastri, D.R. Littman, and J.M. Turner. Requirement for association of p56$^{lck}$ with CD4 in antigen-specific signal transduction in T cells. *Cell* 64:511 (1991).

51. K. Eichmann, J.I. Jonsson, I. Falk, and F. Emmrich. Effective activation of resting mouse T lymphocytes by cross-linking submitogenic concentrations of the T cell antigen receptor with either Lyt-2 or L3T4. *Eur. J. Immunol.* 17:643 (1987).

52. T. Owens, d.S.G.B. Fazekas, and J.F.A.P. Miller. Coaggregation of the T-cell receptor with CD4 and other T-cell surface molecules enhances T-cell activation. *Proc. Natl. Acad. Sci. USA* 84:9209 (1987).

53. T.L. Collins, S. Uniyal, J. Shin, J.L. Strominger, R.S. Mittler, and S.J. Burakoff. p56$^{lck}$ association with CD4 is required for the interaction between CD4 and the TCR/CD3 complex and for optimal antigen stimulation. *J. Immunol.* 148:2159 (1992).

54. S. Volarevic, B.B. Niklinska, C.M. Burns, C.H. June, A.M. Weissman, and J.D. Ashwell. Regulation of TCR signaling by CD45 lacking transmembrane and extracellular domains. *Science* 260:541 (1993).

55. R.R. Hovis, J.A. Donovan, M.A. Musci, D.G. Motto, F.D. Goldman, S.E. Ross, and G.A. Koretzky. Rescue of signaling by a chimeric protein containing the cytoplasmic domain of CD45. *Science* 260:544 (1993).

56. T. Chiba, Y. Nagata, M. Machide, A. Kishi, H. Amanuma, M. Sugiyama, and K. Todokoro. Tyrosine kinase activation through the extracellular domains of cytokine receptors. *Nature* 362:646 (1993).

57. T.R. Hurley, K. Luo, and B.M. Sefton. Activators of protein kinase C induce dissociation of CD4, but not CD8, from $p56^{lck}$. *Science* 245, 407-409 (1989).

58. J.D. Marth, R. Peet, E.G.Krebs, and R.M. Perlmutter. A lymphocyte-specific protein tyrosine kinase gene is rearranged and overexpressed in the murine T cell lymphoma LSTRA. *Cell* 43:393 (1985).

59. J.D. Marth, D.B. Lewis, C.B. Wilson, M.E. Gearn, E.G. Krebs, and R.M. Perlmutter. Regulation of $pp56^{lck}$ during T-cell activation: functional implications for the *src*-like protein tyrosine kinases. *EMBO J.* 9:2727 (1987).

60. W. Swat, L. Ignatowicz, and P. Kisielow. Detection of apoptosis of immature $CD4^+8^+$ thymocytes by flow cytometry. *J. Immunol. Methods* 137:79 (1991).

61. W. Swat, L. Ignatowicz, H. von Boehmer, and P. Kisielow. Clonal deletion of immature $CD4^+8^+$ thymocytes in suspension culture by extrathymic antigen-presenting cells. *Nature* 351:150 (1991).

# CLONING AND CHARACTERIZATION OF NF-$AT_c$ AND NF-$AT_p$: THE CYTOPLASMIC COMPONENTS OF NF-AT

Steffan Ho[1], Luika Timmerman[1], Jeffrey Northrop[1], and Gerald R. Crabtree[1,2]

[1] Department of Pathology
[1,2] Department of Developmental Biology and Pathology
Howard Hughes Medical Institute Rm B211
Stanford University School of Medicine
Stanford, CA 94305-5428

## INTRODUCTION

Signalling through the T cell antigen receptor initiates a complex series of events resulting in the activation of a group of genes (early genes) that are involved in the proliferation of T cells, the development of immune function and a variety of cellular interactions that contribute to immune responses[1,2]. In addition, signalling through the antigen receptor is essential for several transitions in thymic development. These findings raise the question of how such diverse responses are brought about by the actions of a single receptor. Several years ago, we identified nuclear response elements for these signals in the IL-2 promoter/enhancer[3-6]. In the IL-2 genes, two such sites were identified that bound a protein complex that we called NF-AT after the selective expression of this complex in nuclear extracts of activated T cells[5]. The sequence to which this protein bound was shown to direct transcription to activated T cells in the context of a transgenic mouse[7]. Furthermore, transcription directed by the NF-AT binding sites required proper presentation of antigen by MHC matched cells[8]. These finding indicated that this protein might serve as a general terminus for signals coming from the antigen receptor and thereby aid in the elucidation of the signalling pathway carrying information from the cell membrane to the nucleus. Despite many efforts to characterize this protein, it has been elusive; indeed, it has been inferred to be a Pu protein, ets-1, elk-1, a 28 kDa protein and a 57 kDa protein. None of these proteins appear to be actually involved in the biologic activity attributed to this

*Mechanisms of Lymphocyte Activation and Immune Regulation V*
Edited by S. Gupta *et al.*, Plenum Press, New York, 1994

complex. Direct purification of the protein from bovine thymus has revealed that it is composed of two cytosolic components, NF-$AT_c$ and NF-$AT_p$, which bear a vague similarity to a limited region of the rel/Dorsal family, but it is otherwise a pioneer protein[9,10].

## PURIFICATION OF NF-AT PROTEINS BY COMPLEMENTATION WITH HELA CELL DERIVED NF-$AT_N$

An approach to purifying NF-AT was devised based on the studies of Flanagan et al. that indicated that NF-AT was composed of cytoplasmic and nuclear components[11]. The nuclear component was ubiquitous and could be found in the nuclear extracts of PMA-stimulated Hela cells. This component shares many characteristics of AP-1 but a definitive identification has not yet appeared[12]. A biochemical purification was carried out using a complementation assay with Hela cell nuclear extracts to assay the presence of the cytosolic component. After affinity purification, an approximate 10,000-to 50,000-fold purification was obtained. This material was subjected to further purification by HPLC and the fractions showing complementation of NF-AT binding were subjected to proteolytic degradation and sequences obtained from 16 of these proteolytic fragments[9]. These 16 sequences were derived from two proteins that proved to be the products of two separate genes that we call NF-$AT_c$ and NF-$AT_p$[9]. The former protein has the expected characteristics of the cytosolic component of NF-AT originally described and hence will be considered in most detail.

## NF-$AT_c$ AND NF-$AT_p$ ARE DISTANTLY RELATED BY A SHORT SEQUENCE SIMILAR TO THE REL/DORSAL FAMILY OF PROTEINS

The DNA and predicted protein sequence of NF-$AT_c$ and NF-$AT_p$ revealed that they contained a short segment of sequence similar to the DNA-binding and dimerization domains of rel and dorsal[9]. Formally, however, these proteins are probably not rel/dorsal homologues (here used in the strict sense to indicate evolutionary relatedness). Rather they each share about 17 to 20% identity with a short region of rel, dorsal, and NF-kB p50. Outside of this region, they have no similarity. Based on the principals originally proposed by Doolittle for determining the relatedness of proteins[13], we suspect that the rel/dorsal similarity of NF-AT arose by the evolutionary convergence on a similar sequence to serve a similar function. Thus, it is unlikely that NF-$AT_c$ and $_p$ belong to the rel/dorsal gene family.

The DNA-binding and dimerization domains of rel and dorsal have features that are quite distinct from rel and dorsal[14-16]. First, there is a lack of conservation of cysteins: a feature that is felt to be an indication of evolutionary unrelatedness. Second, there are several charge reversals in the DNA-binding and dimerization domain from other rel/dorsal family members. The latter feature indicates that salt bridges may be formed in this region using a strategy that is distinct from that of the rel/dorsal family.

Other features of the sequence are the presence of a short 13 amino acid repeat in the mid-region of the protein. This short repeat is unlike anything found in the data bank, and its function is uncertain. An additional feature of the amino acid sequence is an amino terminus that is very rich in proline, a feature that is not uncommon for a transcription factor.

## THE mRNA FOR NF-AT$_c$ IS EXPRESSED PREDOMINANTLY IN T CELLS WHILE THE mRNA FOR NF-AT$_p$ IS EXPRESSED MORE BROADLY

Using quantitative ribonuclease protection, we studied the expression of NF-AT$_c$ and $_p$ in cell lines and organs of mice[9]. NF-AT$_c$ was expressed in the thymus, to a lesser degree in the spleen and to a much lesser degree in the skin. These are precisely the tissues of the mouse that show NF-AT directed transcription in animals transgenic for a construct in which NF-AT directs transcription of the SV-40 T antigen[7]. However, very long exposures of the films indicated that most tissues of the mouse had some detectable NF-AT$_c$ mRNA, which could either be due to contamination with circulating T cells or to low level expression in all these tissues. NF-AT$_p$ is expressed at somewhat higher levels in the heart and brain than the thymus and can be detected in many tissues[9]. Thus, based on tissue distribution, NF-AT $_c$ appears to be the activity originally defined as the cytosolic component of NF-AT.

## EXPRESSION OF NF-ATc WILL ACTIVATE THE IL-2 PROMOTER IN NON-T CELLS

Based on the tissue distribution of the mRNAs for NF-AT$_c$, one might expect that it could be the tissue-specific component required for the activation of the IL-2 enhancer. To examine this possibility, we expressed NF-ATc in non-T cell lines and found that it gave a 5- to 15-fold activation of the IL-2 enhancer[9]. In contrast, in the Jurkat human T cell line a somewhat larger activation was observed, suggesting that there may still be some additional components of the T cell signalling cascade that contributes to the T cell specificity or that the endogenous NF-AT contributes to the responses seen in T cells. As yet, similar studies have not been carried out for NF-AT$_p$, however, its tissue distribution suggests that it might not be involved at this point. A fragment of the bacterially expressed protein will activate transcription in vitro, but only in the presence of bacterially expressed fos and jun proteins[10].

## A CALCIUM STIMULUS IS NOT REQUIRED FOR NF-AT$_c$-DEPENDENT TRANSCRIPTION

One curious aspect of the function of NF-AT$_c$ revealed in transfection studies is that it does not appear to require calcium for activity[9]. This is not expected since it was originally described to become associated with the nucleus in response to calcium. This stimulus-independent function for activation is also seen with NFκB and AP-1[17-19] and probably relates to the ability of the transfected and over-produced protein to saturate the normal cytoplasmic anchoring mechanism. As yet, however, there is no definitive reason why overexpression of NF-AT$_c$ should result in calcium-independent function.

## MONOCLONAL ANTIBODIES FOR NF-AT$_c$ REACT WITH THE MOST OF THE NF-AT COMPLEX PRESENT IN MURINE THYMOCYTES

Several polyclonal and a total of 8 monoclonal antibodies were raised to NF-AT and used to explore its involvement in the formation of the NF-AT DNA-protein complex[9]. One of the mice immunized made antibodies that bound to the NF-AT DNA-protein complex and supershifted it on native polyacrylamide gels[9]. This mouse was used to make monoclonal antibodies to NF-AT. One of these monoclonals reacts with the native NF-$AT_c$ at a region that is outside the DNA binding and dimerization domain that is shared with NF-$AT_p$ and hence could be used to assay the activity of NF-ATc. Using this assay, most of the NF-AT complex formed from nuclear extracts of thymus cells reacted with the antibody and about 50% of it was "supershifted" to a slower mobility. This effect was specific in that the antibody did not react with the NFκB or AP-1 complex[9] and the antibody did not react with the bare probe. The latter is a common artifact in these types of supershift experiments and is probably due to an interaction with acute phase proteins in the "immune" antisera but not in the "preimmune" antisera. The ability of these acute phase proteins to interact with DNA probably accounts for the apparent ability of an anti-junB antisera to react with the NF-AT complex.

## A DOMINANT NEGATIVE OF NF-$AT_c$ BLOCKS IL-2 PROMOTER FUNCTION IN JURKAT HUMAN T CELLS

The above studies strongly indicated that NF-$AT_c$ is a major component of the NF-AT complex and that it could play a positive role in activating the IL-2 gene in non-T cells. To help determine the contribution of NF-$AT_c$ to the activity of the IL-2 gene enhancer in T cells, we constructed a dominant negative of NF-$AT_c$[9]. This dominant negative was derived from a region of NF-$AT_c$ that was not shared with NF-$AT_p$ and showed no detectable similarity in sequence or predicted secondary structure. This construct was transfected into T cells along with constructs that contained either the whole IL-2 promoter or the NF-AT sites directing transcription of the β–gal reporter gene. The dominant negative blocked IL-2 promoter/enhancer function by more than 95% and blocked NF-AT-directed transcription by more than 90%, strongly indicating that NF-$AT_c$ is the major contributor to NF-AT function on the IL-2 gene. Although the NF-$AT_p$ clone has not been reported to have activity after transfection into T cells and it is not known if a dominant negative can be constructed from this cDNA, these studies do not rule out a role for NF-$AT_p$ and it may function in the NF-AT complex in a fashion that is not in competition with NF-$AT_c$.

## COULD NF-$AT_c$ BE A SUBSTRATE FOR CALCINEURIN?

Our finding that NF-ATc associated with the nucleus quickly after stimulation with calcium ionophore and that cyclosporin A blocked this nuclear association strongly indicated that calcineurin was essential for some aspect of the nuclear association. This led us to speculate that NF-$AT_c$ could be the direct substrate for calcineurin[20]. We and others have found that very high, unphysiologic concentrations of calcineurin will alter the mobility of NF-AT on a polyacrylamide gel, however, the exact subunit that is affected is not known and it is not known if this result actually corresponds to a dephosphorylation. However, any phosphatase will also alter the mobility of the complex and these results are difficult to fully interpret. Identifying NF-AT as a functional substrate of calcineurin will require the following steps: 1) mapping the sites of phosphorylation in NF-AT, 2) mapping the sites of dephosphorylation in vivo and in demonstrating that the are identical to the site dephosphorylated in vitro, 3) mutation of these sites of dephosphorylation and 4) demonstration that the mutated protein is defective in nuclear assoication or some other activity.

## THE POSSIBLE IDENTITY OF NF-AT$_n$ WITH AP-1

Several studies have indicated that the nuclear component of NF-AT might be a fos-jun complex. Present evidence indicates that overexpression of c-fos, fra, c-jun or jun D will replace PMA for the activation of NF-AT transcription but not IL-2 directed transcription[21,22]. In addition, the NF-AT complex can be reconstituted with fos or jun in a gel mobility shift assay[12]. Finally, antisera to fos or jun will often interfere with the formation of the NF-AT complex and in some cases will produce a reduction in the NF-AT complex accompanied by the formation of a supershift[21]. These data are all consistent with a role for AP-1 in NF-AT-dependent transcription, however, definitive evidence that AP-1 is involved has not yet been presented and several pieces of data suggest that other proteins may serve this function. For example, while it is possible to reconstitute the NF-AT complex with fos or jun proteins, it usually requires from 10- to 100-times as much as to reconstitute an AP-1 gel shift, leaving one to wonder if AP-1 ever achieves this concentration within the cell. Further, several of the supershifts reported, as for jun B, are artifacts due to the interaction of the antisera with the DNA and do not require the presence of a nuclear extract. Finally, AP-1 derived peptides have not appeared in any of the sequences found in purified NF-AT. This may be due to the fact that NF-AT has been purified by a reconstitution assay[9] and hence may not be accompanied by associated proteins. While the evidence for the involvement of AP-1 is good, proof in biology is elusive and more definitive evidence will require the purification of the nuclear subunit. The importance of a separate purification of NF-AT$_n$ is underlined by the observation that this subunit is developmentally regulated[22], and a complacent approach to this problem may lead to a misunderstanding of the factors underlying thymic development.

## SUMMARY

Present evidence indicates a pathway of signal transmission in T cells that is outlined in figure 1. The elevation in intracellular calcium that is induced by interactions at the antigen receptor leads to the activation of the calcium-dependent phosphatase calcineurin. This in turn leads to the nuclear association of the cytosolic component of NF-AT$_c$. The activation of calcineurin and the nuclear import of NF-AT$_c$ can both be blocked by cyclosporin A or FK506 in complex with their respective immunophilins. Once in the nucleus, NF-AT$_c$ interacts with NF-AT$_n$ to form an active transcriptional complex. NF-AT$_n$ is a ubiquitous protein, can be synthesized in response to PMA, and has many similarities to AP-1. The mechanism by which NF-AT$_c$ enters the nucleus is unknown, and although it appears to require calcineurin, NF-AT$_c$ has not yet been shown to be an in vivo substrate of calcineurin. Alternative mechanisms include the possibility that NF-AT$_c$ operates on some cytoplasmic anchor or that other proteins that are controlled by calcineurin carry out the nuclear import of NF-AT$_c$. Although NF-AT$_p$ copurifies with NF-ATc, there is as yet no understanding of how NF-AT$_p$ is functioning in vivo. Now that these proteins are purified and cloned, the major goals will be to understand their role and the roles of other family members in thymic development.

## REFERENCES


1. G.R. Crabtree. Contingent genetic regulatory events in T lymphocyte activation. *Science* 243:355 (1989).
2. A. Weiss. T cell antigen receptor signal transduction: a tale of tails and cytoplasmic protein-tyrosine kinases. *Cell* 73:209 (1993).
3. U. Siebenlist, D.B. Durand, P. Bressler, N.J. Holbrook, C.A. Norris, M. Kamoun, J.A. Kant, and G.R. Crabtree. Promoter region of the IL-2 gene undergoes chromatin structure changes and confers inducibility on chloramphenicol acetyltransferase gene during activation of T cells. *Mol. Cell. Biol.* 6:3042 (1986).
4. D.B. Durand, M.R. Bush, J.G. Morgan, A. Weiss, and G.R. Crabtree. A 275 bp fragment at the 5' end of the IL-2 gene enhances expression from a heterologous promoter in response to signals from the T cell antigen receptor. *J. Exp Med.* 165:395 (1987).
5. J.-P. Shaw, P.J Utz, D.B. Durand, J.J. Toole, E.A. Emmel, and G.R. Crabtree. Identification of a putative regulator of early T cell activation genes. *Science* 241:202 (1988).
6. D.B. Durand, J.-P. Shaw, M.R. Bush, R.E. Replogle, R. Belageje, and G.R. Crabtree. Characterization of antigen receptor response elements within the interleukin 2 enhancer. *Mol. Cell Biol.* 8:1715 (1988).
7. C.L. Verweij, C. Guidos, and G.R. Crabtree. Cell type specificity and activation requirements for NFAT-1 (nuclear factor of activated T-cells) transcriptional activity determined by a new method using transgenic mice to assay transcriptional activity of an individual nuclear factor. *J. Biol. Chem.* 265:5788 (1990).
8. J. Karttunen, and N. Shastri. Measurement of ligand-induced activation in single viable T cells using the *lacZ* reporter gene. *Proc. Natl. Acad. Sci. USA* 88:3972 (1991).
9. J.P. Northrop, S.N. Ho, D.J. Thomas, L. Chen, G.P. Nolan, A. Admon, and G.R. Crabtree. Signaling pathways for T cell activation converge on NF-ATc, a *Dorsal/Rel* homologue. *Nature,* in press (1994).
10. P.G. McCaffrey, C. Luo, T.K. Kerppola, J. Jain, T.M. Badalian, A.M. Ho, E. Burgeon, W.S. Lane, J.N. Lambert, T. Curran, G.L. Verdine, A. Rao, and P.G. Hogan. Isolation of the cyclosporin-sensitive T cell transcription factor NFATp. *Science* 262:750 (1993).
11. W.F. Flanagan, B. Corthesy, R.J. Bram, and G.R. Crabtree. Nuclear association of a T-cell transcription factor blocked by FK-506 and cyclosporin A. *Nature* 352:803 (1991).
12. J.P. Northrop, K.S. Ullman, and G.R. Crabtree. Characterization of the nuclear and cytoplasmic components of the lymphoid-specific nuclear factor of activated T cells (NF-AT) complex. *J. Biol. Chem.* 268:2917 (1993).
13. R.F. Doolittle. Similar amino acid sequences: chance or common ancestry. *Science* 214:149 (1981).
14. S. Ghosh, A.M. Gifford, L.R. Riviere, P. Tempst, G.P. Nolan, and D. Baltimore. Cloning of the p50 DNA binding subunit of NF-KB: Homology to rel and dorsal. *Cell* 62:1019 (1990).

15. G.P. Nolan, S. Ghosh, H.-C. Liou, P. Tempst, and D. Baltimore. DNA binding and IkB inhibition of the cloned p65 subunit of NF-κB, a *rel*-related polypeptide. *Cell* 64:961 (1991).
16. G.P. Nolan, and D. Baltimore. The inhibitory ankyrin and activator Rel proteins. *Current Biology, Ltd*. 2:211 (1992).
17. H.C. Liou, and D. Baltimore. Regulation of the NF-kappa B/rel transcription factor and I kappa B inhibitor system. *Curr. Opin. Cell Biol.* 5:477 (1993).
18. R. Chiu, W.J. Boyle, J. Meek, T. Smeal, T. Hunter, and M. Karin, M. The c-Fos protein interacts with c-Jun/AP-1 to stimulate transcription of AP-1 responsive genes. *Cell* 54:541 (1988).
19. T.J. Bos, D. Bohmann, H. Tsuchie, R. Tjian and P.K. Vogt. v-jun encodes a nuclear protein with enhancer binding properties of AP-1. *Cell* 52:705 (1988).
20. S.L. Schreiber, and G.R. Crabtree. The mechanism of action of cyclosporin A and FK506. *Immunology Today* 13:136 (1992).
21. J. Jain, P.G. McCaffrey, V.E. Valge-Archer, and A. Rao. Nuclear factor of activated T cells contains Fos and Jun. *Nature* 356:801 (1992).
22. D. Chen, and E.V. Rothenberg. Molecular basis for developmental changes in interleukin-2 gene inducibility. *Mol. Cell. Biol.* 13:228 (1993).

# QUANTITATIVE ASPECTS OF RECEPTOR AGGREGATION

Henry Metzger,[1] Byron Goldstein,[2] Ute Kent,[1] Su-Yau Mao,[1] Clara Pribluda,[1] Victor Pribluda,[1] Carla Wofsy [3] and Toshiyuki Yamashita[1]

[1] Arthritis & Rheumatism Branch, National Institute of Arthritis and Musculoskeletal and Skin Diseases, National Institutes of Health, Bethesda MD 20892

[2] Theoretical Biology and Biophysics, Los Alamos National Laboratory Los Alamos, NM 87545

[3] Department of Mathematics and Statistics, University of New Mexico Albuquerque, NM 87131

## INTRODUCTION

The critical role aggregation plays in antibody-mediated activation of cellular responses is now familiar to immunologists. The experimental evidence that led to that understanding has been recounted on several occasions[1,2] and the interpretation of the data has stood up well to the newer findings particularly as they relate to cell receptors. It is likewise well known that aggregation is also a widely used mechanism by systems outside those central to the immune response[3]. In this brief review we wish to focus on what progress has been made in extending our knowledge about some quantitative aspects of the aggregation mechanism.

## SIZE OF AGGREGATES

For some cellular receptors the initial size of the aggregate that forms in response to the binding of ligand is determined by the ligand itself. One example is the receptor for human growth hormone in which identical binding sites on two receptors react with epitopes on two

*Mechanisms of Lymphocyte Activation and Immune Regulation V*
Edited by S. Gupta *et al.*, Plenum Press, New York, 1994

discrete regions of the monomeric hormone[4,5]; another is the receptor for platelet-derived growth factor in which two identical or homologous receptors become linked by binding a dimeric ligand[6].

In other instances the size of the aggregate is determined by the changes in the receptor. Thus, in the case of the receptor for epidermal growth factor, binding of the monomeric ligand appears to induce a conformational change in the extracellular domain of the monomeric receptor leading to its dimerization[7].

In the case of receptors that play central roles in the immune system, the initial size of the aggregate is not under similar tight control. For example, the soluble antigens that interact with the antigen receptors on B lymphocytes may have a variable number of repeating epitopes. Similarly an antigen with multiple different epitopes can interact with a variable number of antibody molecules and the complex can in turn react with a variable multiple of Fc receptors. Nevertheless, as has been demonstrated in the case of the receptor with high affinity for IgE (FcεRI), in these systems also, it appears that the cells possess a mechanism that permits them to distinguish as small a cluster as a dimer from monomeric receptors[8,9].

## TEMPORAL CONSIDERATIONS

For over 70 years immunochemists have utilized haptens and hapten conjugates to analyze antibody specificity. Subsequently, such materials also provided powerful tools to analyze the role of aggregation in initiating a variety of immunological phenomena. Indeed, it was through the use of univalent and multivalent chemically defined ligands that 35 years ago, Ishizaka and Campbell provided the first evidence in any biological system of the importance of aggregation[10]. The experimental model they examined is similar to the IgE-mast cell system we have been studying. In the latter, hapten specific IgE antibodies bound to FcεRI on the surface of mast cells will mediate a virtually instantaneous cellular response when exposed to a multivalent hapten conjugate and this response is abruptly halted if an excess of univalent hapten is added at any time during the course of the reaction[11-14].

The rapid cessation of the response could be explained by two principal alternative mechanisms: If the hapten stops the response simply by preventing the formation of new aggregates it raises the possibility that receptors clustered by the aggregated IgE can have only a brief active lifetime[13,15]. Alternatively, if hapten also or largely inhibits the response because it dissociates previously clustered receptors, the results would suggest that the aggregated receptors can remain active for relatively long periods.

We recently attempted to discriminate between these alternatives by using oligomers of IgE stabilized by covalent cross links. By aggregating the receptors directly, such oligomers stimulate cellular responses in the absence of added antigen[8,9]. Because of the slow rate of dissociation of IgE from FcεRI, once individual surface receptors become bound to the oligomers (e.g. dimers or trimers) they will remain clustered for long periods. If excess

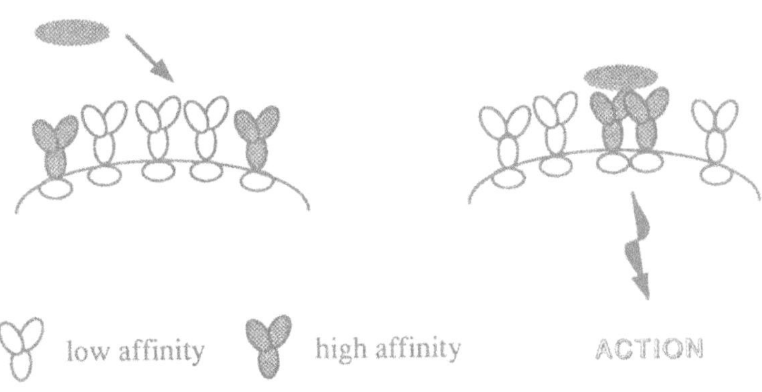


Fig.1 Preferential signaling by receptors bearing IgE with high affinity for antigen

monomeric IgE is added to the cell shortly after the addition of oligomer, no new receptors can be recruited into clusters. In our new studies, we observed that the stably clustered receptors continued to generate transmembrane signals for prolonged periods; in related experiments we obtained evidence that even stably clustered receptors are in a dynamic equilibrium with respect to the phosphorylation of their tyrosine side-chains[16].

We surmise that under more natural conditions, any individual receptor-bound IgE is associating and dissociating from the individual epitopes on a multivalent antigen at rates that reflect the intrinsic binding properties of the IgE's combining sites. Therefore the receptors are similarly in a dynamic equilibrium between being clustered and monomeric. Together, the data on the rapid inhibition of signaling by hapten and the prolonged signaling by oligomers suggest that *the intensity of signal generation is temporally linked to the state of aggregation of individual IgE receptors*. This characteristic makes the system responsive to affinity because receptors associated with an IgE whose antibody combining sites bind the ligand with a high intrinsic affinity will remain clustered preferentially and therefore will continue to signal preferentially (Fig. 1).

A system with characteristics similar to those of the FcεRI receptors would be particularly advantageous to the central elements of the immune system. The similarity in the overall structural and functional characteristics of the "multichain immune response receptors"[17] and the fact that several of these may actually share a critical signaling subunit[18,19] makes it reasonable to extrapolate our results on the active lifetime of the clustered FcεRI receptors, to antigen mediated phenomena with B and T cells.

For receptors such as the antigen receptors of B lymphocytes, in which the diversity in the intrinsic affinity of the ligand binding sites is clonally segregated, cluster-responsive signaling would provide an effective means for differential stimulation (Fig. 2). Because of multivalent binding, those cells bearing high affinity receptors will not necessarily bind much more antigen than cells with receptors having a lower intrinsic affinity[20]. Likewise, therefore, even the weakly interacting receptors will be clustered. However, on those cells bearing high affinity receptors, individual receptors would be expected to occupy a clustered state for longer periods[21]. If their signaling lifetime, like that of FcεRI, is proportional to the length of time they are clustered, it follows that *the intensity of signaling is directly related to the intrinsic affinity of the antigen receptor for the antigen* [16,21].

These considerations may likewise be relevant to an interesting phenomenon described for the activation of T lymphocytes. It is well known that the antigen receptors on such cells are activated by "antigen-presenting cells" bearing antigenic peptides bound in the groove at the distal end of proteins coded by the major histocompatibility complex (MHC). The currently accepted paradigm proposes that when the match between the components on the two cells is such that the antigen receptors on the T cell are efficiently clustered, the cells will be stimulated. As expected, when an antigenic peptide is mutated, the variant may fail to stimulate the T cell; unexpectedly, in some instances it acts as an inhibitor of the stimulation induced by the wild-type peptide[22]. Inhibition can also be observed with the wild-type peptide by introducing a small variation in the presenting MHC protein[23]. Various simple explanations can be ruled out. For example, the inhibition by the variant peptide does not simply result from the displacement by the analogue, of the wild-type antigen on the presenting cell[22].

We suggest that these experimental results can be readily explained if, as is true for FcεRI, *the signaling lifetime of an individual T cell antigen receptor is coterminous with its clustered state*. Thus, we assume that the intrinsic affinity of the T cell antigen receptor for the variant antigen-MHC complex is less than that for the wild-type complex. If so, the presence of both wild-type and variant antigen-MHC complexes on the same presenting cells could actually reduce the time an individual T cell receptor reacting with the wild-type complex is in a clustered state. Some of the time it will be clustered with receptors interacting

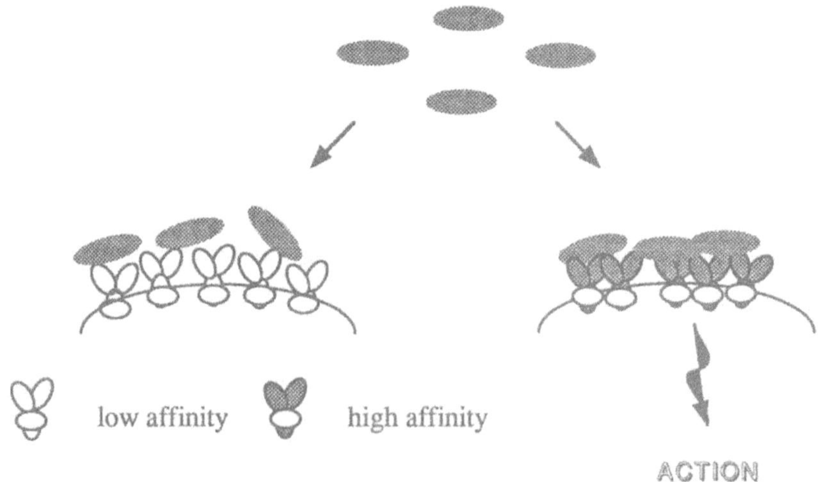


Fig.2 Preferential triggering of B cells bearing high affinity antigen receptors

weakly with the variant presenting complexes, and such a mixed cluster would be more transient than a homogenous wild-type cluster. The consequences of these variant presenting complexes are much like those produced by the hapten in the experiments on mast cells referred to above, even though the mechanism is somewhat different.

## NATURE OF AGGREGATED SPECIES

In order to pursue these considerations further it is necessary to understand the biochemical nature of what we have referred to as the "active" state of the receptor and of the proximal "signals" it generates. Those working in this field know that a large amount of new information is being rapidly obtained, but that even for the most completely studied and structurally simplest receptors, such as the single chained, kinase-encoding receptors for various "growth factors", a complete model remains to be developed. So far, the information that is being acquired is largely qualitative: Critical regions of the receptors are beginning to be discerned, the "receptor-associated" molecules with which they interact are being identified, and some of the early covalent changes such as phosphorylations and dephosphorylations are being mapped.

In the system with which we work, it is now amply documented that in response to aggregation of FcεRI, both the β and γ subunits of the receptor as well as a variety of other cellular components become rapidly phosphorylated--prominently but not exclusively on tyrosines[24,25]. In addition, there is beginning evidence that src-like kinases may be physically associated with the receptors: p53/p56$^{lyn}$ and pp60$^{c\text{-}src}$ in the rat mucosal mast cell line RBL, p62$^{c\text{-}yes}$ in the murine PT18 line[26]; evidence for involvement of p72$^{syk}$ is also developing[27].

The quantitative relationship of these components to the receptors is difficult to assess because the more definitive approaches require isolation of the solubilized receptors and the methods necessary to purify the receptors may disrupt the interactions one is attempting to quantitate. Previously, the method of chemical crosslinking had helped us to identify the β and γ subunits of FcεRI although no other substantial amounts of other components were observed. Recently we have once again applied this approach. Our initial results will be reported in detail elsewhere (S.-Y.Mao, T. Yamashita, H. Metzger, manuscript in preparation). Here we simply summarize our principal findings:

1. Any components that are physiologically associated before or after aggregation, interact with the receptor much less firmly than the receptor's subunits do with each other.

2. After chemical crosslinking multiple components are found to be covalently attached to the receptor in significantly greater amounts than is likely to have resulted from non-specific crosslinking. Thus, after reaction with crosslinking reagent, the increase in receptor-associated [$^{35}$S]-cysteine-labeled proteins and the p53/56$^{lyn}$ kinase exceeded almost by an

order of magnitude the increase observed with a control protein (other receptors) known to be independent of the receptor. Several other proteins whose association with the receptor is unknown, fail to become covalently linked to the receptor. Interestingly, the increase in the association of p53/p56$^{lyn}$ with the FcεRI after chemical crosslinking was quantitatively similar to the increase observed for other components. The most obvious explanation is that the chemical crosslinker stabilized the association of the receptor with a discrete, multi-component "signaling particle" [3,28], a possibility we are currently exploring.

3. If such a "signaling particle" exists, our data suggest its association with FcεRI is not substantially enhanced by aggregation of the receptor, but we cannot exclude the possibility that aggregation promotes the association or dissociation of larger amounts of particular non-receptor components. Eisman and Bolen's data suggested to them that p53/p56$^{lyn}$ is pre-associated with FcεRI[26], whereas Benhamou et al. conclude that p72$^{syk}$ becomes associated only after the receptors are aggregated and phosphorylated[27]. The findings of both groups as well as our own are not necessarily inconsistent.

4. The phosphorylation of the receptor-associated components is confined to those associated with the receptor being aggregated. As with the phosphorylation of the receptors themselves[14,29], there appears to be no "bystander" effect.

## TOWARDS IN VITRO RECONSTITUTION

Reconstitution in vitro of the initial consequences of aggregation will provide an important approach to a complete understanding of how the receptors function. A critical first step is to determine whether some of the early phenomena can be reproduced on broken cell preparations. We have recently reported on the first successful results[29]. In those experiments cell-free particulates of IgE-sensitized cells could be shown to undergo phosphorylation of receptor tyrosines when the receptors were aggregated in vitro. Recently we have pursued the goal of reconstitution by focussing on the kinase activity associated with immunoprecipitated receptors[26,28]. The initial studies (V.Pribluda, C. Pribluda and H. Metzger, manuscript in preparation) have clarified some of the critical factors that affect the recovery of activity and characterized the tyrosine kinase activity that is associated with the FcεRI in greater quantitative detail.

Parallel studies that have focussed on the p53/p56$^{lyn}$ (T.Yamashita, S.-Y. Mao, and H. Metzger, work in progress) and have revealed the following:

1. In resting cells, under conditions where much of p53/p56$^{lyn}$ and FcεRI have reacted with a chemical crosslinking reagent, only about 2% of each of these two entities are crosslinked to each other. Then if they are linked to each other in a 1:1 ratio, the cells must contain roughly equal numbers of molecules of FcεRI and p53/p56$^{lyn}$. It also suggests that only a small fraction of the receptors may be pre-associated with the p53/p56$^{lyn}$ kinase

although we cannot exclude the possibility that this reflects a methodological problem rather than reality.

2. When the FcεRI are substantially aggregated, the percent of receptors associated with p53/p56$^{lyn}$ increases several fold.

3. When incubated with ATP under permissive conditions, immunoprecipitates of such receptors aggregated and cross-linked in vivo show a large increase in protein phosphorylation in comparison to immunoprecipitates (isolated under identical conditions) of cells whose FcεRI had not been aggregated in vivo or had been aggregated but not cross-linked.

4. Interestingly, if the FcεRI are aggregated in vivo under conditions where kinases are inhibited, the aggregates when tested in vitro behave as resting (unaggregated) receptors.

## CONCLUDING REMARKS

Studies by ourselves and others have demonstrated that the FcεRI, like other members of the family of multi-subunit immune response receptors, are likely associated with other, weakly interacting, cellular components. Defining the qualitative and quantitative characteristics of these interactions remains a formidable challenge. One of the earliest consequences of the aggregation of the receptors is phosphorylation of the receptor and some of the associated components. It appears that at least small aggregates can undergo repeated rounds of phosphorylation and dephosphorylation but the exact mechanism that accounts for these changes remains to be defined. An important consequence of these characteristics is that signaling appears to be directly related to the number and persistence of clustered receptors. Since clustering should reflect the intrinsic affinity between the receptor (or its bound immunoglobulin) and antigen, it is apparent that the mechanism is ideally suited to be differentially responsive to different ligand-receptor affinities. Thus we are nearing a molecular explanation of antigen directed activation of cells mediated by the multi-subunit immune response receptors, and can therefore explore rationally whether these responses can be therapeutically manipulated.

## ACKNOWLEDGEMENTS

Partly supported by the N.I.H.Grant GM35556 (B.G.), and N.S.F. Grant DMS 8911623 (C.W.) and by the U.S. Department of Energy.

## REFERENCES

1. H. Metzger. The effect of antigen binding on the properties of antibody. *Adv.Immunol.* 28:169 (1974).

2. H. Metzger. Transmembrane signaling: The joy of aggregation. *J.Immunol.* 149:1477 (1992).

3. A. Ullrich and J. Schlessinger. Signal transduction by receptors with tyrosine kinase activity. *Cell* 61:203 (1990).

4. B.C. Cunningham, M. Ultsch, A.M. De Vos, M.G. Mulkerrin, K.R. Clauser and J.A. Wells. Dimerization of the extracellular domain of the human growth hormone receptor by a single hormone molecule. *Science* 254:821 (1991).

5. A.M. De Vos, M. Ultsch and A.A. Kossiakoff. Human growth hormone and extracellular domain of its receptor: crystal structure of the complex. *Science* 255:306 (1992).

6. W. Chao and M.S. Olson. Platelet-activating factor: Receptors and signal transduction. *Biochem.J.* 292:617 (1993).

7. J. Schlessinger. Signal transduction by allosteric receptor oligomerization.*TIBS* 13:442 (1988).

8. D.M. Segal, J.D. Taurog and H. Metzger. Dimeric immunoglobulin E serves as a unit signal for mast cell degranulation. *Proc.Natl.Acad.Sci.USA* 74:2993 (1977).

9. C. Fewtrell and H. Metzger. Larger oligomers of IgE are more effective than dimers in stimulating rat basophilic leukemia cells. *J.Immunol.* 125:701 (1980).

10. K. Ishizaka and D.H. Campbell. Biologic activity of soluble antigen-antibody complexes IV. The inhibition of the skin reactivity of soluble complexes and the PCA reaction by heterologous complexes. *J.Immunol.* 83:116 (1959).

11. Fewtrell, C. Activation and desensitization of receptors for IgE on tumor basophils. In: *Calcium in Biological Systems*, edited by Rubin, R.P., Weiss, G.P. and Putney, J.W.,Jr. New York: Plenum Publishing Corporation, 1985, p. 129-135.

12. A. Kagey-Sobotka, D.W. MacGlashan and L.M. Lichtenstein. Role of receptor aggregation in triggering IgE-mediated reactions. *Fed.Proc.* 41:12 (1982).

13. J.M. Oliver, J. Seagrave, R.F. Stump, J.R. Pfeiffer and G.G. Deanin. Signal transduction and cellular response in RBL-2H3 mast cells. *Prog.Allergy* 42:185 (1988).

14. R. Paolini, M.-H. Jouvin and J.-P. Kinet. Phosphorylation and dephosphorylation of the high-affinity receptor for immunoglobulin E immediately after receptor engagement and disengagement. *Nature* 353:855 (1991).

15. D. Holowka and B. Baird. Structure and function of the high-affinity receptor for immunoglobulin E. *Cell.Molec.Mech.Inflam.* 1:173 (1990).

16. U.M. Kent, S.-Y. Mao, C. Wofsy, B. Goldstein, S. Ross and H. Metzger. Dynamics of signal transduction after aggregation of cell-surface receptors: Studies on the Type I receptor for IgE . *Proc.Natl.Acad.Sci.USA* 91:3087 (1994).

17. A.D. Keegan and W.E. Paul. Multichain immune recognition receptors: Similarities in structure and signaling pathways. *Immunol.Today* 13:63 (1992).

18. D.G. Orloff, C. Ra, S.J. Frank, R.D. Klausner and J.-P. Kinet. Family of disulfide-linked dimers containing the zeta and eta chains of the T-cell receptor and the gamma chain of Fc receptors. *Nature* 347:189 (1990).

19. D. Qian, A.I. Sperling, D.W. Lancki, Y. Tatsumi, T.A. Barrett, J.A. Bluestone and F.W. Fitch. The γ chain of the high affinity receptor for IgE is a major functional subunit of the T-cell antigen receptor complex in γδ T lymphocytes. *Proc.Natl.Acad.Sci.USA* 90:11875 (1993).

20. M.R. Jarvis and E.W. Voss,Jr.. Consequences of avidity in lymphocyte receptor -multivalent antigen binding in affinity maturation. *Mol.Immunol.* 19:1063 (1982).

21. H. Metzger. A comment on the "Speculation" of Jarvis and Voss [letter]. *Mol.Immunol.* 19:1071 (1982).

22. M.T. De Magistris, J. Alexander, M. Coggeshall, A. Altman, F.C. Gaeta, H.M. Grey and A. Sette. Antigen analog-major histocompatibility complexes act as antagonists of the T cell receptor. *Cell* 68:625 (1992).

23. L. Racioppi, F. Ronchese, L.A. Matis and R.N. Germain. Peptide-major histocompatibility complex class II complexes with mixed agonist / antagonist properties provide evidence for ligand-related differences in T cell receptor -dependent intracellular signaling. *J.Exp.Med.* 177:1047 (1993).

24. M. Benhamou and R.P. Siraganian. Protein tyrosine phosphorylation: An essential component of FcεRI signaling. *Immunol.Today* 13:195 (1992).

25. M.A. Beaven and H. Metzger. Signal transduction by Fc receptors: The FcεRI case. *Immunol.Today* 14:222 (1993).

26. E. Eiseman and J.B. Bolen. Engagement of the high-affinity IgE receptor activates *src* protein-related tyrosine kinases. *Nature* 355:78 (1992).

27. M. Benhamou, N.J.P. Ryba, H. Kihara and R.P. Siraganian. Protein-tyrosine kinase p72syk in high affinity IgE receptor signaling. *J.Biol.Chem.* 268:23318 (1993).

28. R. Paolini, R. Numerof and J.-P. Kinet. Phosphorylation/dephosphorylation of high-affinity IgE receptors: A mechanism for coupling/uncoupling a large signaling complex. *Proc.Natl.Acad.Sci.USA* 89:10733 (1992).

29. V.S. Pribluda and H. Metzger. Transmembrane signaling by the high affinity IgE receptor on membrane preparations. *Proc.Natl.Acad.Sci.USA* 89:11446 (1992).

# Fc RECEPTOR SIGNALING

David Brooks and Jeffrey V. Ravetch

Dewitt Wallace Research Laboratory
Sloan-Kettering Institute
1275 York Avenue
New York, NY 10021

## INTRODUCTION

Immune complex cross-linking of cell surface receptors for immunoglobulin (Ig) triggers pleiotropic cellular events that underpin a wide variety of immune responses. In this fashion receptors for immunoglobulin form a molecular bridge between the humoral and cellular immune responses. Elucidation of the primary structure of receptors for the Fc domain of Ig (FcγRs) has provided invaluable insight into their contribution to immune cell regulation. This revelation has been made possible as the result of cloning of both cDNAs and genes that encode this diverse family of molecules. Over the past decade a nearly complete "dissection" of the molecules that mediate this binding of immune complexes to cells has been accomplished. Current efforts focus on putting these pieces back into an organismic context and determining their roles in systemic processes such as inflammation, autoimmunity, allergy and development. The genetic map of most FcγRs will soon be complete providing a valuable tool to evaluating involvement of these genes in karyotypic instability and landmarks in the search for linked human genes/conditions.

Multiple lines of evidence established the existence of three classes of FcRs for IgG (FcγRs): I, II and III [1]. FcγRI binds monomeric Ig and is thereby defined as high affinity while cells bearing receptors of class II or III only manifest avidity for multimeric immune complexes and are therefore referred to as low affinity receptors. While each class is believed to be encoded by a single gene in mice, further diversity exists in man where multiple genes encode FcγRs of each class. Three subclasses, known as A, B and C are known in both FcγRI and FcγRII. FcγRIII is encoded by two genes, A and B. Therefore, a total of eight subclasses of human FcγRs are recognized on the basis of their ligand binding polypeptide chains as shown in Fig. 1.

Adding to the diversity of the FcγR gene family and the opportunity for pleiotropic cellular responses to the binding of immune complexes is formation of multimeric receptor complexes. An emerging theme of cell surface receptors that is in evidence here is the existence of hetero-oligomeric receptor complexes at the plasma membrane. Unlike growth factor receptors where a catalytic domain is fused to the ligand binding domain, members of each class of FcγR exist as hetero-oligomeric

*Mechanisms of Lymphocyte Activation and Immune Regulation V*
Edited by S. Gupta *et al.*, Plenum Press, New York, 1994

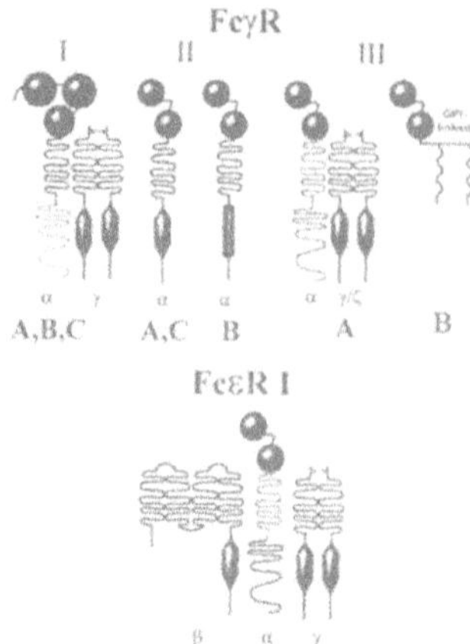


**Figure 1.** Schematic representation of FcRs that are members of the Ig super family (IgSF) and their associated subunits. IgSF proteins are shown with balls to symbolize their Ig-like domains. FcR ligand binding chains have two or three (solely FcγRI) of these domains. With the exception of FcγRIIIB, which is GPI-linked, all receptors are illustrated in their transmembrane forms. Three FcγRs: IIA, IIB and IIIB are not known to have associated subunits. The cytoplasmic tails of FcγRIIA and IIB contain functional domains involved in signal transduction. The hexagonal symbol common to several receptors represents the antigen receptor activation motif (ARAM). The cylindrical symbol in FcγRIIB represents the motif capable of down modulating BCR signalling. In addition to the ligand binding (α) subunit and the signal transducing γ subunit, FcεRI has a novel β subunit that is unique among FcRs.

complexes. Two related proteins that were originally discovered in the context of other receptors have been found to associate with these FcγRs. In particular, the gamma subunit discovered as a subunit of the FcεRI and the zeta subunit discovered in the T cell receptor both associate with FcγRs as illustrated in Fig. 2. These subunits play critical roles in the cases of FcγRI and FcγRIIIA where they not only contribute to protein trafficking through the cell but also in signal transduction as will be discussed below.

The FcγRs discussed above and the high affinity IgE receptor, FcεRI, are all members of the immunoglobulin super family (IgSF). The paradigmatic feature of the

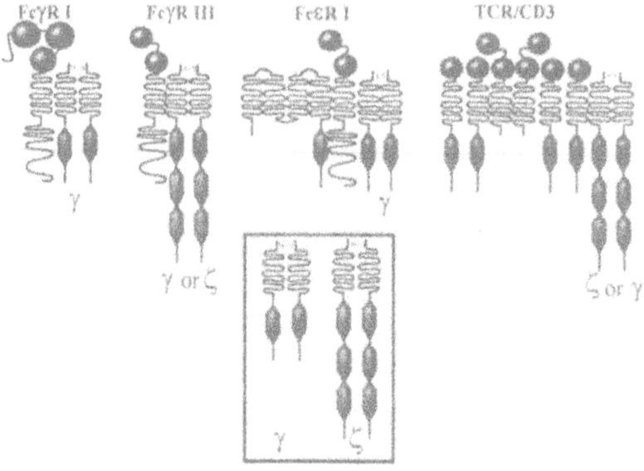


**Figure 2.** The subunit composition of 4 classes of antigen receptors is shown schematically. FcγRI has been shown to associate with γ at the cell surface. FcγRIII depends on γ and/or ζ to get to the cell surface and to transduce signals. The composition of this disufide-linked dimer varies depending on cell type. FcεRI requires a γ dimer to be expressed at the cell surface and for signal transduction. Like FcγRIII, the TCR/CD3 complex must associate with homo or heterodimers of γ and/or ζ. Shown in the box is a schematic of the γ or ζ structure indicating that the activation domain found in the cytoplasmic tail of γ is found in triplicate in ζ as discussed in the text.

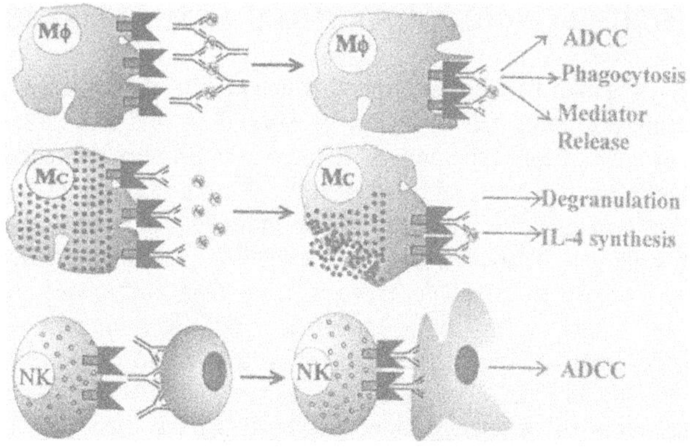


**Figure 3.** The role of FcRs in immune cell activation is highlighted. Hallmark functions of immune effector cells that are elicited by FcR crosslinking are illustrated on the right side of the figure. Macrophages (top), mast cells (Mc, middle) and natural killer cells (NK, bottom) are illustrated but other cells including granulocytes are also activated by FcRs. Antibody dependent cellular cytotoxicity (ADCC) of IgG coated targets is mediated by macrophages, NK cells and neutrophils. Macrophages and monocytes are capable of phagocytosis of IgG opsonized particles. Many mediators of inflammation such as cytokines, prostaglandins, leukotrienes, activated oxygen species and histamine are released by various cells in response to multivalent antigen binding to FcRs.

IgSF is an extracellular Ig-like domain structure that is stabilized by a disulfide bond (symbolized by ball in Fig. 1). Almost all members of the IgSF have been shown to play a role in cell surface recognition events. The FcR subfamily is characterized by the C2 set, Ig-like domains in extracellular portion of the receptor[2] that is responsible for binding the Fc portion of their respective Ig ligand(s). The conserved domain structure of FcγRs, which correlates perfectly with exon structure of the genes, is the basis on which these genes are considered to be a subfamily of the IgSF and are believed to have evolved by sequential rounds of gene duplication and divergence. With exceptions of FcγRIIIB and "soluble FcγRs", all known FcRs are transmembrane proteins. FcγRIIIB is phosphatidylinositol-glycan linked to the membrane and is unique to man; no rodent homolog is known. In contrast to the conserved extracellular structure of FcγRs, considerable diversity characterizes their intracellular domains. This led to recognition that a common extracellular structure and function has been retained while being coupled to divergent membrane anchorage and intracellular domains capable of mediating pleiotropic signals in response to a common stimulus.

## FcγRs PLAY IMPORTANT ROLES IN IMMUNE CELL FUNCTIONS

The structure of FcγRs discussed above includes the common element of an activation domain that is shared by multiple antigen receptors. This domain of the FcγRs is involved in coupling antigen binding to cell signalling events that feature protein phosphorylation. Immune cells such as macrophages, mast cells and natural killer cells are activated by FcγR crosslinking to mediate a number of important functions as shown in Fig. 3. Activated macrophages and neutrophils mediate antibody dependent cellular cytotoxicity (ADCC), phagocytosis and release of inflammatory mediators such as cytokines and prostaglandins. A unique example where a single class of FcγR is expressed on a given cell type and can be assigned a specific function is illustrated by NK cell ADCC of IgG coated targets. In addition to the FcεRI on mast cells, FcγRs can stimulate IL-4 secretion and degranulation leading to histamine release.

## FcγRI TRANSCRIPTS INDICATE MULTIPLE GENES EXIST IN MAN

Each investigation of the transcripts that encode the high affinity IgG receptor has revealed a notable degree of heterogeneity; between three and five distinct cDNAs have been reported by each of four laboratories. Diversity in primary sequence and alternative splicing of three exons accounts for this diversity[3,4,5,6]. The finding of three polymorphic cDNAs in a library of mononuclear cells of one individual led one group to suggest that more than one gene was likely to encode these cDNAs[3]. Southern analysis of genomic DNA led these authors to conclude that no more than two genes exist. Our data, substantiated by others[7,5], has now established that a minimum of three highly homologous genes exist. These genes are referred to as IA, IB and IC; however, as discussed below only FcγRIA encodes the consensus high affinity IgG receptor.

The first report of a genetic clone of FcγRI set the precedent for what is now thought of as the consensus high affinity IgG receptor, i.e. FcγRIA[3]. This is the type I transmembrane glycoprotein with three Ig-like extracellular domains, a 21 amino acid membrane spanning domain and a novel, highly charged cytoplasmic tail. The existence of a third Ig-like domain in this receptor is unique among FcγRs. It is interesting that while each of the first two Ig-like domains is most homologous to their respective counterparts in low affinity FcγRs, the third domain shows considerably less homology to known FcRs[3]. The functional properties associated with this EC3 domain reviewed below substantiate that it is unique to, and instrumental in, the high affinity IgG receptor.

Transcripts encoding FcγRIB or IC have two notable features. Full length transcripts that encode all six exons (see Results) of each gene have been reported that include in frame stop codons in the third extracellular (EC3) exon . FcγRIB transcripts have been described that lack this exon presumably as a result of alternative splicing[4,5,6]. While the cDNAs encoding these transcripts were generated by RT coupled PCR, the transcripts have been detected by RNase protection[5] and northern analysis[6] as well; therefore, these are not simply rare abortive spliced mRNAs. In fact, such an $EC_3^-$ FcγRIB cDNA encodes a low affinity FcγR that no longer binds monomeric IgG but can still bind multivalent ligand[6]. To further complicate the pattern of gene expression in this family two groups have reported FcγRI transcripts that lack EC1 both with[4] and without EC3[4,6]. Northern analysis has revealed a small, faint 1kb mRNA that may correspond to the cDNA lacking both EC1 and EC3[6]. However, the function of a protein with a single EC domain is entirely speculative. It must be considered that genetic drift in this reiterated gene family contributes to the heterogeneity of gene products observed. Indeed some members of the family may be drifting toward pseudogene status. In summary, no evidence exists for consensus transcripts from either FcγRIB or IC genes. FcγRIB is capable of encoding an alternatively spliced low affinity IgG receptor.

## GENOMIC LOCI ENCODING FcRs

While members of the IgSF are encoded by genes that are believed to have evolved from a common ancestor, the loci are spread throughout the human genome in present-day man. A few conserved linkage groups of homologous genes have been retained and have prompted speculation that this arrangement is the result of recent duplication events. One such linkage group is the IgSF FcRs; all the genes encoding these receptors are found on human chromosome 1. The only reported locus of human FcγRI was on 1q21[8]. Both in situ hybridization and Southern analysis of rodent-human hybrid cell lines have been used to localize FcγRII and FcγRIII genes to bands 1q22-24. This chromosomal region also contains the gene, FCRG, for the gamma subunit that is shared by no less than three FcRs including the alpha subunit of the FcεRI whose gene maps nearby on 1q22 .

## DISTRIBUTION OF FcR EXPRESSION

FcγRs are expressed on a variety of cell types. All hematopoietic lineages except for RBCs have at least one type of FcγR on their cell surface. Cells constituting the reticuloendothelial system throughout the body are $FcγR^+$. In addition, endothelial and placental epithelium FcγR expression has been reported.

## STRUCTURE: FUNCTION RELATIONSHIPS

### Structure: Function Correlations of FcγRI

As emphasized above the hallmark feature of FcγRIA is its high affinity for IgG ($K_a = 10^8$-$10^9$ $M^{-1}$). This property is dependent on the presence of a third Ig-like domain that is unique to this receptor. To emphasize this point one may consider that the high affinity IgE receptor has only two Ig-like domains. Receptors lacking this third domain lose specificity and affinity for ligand suggesting that it is essential to the higher affinity of this receptor relative to FcγRII/III[19,6]. The fact that receptors lacking the EC3 domain exhibit characteristics of a low affinity receptor demonstrates that the first two EC domains do contribute significantly to binding Fc domains of IgGs. This is substantiated by the finding that these exons cannot be substituted by their counterparts in the murine FcγRII. Replacing EC1 and 2 domains of moFcγRI with those of FcγRII not only fails to maintain high affinity binding it alters ligand specificity[19]. Surprisingly, all six CD64 mAbs tested failed to detect this variant FcγRIb1 receptor (as do CD16 and CD32). This degree of epitope loss/modification suggests that a more extensive conformational change may occur in the absence of EC3 than simple loss of epitopes encoded by that domain. Together these data point to a strong interaction between extracellular domains of FcγRI that is required for binding monomeric IgG of the appropriate subtype.

Recent evidence suggests that FcγRI is found associated with the FcR gamma subunit at the cell surface as diagrammed in Fig. 1. Strong evidence suggesting that gamma plays a role in signal transduction by antigen receptors has been reviewed[20]. Evidence of physical association between FcγRI and gamma in mononuclear and transfected cells has been reported[21]. Direct functional evidence to support a role for gamma in signal transduction by this complex has not been reported. However, a very interesting series of studies suggests that this subunit is critical to FcγRI function in mice that are deficient in the gamma subunit as a result of targeted disruption.

FcγRI is unique among FcRs in that a strong association with the cytoskeleton has been demonstrated[22]. In particular, all FcγRI on U937 cells was found to be associated with actin-binding protein (ABP). This dimeric protein, which cross links actin filaments, serves to couple the cell membrane to the cytoskeleton through this high avidity ($K_d = 3 \times 10^{-8}$ M) association. Ligand or mAb to FcγRI as well as MAb to ABP could coimmunoprecipitate these proteins from COS cells transfected with FcγRI. In a finding that has intriguing implications for phagocyte cell biology, IgG was found to rapidly decrease the avidity of FcγRI for ABP in whole cells or extracts thereof.

### FcγRIIA/C

Structure-function studies on FcγRs have mapped some important domains within the polypeptide chains. In the case of FcγRIIA three functions have been investigated in some detail: IgG binding, phagocytosis and signal transduction. A series of studies identified polymorphisms of FcγRIIA in human cells by virtue of differential binding of huIgG2 (or moIgG1). Isolation of FcγRIIA cDNAs from high and low responder cell lines identified a single amino acid difference that determines this polymorphism[9]. $Arg^{131}$ in the second extracellular domain results in significantly lower avidity for huIgG2 than $His^{131}$. An additional difference was noted between reactivity of these allelic forms with CD32 mAb 41H16, the former receptor is recognized while the latter is not[10,11]. It is curious that these avidity differences are reciprocal for moIgG1[11]. This polymorphism is of unknown clinical significance in human populations. It is noteworthy that cell types expressing FcγRIIA all have redundant FcγR expression and thus inability of FcγRIIA to bind a particular IgG subtype may not be overtly detrimental.

A hallmark function of FcγRs is phagocytosis. Initial assays of phagocytosis relied on the readily visualized ingestion of RBCs; an extension of rosetting assays that were used to detect FcγRs. To isolate the confounding effect of other FcγRs, cDNAs under the control of promiscuous promoters were transferred into fibroblasts or epithelial cell lines. FcγRIIA-dependent rosetting and phagocytosis was observed that could be reduced by more than 90% after preincubation with CD32 mAb and was shown to depend on actin polymerization[12]. In deletion studies removal of all but one of the 76 amino acid cytoplasmic domain decreased phagocytosis[13]. Deletions retaining as little as the 16 amino acids just inside the membrane were phagocytic and this level gradually increased in receptors retaining 28, 46 and 62 amino acids[13]. This analysis failed to generate a discrete localization of a "phagocytosis determinant" yet demonstrated that as few as 16 membrane proximal residues of the cytoplasmic tail are sufficient.

A putative activation motif has been identified in the cytoplasmic tail of FcγRIIA/C. This motif is believed to confer physical association with PTKs and the ability to trigger a tyrosine phosphorylation cascade on FcγRIIA crosslinking.

## STRUCTURE: FUNCTION RELATIONSHIPS OF Fcγ RIIB

Human FcγRIIB2 has been shown to mediate phagocytosis of IgG opsonized RBCs in epithelial cells. In parallel studies with those described above for huFcγRIIA, deletion analysis showed a dependence of this function on the cytoplasmic tail of FcγRIIB2. However, as little as the 13 membrane proximal amino acids of FcγRIIB2 were sufficient to observe phagocytosis[13]. In complete agreement with studies on FcγRIIA above, deletions of most of the cytoplasmic domain progressively decrease phagocytosis and subsequent phagosome maturation into lysosomes. The clearest result of these experiments is that the membrane proximal 13 or 16 residues of FcγRIIB or IIA respectively are sufficient to promote phagocytosis.

Studies on mouse FcγRIIB2 support a phagocytic function of this receptor. Immune complexes can be endocytosed through coated pits in murine FcγRIIB2 transfected fibroblasts. The 47 amino acid insertion in the cytoplasmic tail encoded by the alternatively spliced exon found in FcγRIIB1 abrogates immune complex concentration in coated pits. A cytoplasmic minus mutant of this receptor behaves like FcγRIIB1 in exhibiting inefficient, coated pit-independent internalization of immune complexes[14]. Subsequent studies reaffirmed these results in an FcγR-deficient mouse B lymphocyte cell line[15]. Moreover, only when this cell line was transfected with the endocytosis competent FcγRIIB2 was it transformed into an efficient antigen presenting cell for IgG bound antigen. Therefore, murine FcγRIIB2 is sufficient to confer the ability to endocytose immune complexes on fibroblasts and to mediate antigen delivery to processing compartments within B lymphocytes.

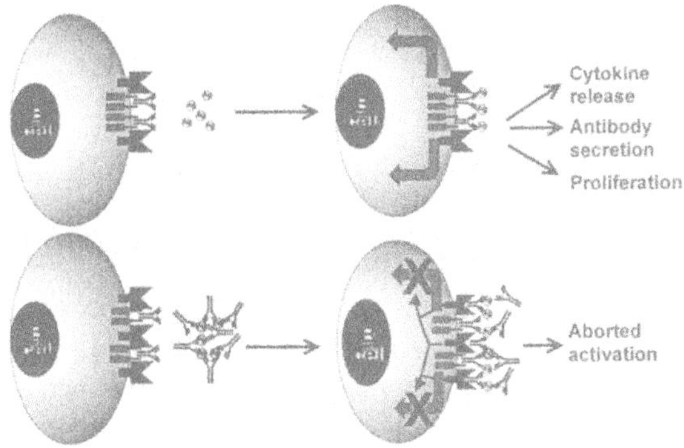


**Figure 4.** Repression of B lymphocyte activation by FcγRIIB1. B lymphocyte activation through the B cell receptor is depicted at top. Soluble antigen crosslinks the B cell receptor leading to transduction of signals that result in cell activation. Features of the activated B lymphocyte include cytokine and antibody release and proliferation. Bottom: IgG complexed antigen crosslinks FcγRIIB1 as well as the BCR.Signal transduction events such as calcium influx are inhibited and cell activation is abrogated.

As illustrated in Fig. 4, FcγRIIB modulates signalling through the B cell receptor (BCR) surface Ig complex. Crosslinking the BCR increases inositol phosphates and free calcium as well as activation of phosphorylation cascades which result in B cell differentiation and proliferation[16]. However, if FcγRII is crosslinked to the BCR an attenuation of calcium influx across the plasma membrane is observed[17]. In this manner, coengagement of FcγRIIB mediates a dominant inhibitory signal that abrogates B cell activation by the BCR[18]. Calcium influx seems to be selectively inhibited as patterns of protein phosphorylation after BCR activation are unchanged[17]. The region of muFcγRIIB responsible for this ability to modulate signal transduction by the BCR has been mapped. Initial studies determined that a 13 amino acid stretch, of which 12 residues are conserved between human and murine receptors, is necessary and sufficient to mediate this function[17]. Subsequent studies demonstrated phosphorylation of a specific tyrosine (#309 in the murine receptor) in this 13 amino acid region upon BCR activation[17]. Investigation of the contribution of this residue, and a 65kDa protein that coprecipitates with FcγRIIB1, are in progress. These studies provide insight into the function of FcγRIIB and may inform the mechanisms by which IgG or anti-idiotype Abs induce tolerance or modify the B cell repertoire.

## STRUCTURE-FUNCTION RELATIONSHIPS FOR FcγRIII[24]

The structural heterogeneity of Fc receptors for IgG, in which the IgG immune complex interacts with a diverse array of related receptors, and the overlapping pattern of FcR expression on effector cells have precluded detailed determination of the specific functions of individual receptors in mediating effector responses in vivo. We have created a mouse strain genetically deficient in the γ subunit by homologous recombination in ES cells in order to determine the role os FcγRIII and FcεRI in effector responses to IgG and IgE, respectively. Selective ablation of this chain has indeed resulted in the loss of these receptors on NK cells, macrophages, and mast cells. The functional deficit, however, is more pronounced than would have been predicted from in vitro reconstitution studies alone. Inflammatory macrophages are unable to mediate phagocytosis through

FcγRI, FcγRII, or FcγRIII, indicating an unexpected pleiotropic role of this subunit. These mice thus provide the first clear mutations with which we can determine the distinct roles of these structurally similar receptors in mediating effector responses in host defense.

## PHAGOCYTIC ACTIVITY IS ABSENT IN γ-DEFICIENT MICE

FcγRIII, in common with FcγRI and II, mediated ADCC, phagocytosis, release of inflammatory mediators, and degranulation when cross-linked with antigen-antibody complexes. Since the deletion of the γ chain resulted in the marked reduction or total loss of FcγRIII, functional characterization of the macrophage, NK, and mast cells from these γ-deficient animals were studied to determine the contribution of FcγRIII to these physiological responses to IgG immune complexes.

FcR-mediated phagocytosis was assessed by the ability of TEMs to internalize sheep red blood cells (SRBCs) opsonized with IgG. IgG2a-opsonized RBCs are bound and internalized preferentially by the high-affinity FcγRI, while IgG1 and 2b are only bound and internalized by the low-affinity receptors[24]. Macrophages from wild-type (+/+) mice displayed robust binding of IgG-opsonized SRBCs of G1 and 2a subclasses. Heterozygous mice displayed identical binding, and IgG2b binding was comparable to IgG1 in all cases. These bound opsonized particles were efficiently internalized. Macrophages from homozygous mutant (-/-) mice bind IgG1-opsonized SRBCs, owing to retained expression of FcγRII. This binding is completely blocked by Mab 2.4G2. These same macrophages failed to demonstrate IgG2a binding, indicating an unexpected loss of FcγRI binding activity. This is not the result of loss of FcγRI α chain expression,

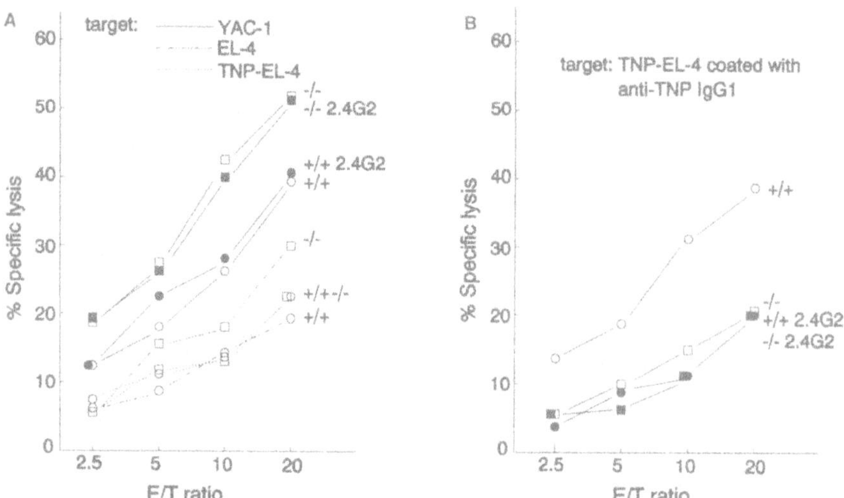


**Figure 5.** Natural Killing and ADCC activities of Splenic NK Cells from Wild-type and $FcR\gamma^{nl}$ Mice. Splenic NK cells generated after in vitro culture in the presence of exogenous IL-2 were tested for lytic activity against a set of targets (1) including the NK-sensitive target YAC-1, the thymoma EL-4, and the TNP-derivatized EL-4. (B) ADCC activities were tested on the following targets: TNP-derivatized EL-4 cells coated with an anti-TNP IgG1 antibody, or TNP-derivatized, anti-TNP-coated EL-4 cells in the presence of the anti-FcγRII/III antibody 2.4G2.

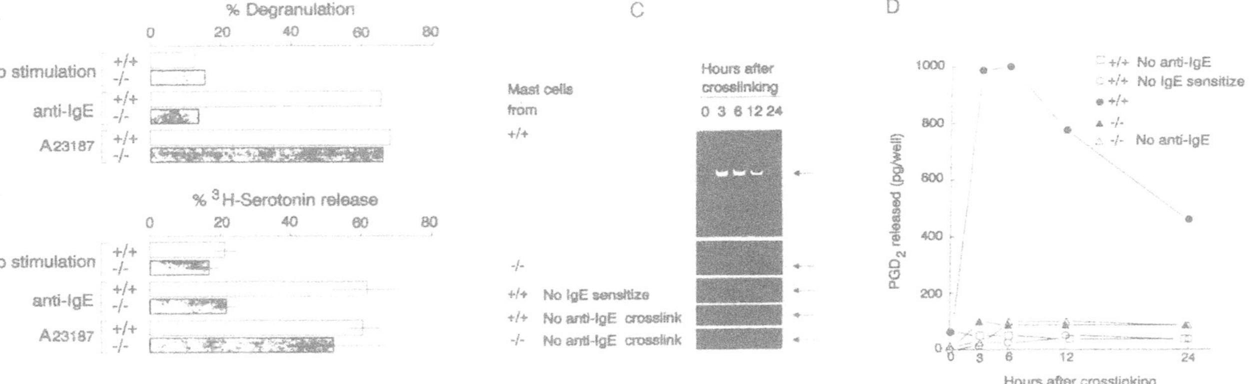


**Figure 6.** Functional Characterization of Mast Cells from Wild-Type and FcRγ$^{nl}$ Mice. (A) Mast cell degranulation. Bone marrow-derived mast cells were presensitized with monoclonal murine IgE and subsequently degranulated by cross-linking with murine anti-IgE antibody. Maximal degranulation was determined using ionophore A23187. (B) Mast cell serotonin release. Bone marrow-derived mast cells were preincubated with [$^{3}$H]-serotonin and then incubated with monoclonal murine IgE. [$^{3}$H]-serotonin released into the supernatant was quantitated after exposure to monoclonal anti-IgE or to ionophore A23187. (C) RT-PCR analysis of IL-4 mRNA after cross-linking of IgE on mast cells. Mast cells were prepared and sensitized as above, and cells were collected and subjected to RT-PCR analysis for IL-4 mRNA. (D) Prostaglandin D2 release from mast cells are cross-linking of IgE. Bone marrow-derived mast cells were sensitized overnight with mouse monoclonal IgE. The cells were stimulated as above, and culture supernatants were collected at various time intervals. Amount of prostaglandin D2 released in the supernatant was determined by radioimmunoassay.

since these macrophages express equivalent levels of mRNA for this chain when compared with wild-type mice. Rather, it indicates a functional dependence of the high-affinity FcγRI for the γ chain either for facilitating surface expression or ligand binding for this receptor. Despite the efficient binding of SRBCs to -/- macrophages through FcγRII, these cells fail to internalize such opsonized particles, indicating a more global defect in FcR-mediated phagocytosis. Deletion of γ chain thus has a pleiotropic effect on macrophage FcR-mediated ligand binding and phagocytosis beyond what might be expected from the loss of FcγRIII.

### ADCC Activity, but Not Natural Killing, in NK Cells Was Severely Abrogated in FcRγ$^{nl}$ Mice

Natural killing activity, as measured against the YAC-1 tumor target, was normal for wild-type and mutant NK cells either freshly isolated from the spleen or purified over glass wool and cultured in IL-2 for 7 days (Figure 5A). The EL-4 target was less sensitive to NK lysis in both wild-type and mutant cell preparations. Deletion of γ chain thus has no effect on natural killing of tumor targets, consistent with the evidence that receptors other than FcγRIII mediate this process. ADCC activity against TNP-derivatized and anti-TNP IgG1-coated EL-4 target, however, was markedly diminished in NK cells from γ-deficient mice, while wild-type NK cells showed clear ADCC activity against the same target cells; this ADCC activity was completely blocked by the Mab2.4G2 (Figure 5B). These results indicate that ADCC activity is almost totally lost in NK cells in these mutant mice, owing to loss of functional FcγRIII expression.

### Characterization of Mast Cell Functions

Mast cells were first sensitized with monoclonal mouse IgE and then triggered with monoclonal anti-mouse IgE antibody. Degranulation, [$^{3}$H]-serotonin release, Il-4 production, and prostaglandin D2 release of these cells were measured for wild-type as well as γ-deficient mice (Figure 6). As expected, degranulation and serotonin release from γ-deficient mast cells were only negligible above background, while +/+ mast cells responded well to IgE crosslinking (Figures 6A and 6B). Similarly, the amount of an IL-4 specific 267 bp cDNA fragment was markedly increased after stimulation of mast cells from wild-type mice (Figure 6C). In contrast, mast cells from γ-/-deficient mice showed apparently no response to cross-linking stimulation. Finally, prostaglandin D2 release into the culture supernatant was significantly reduced in mutant mast cells, as compared with wild-type cells (Figure 6D). Thus, by several criteria, assessing early and late activation responses of mast cells to IgE cross-linking, the mutant mice fail to respond.

### IgE-Mediated Anaphylaxis is Absent in FcRγ$^{nl}$ Mice

IgE-mediated anaphylaxis is thought to be dependent upon FcεRI triggering of mast cells, although the role of other cell types and other IgE binding molecules to this in vivo response is debated. To address this question, we challenged mutant and wild-type mice in a passive cutaneous anaphylaxis (PCA) model. The characteristic increase in vascular permeability triggered by mast cell degranulation is readily visualized by Evans blue extravasation. Wild-type and heterozygous mice mount a prompt PCA reaction in response to IgE cross-linking. In contrast, homozygous deficient animals are unable to mount a PCA reaction.

In marked contrast with the defect in innate immunity resulting from FcR perturbation, T cell development appears to be grossly normal. The same populations of cells were observed in thymocytes and peripheral T cells from 2-week- or 10-week-old mice in the γ-deficient animals and their wild-type littermates. This chain, while found associated with TCR-CD3 in some populations of T cells, is clearly not critical for normal T cell development, in contrast with what has been reported in ζ-deficient animals. This preliminary analysis of T cell populations in these mutant mice further suggests that the dominant $Fc\gamma RIII^+$ population of early thymocytes (days 14.5 through 16.5) is not essential for progression through the normal T cell developmental program. Whether specific defects in T cell populations that appear to use the γ chain exclusively are found in these animals remains to be determined.

These mice therefore will enable studies to be performed evaluating the contribution of IgG- and IgE-triggered effector responses to a variety of pathogens and will aid in further defining the role of this subunit in both T cell and effector cell pathways.

## REFERENCES


1. J.V. Ravetch and J.-P. Kinet, *A. Rev. Immunol.* 9:457 (1991).
2. A.F. Williams and A.N. Barclay, *A. Rev. Immunol.* 6:381 (1988).
3. J.M. Allen and B. Seed, *Science* 243:378 (1989).
4. M.A. Chaikin, R.W. Sweet, D.R. Sylvester, and D.J. Bergsma, Gene 104:285 (1991).
5. L.K. Ernst, J.G.J. van de Winkel, I.-M. Chiu, and C.L. Anderson, *J. Biol. Chem.* 267:15692 (1992).
6. A.J. Porges, et al., *J. Clin. Invest.* 90:2102 (1992).
7. J.G. J. van de Winkel, L.K. Ernst, C.L. Anderson, and I.-M. Chiu,, *J. Biol. Chem.* 266:13449 (1991).
8. E. Dietzsch, N. Osman, I.F. McKenzie, O.M. Garson, and P.M. Hogarth, *Immunogenetics* 38:307 (1993).
9. M.R. Clark, S.B. Clarkson, P.A. Ory, N. Stollman, and I.M. Goldstein, *J. Immun.* 143:1731 (1989).
10. P.A. Warmerdam, J.G. van de Winkel, A. Vlug, N.A. Westerdaal, and P.J. Capel, *J. Immun.* 147:1338 (1991).
11. B.J. Tate, E. Witort, I.F. McKenzie, and P.M. Hogarth, *Immunol. & Cell Bio.* 70:79 (1992).
12. W.B. Tuijnman, *Blood* 79:1651 (1992).
13. W. Engelhardt, *Eur. J. Immunol.* 21:2227 (1991).
14. H.M. Miettinen, *Cell* 58:317 (1989).
15. S. Amigorena, *Science* 256:1808 (1992).
16. J.C. Cambier, *Immunol. Rev.* 95:37 (1987).
17. T. Muta, T. Kurosaki, M. Nussenzweig, and J. V. Ravetch, *Nature* 368:70 (1994).
18. N.E. Phillips, and D.C. Parker, *J. Immunol.* 132:627 (1984).
19. P.M. Hogarth, M.D. Hulett, and N. Osman, *Immunol. Res.* 11:217 (1992).
20. A. Weiss, and D.R. Littman, *Cell* 76:263 (1994).
21. L.K. Ernst, A.-M. Duchemin, and C.L. Anderson, *Proc. Natl. Acad. Sci. USA* 90:6023 (1993).
22. Y. Ohta, T.P. Stossel, and J.H. Hartwig, *Cell* 67:275 (1991).
23. R.L. Weinshank, A.D. Luster and J. V. Ravetch, *J. Exp. Med.* 163:1909 (1988).
24. T. Takai, et al.,*Cell* 76:519 (1994).

# PROXIMAL SIGNALS AND THE CONTROL OF S-PHASE ENTRY IN INTERLEUKIN-2-STIMULATED T LYMPHOCYTES

Robert T. Abraham, Larry M. Karnitz, Leigh Ann Burns, and Gregory J. Brunn

Departments of Immunology and Pharmacology
Mayo Clinic/Foundation
Rochester, MN 55905

## INTRODUCTION

Stimulation of resting, $G_0$-phase T lymphocytes with antigenic peptides presented in the context of self-MHC triggers a pleiotropic activation program that culminates in cell-cycle entry and the expression of high-affinity receptors for T-cell-derived growth factors. The principal growth and differentiation factor for antigen-activated T lymphocytes is the T-cell-derived cytokine, interleukin-2 (IL-2). Binding of IL-2 to the high-affinity IL-2 receptor (IL-2R) drives the progression of activated, $G_1$-phase T cells into S-phase and, ultimately, mitosis. The intracellular pathways through which IL-2R occupancy provokes this cell-cycle progression response remains an area of intense interest in the field of T-cell biology.

The high-affinity ($K_d$ = $10^{-11}$ M) IL-2R is multi-subunit complex comprised of α, β, and γ polypeptides. The α chain alone binds IL-2 with low affinity ($K_d$ = $10^{-7}$ M), and appears incapable of transmitting IL-2-mediated regulatory signals. In contrast, the βγ heterodimer displays an intermediate ligand-binding affinity ($K_d$ = $10^{-9}$ M), and apparently constitutes the critical signal-transducing component of the high-affinity receptor complex. On the basis of conserved structural motifs in their extracellular domains, both the β and γ subunits have been grouped into the newly-defined superfamily of cytokine receptors[1]. Like the other members of this receptor superfamily, the intracytoplasmic domains of the β and γ subunits contain no recognizable catalytic motifs suggestive of direct roles in second messenger generation. In particular, the cytoplasmic tails of the β and γ subunits do not function as ligand-activated protein tyrosine kinases (PTKs) unlike a number of other growth factor receptors. Nonetheless, IL-2R occupancy provokes a rapid increase in the tyrosine phosphorylation of several intracellular proteins in activated T lymphocytes. The latter observation strongly suggests that the ligand-bound IL-2R regulates the activity of one or more cytoplasmic PTKs in T lymphocytes. The most often-cited candidates are two members of the *src* PTK family, Lck and Fyn. Indeed, evidence from several

*Mechanisms of Lymphocyte Activation and Immune Regulation V*
Edited by S. Gupta *et al.*, Plenum Press, New York, 1994

laboratories suggests that both Lck and Fyn are physically and functionally coupled to the IL-2R via the cytoplasmic domain of the β-subunit. Whether these *src*-related PTKs subserve distinct as well as overlapping functions during signal transduction through the IL-2R has not been elucidated.

The proximal signals initiated by IL-2R ligation ultimately converge on the biochemical machinery that controls the cell-cycle clock in T lymphocytes. The decision to replicate DNA commits the IL-2-stimulated T cell to execute the remaining phases of the cell cycle, even in the absence of continued growth factor exposure[2]. Clearly, if cell-cycle progression is to occur in an orderly and appropriate fashion, the process of S-phase commitment must be tightly regulated. This intrinsic mechanism of cell-cycle regulation, the general workings of which have been highly conserved throughout eukaryotic evolution, consists of the $G_1$ cyclins,the associated cyclin-dependent kinases (cdks), and a series of regulatory phosphatases and kinases. Although many of the early intracellular responses triggered by IL-2 appear relatively transient, these events provoke a series of delayed responses that orchestrate the sequential activation of distinct complexes of cyclins and cdks[3,4]. The phosphorylation of specific substrates by these cyclin-cdk complexes presumably sets in motion the cellular responses required for competence to enter S phase. Based on the available data, a plausible model suggests that growth factor stimulation induces the expression of genes encoding the D-type cyclins (T cells express cyclins $D_2$ and $D_3$) in early to mid- $G_1$-phase[4]. The newly-expressed cyclin D's associate stoichiometrically with specific cdk partners (cdk2, cdk4, and cdk5) to generate active complexes, which presumably catalyze phosphorylation events required for cell-cycle progression into late $G_1$-phase. The subsequent appearance of active cyclin E-cdk2 complexes, in turn, drives the IL-2-stimulated T cell toward and possibly through the $G_1$-S boundary. A complete understanding of the molecular basis of T-cell growth regulation will clearly require a thorough analysis of the interface between proximal tyrosine phosphorylation signals and the cyclin-cdk machinery responsible for orchestrating $G_1$-phase transit and S-phase commitment in IL-2-stimulated T lymphocytes.

This chapter will summarize recent studies from our laboratory concerning the mechanisms underlying both proximal signal initiation by the ligand-bound IL-2R, and the regulation of T-cell entry into S phase by IL-2. The former discussion will focus on the roles of *src*-family PTKs in coupling the IL-2R to the phosphatidylinositol 3-kinase (PI 3-K)- and Ras-dependent signaling cascades in T cells. Subsequently, we will examine more distal events related to the fundamental process of S-phase commitment induced by IL-2 stimulation. These studies will focus on the potent immunosuppressant, rapamycin, a novel pharmacologic probe of the cell-cycle control machinery engaged by IL-2-dependent regulatory signals in $G_1$-phase T cells.

## ROLE OF FYN IN IL-2-INDUCED PI 3-K ACTIVATION

PI 3-K is a heterodimeric enzyme containing an 85 kD regulatory subunit and a 110 kD catalytic subunit. This enzyme catalyzes the phosphorylation of the D-3 hydroxyl groups of the *myo*-inositol rings of phosphatidylinositol (PI), PI-4-monophosphate, and PI-4,5-bisphosphate. Unlike the parent phosphoinositides, the resulting D-3 phosphorylated metabolites do not serve as precursors in the canonical phospholipase C-dependent second messenger pathway. The actual role of PI 3-K and its metabolic products in cell signaling remains obscure. Nonetheless, a substantial body of data implicates this enzyme as an essential downstream component of the mitogenic signaling cascades engaged by numerous growth factor receptors, and in

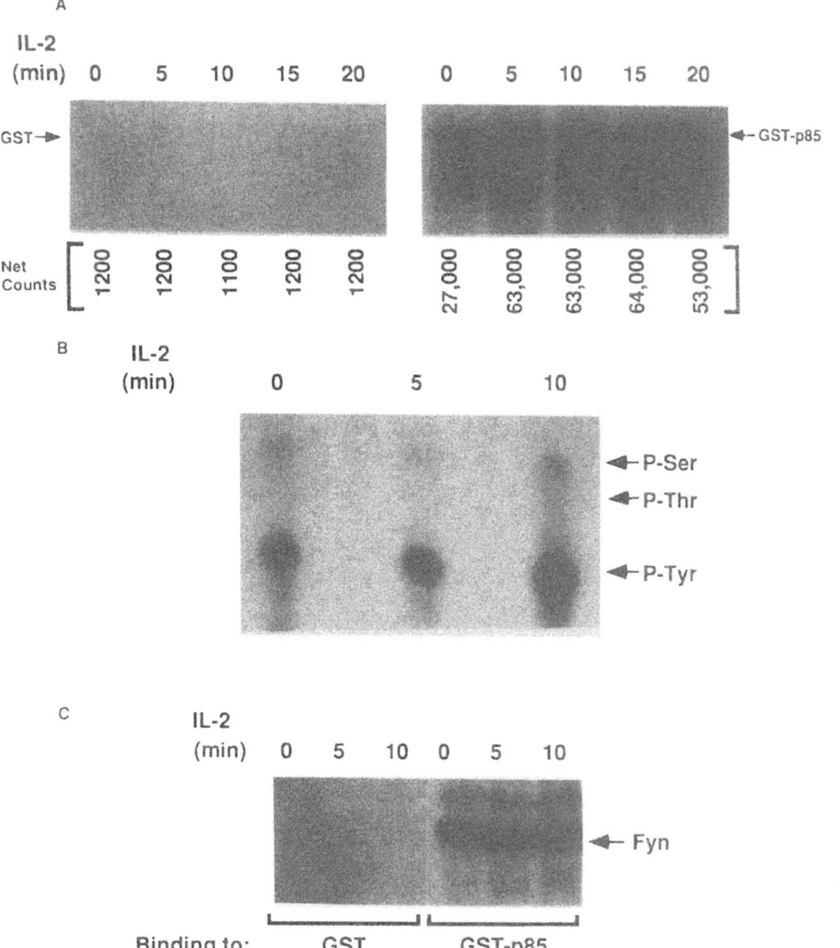


**Figure 1. Association of an IL-2-inducible PTK activity with recombinant p85.** (A) Factor-deprived CTLL-2 cells were stimulated with IL-2 for the indicated times. The cells were lysed, and detergent-soluble proteins were incubated with GSH-agarose bound GST only or GST-p85 (amino acids 158-274). The precipitates were washed and phosphorylation reactions were performed in the presence of [$\gamma$-$^{32}$P]ATP. Precipitated proteins were separated by SDS-PAGE and radioactivity (net counts) incorporated into the GST-p85 band was quantitated over a 30 minute data acquisition period with an Ambis imaging system. (B) Phosphoaminoacid analysis of in vitro-phosphorylated GST-p85. (C) Association of Fyn with GST-p85. Protein precipitates prepared as indicated in panel A were disrupted with SDS, and the solubilized proteins were reprecipitated with Fyn-specific antiserum. The $^{32}$P-labeled Fyn band is indicated with the arrow.

cellular transformation by certain oncogene products, including the polyomavirus middle-sized tumor antigen (MTAg)[5,6].

Previous studies have shown that IL-2 stimulation induces the activation of PI 3-K in the IL-2-dependent cytotoxic T cell line, CTLL-2[7]. The appearance of increased levels of PI 3-K activity in anti-phosphotyrosine (anti-pY) antibody immunoprecipitates from these cells suggested that this inducible activity reflected the tyrosine phosphorylation of either PI 3-K itself or a tightly associated protein. In more recent work, we have observed that IL-2 exposure induces a time-dependent increase in the tyrosine phosphorylation of the p85 subunit of PI 3-K. The IL-2-dependent increase in p85 phosphorylation paralleled the increase in anti-pY antibody-bound phospholipid kinase activity. The first evidence supporting a functional role of the *src*-related PTK, Fyn, in this response was provided by the finding that anti-Fyn antibodies co-precipitated substantial amounts of PI 3-K activity from detergent extracts of CTLL-2 cells[7]. The association of Fyn with PI 3-K activity was not dependent on cellular stimulation with IL-2. Moreover, transfection of CTLL-2 cells with a polyomavirus expression plasmid yielded stable sub-clones that displayed constitutive elevations in Fyn (but not Lck) kinase activity, accompanied by 3-6-fold increases in the levels of anti-pY antibody-bound PI 3-K activity[8]. Collectively, these results were consistent with the notion that the linkage between IL-2R occupancy and PI 3-K activation might be mediated through the Fyn tyrosine kinase.

To more fully understand the roles of Fyn and p85 in IL-2-induced PI 3-K activation, we generated a series of plasmid constructs containing Fyn or p85 coding sequences fused to glutathione S-transferase (GST). After expression in *Escherichia* coli, the GST fusion proteins were purified by chromatography over glutathione (GSH)-agarose. In initial studies, factor-deprived CTLL-2 cells were stimulated for various times with IL-2, the cells were lysed in detergent-containing buffer, and extracts were incubated with GSH-agarose beads coupled to GST alone or GST-p85. The resulting precipitates were washed and assayed for associated protein kinase activity(s) by incubating with $Mg^{2+}$-[$\gamma$-$^{32}$P]ATP. IL-2 stimulation induced a rapid increase (2- to 4-fold in repeat experiments) in the phosphorylation of the GST-p85 fusion protein (Fig. 1A). In contrast, GST alone precipitated virtually no detectable protein kinase activity from parallel cell lysates (data not shown). Phosphoamino acid analysis demonstrated that GST-p85 was phosphorylated primarily on tyrosyl residues, indicating that an IL-2-regulated PTK was bound to this fusion protein (Fig. 1B).

To determine whether Fyn was present in the GST-p85 precipitates, protein complexes were disrupted by heating in the presence of SDS, and the solubilized proteins were re-precipitated with Fyn-specific antibodies. Radiolabeled Fyn was readily detected in GST-p85 precipitates from both unstimulated and IL-2-stimulated T cells (Fig. 1C), which provides indirect support for the model that the tyrosine phosphorylation of p85 in intact CTLL-2 cells is mediated, at least in part, through an increase in the specific activity of the constitutively-bound Fyn kinase. Obviously, this model predicts that IL-2R occupancy leads to activation of intracellular Fyn. This issue was addressed by transfection of a previously described, Lck-negative CTLL-2 cell line[8] with a plasmid expression vector containing full-length Fyn bearing a C-terminal epitope tag recognized by the KT3 monoclonal antibody[9]. CTLL-2 cell sub-clones expressing the epitope-tagged Fyn construct were deprived of growth factors for 12 h, and then were restimulated with IL-2 for various times. Detergent-soluble proteins were immunoprecipitated with KT3 antibodies, and the immunoprecipates were assayed for protein kinase activities. These studies documented that IL-2 stimulation induced a reproducible, albeit relatively low-level, increase in the catalytic activity of KT3-tagged Fyn in CTLL-2 cells (Fig. 2). The rapid time course of Fyn activation induced by IL-2 was entirely consistent with the postulated role of this *src*-family kinase in the IL-2-dependent phosphorylation of PI 3-K.

The physical basis of the Fyn - PI 3-K interaction was examined in greater detail using isolated Fyn sub-domains as potential ligands for PI 3-K present in CTLL-2 cell lysates. GST fusion proteins containing Fyn sequences corresponding to the unique amino terminus (NH), the *src*-homology 3(SH3) domain, the SH2 domain, or the tandem NH-SH3-SH2 regions were expressed in *E.* coli (Fig. 3A), and the purified fusion proteins were immobilized on GSH-agarose beads. Lysates from factor-deprived

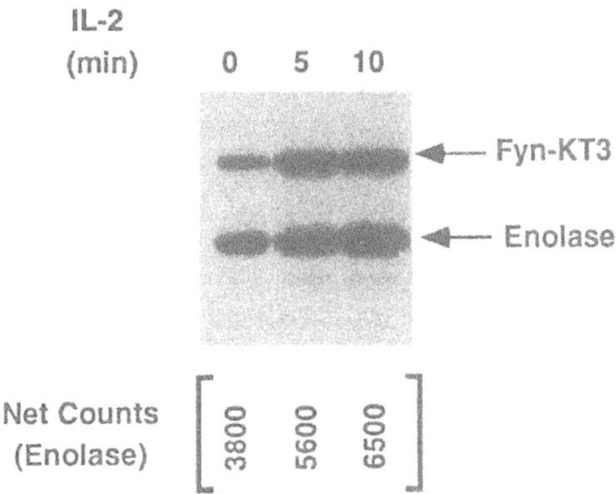


**Figure 2. IL-2-induced activation of epitope-tagged Fyn.** CTLL-2 cell transfectants expressing full-length Fyn bearing a C-terminal KT3-epitope tag were stimulated for the indicated times with IL-2. KT3 antibody immunoprecipitates were phosphorylated in vitro in the presence of [$\gamma$-$^{32}$]ATP and the exogenous PTK substrate, acid-denatured enolase. Radioactivity (net counts) incorporated into the enolase band was quantitated over a 30 min data acquisition period with an Ambis imaging system.

or IL-2-stimulated CTLL-2 cells were precipitated with the fusion protein-coupled beads, and the bound proteins were assayed for PI 3-K activity. Both the SH2 and the SH3 domains of Fyn independently precipitated significant amounts of PI 3-K activity from the CTLL-2 cell extracts (Fig. 3B). The PI 3-K binding activities of the isolated Fyn SH2 and SH3 domains were approximately additive, based on the results obtained with the construct containing the Fyn amino terminus through SH2 domain. In contrast, a similar fusion protein containing the amino terminus through the SH2 domain of Lck consistently precipitated much lower levels of PI 3-K activity from CTLL-2 cell extracts, which suggests that PI 3-K interacts preferentially with Fyn in this experimental system. Immunoblot analyses subsequently revealed that the level of PI 3-K precipitated by the various fusion proteins was well-correlated with the amount of p85 present in these precipitates.

The above results prompted speculation that the interaction of Fyn with PI 3-K might be mediated through the p85 subunit. Whereas the SH2 domain of Fyn recognizes specific target sequences containing phosphotyrosine, recent studies have shown that SH3 domains bind with high-affinity to distinct, proline-rich motifs[10]. Examination of the p85 coding sequence revealed two putative SH3 domain- binding regions located at amino acid residues 84-96 (PPTPKPRPPRPLP) and 303-314 (APALPPKPPKP). To determine whether Fyn was capable of interacting directly with p85 via its SH3 or SH2 domains, the Fyn fusion proteins described above were coupled covalently to cyanogen bromide-activated Sepharose beads. The immobilized Fyn SH3 domain, and to a lesser extent the SH2 domain, specifically precipitated recombinant p85 purified from bacteria (results not shown). Thus, these results suggest that the SH3 domain of Fyn plays an important role in the recruitment of a putative substrate, the PI 3-K heterodimer, to the plasma membrane, and, possibly, into the signal-transducing complexes induced by IL-2R receptor ligation. The IL-2-dependent increase in PI 3-K activity reflects, at least in part, the tyrosine phosphorylation of p85, although we cannot rule out an allosteric effect of the Fyn SH3 domain on this enzyme. In a broader context, these results suggest that the ligand-binding specificities of the Fyn and Lck SH3 domains may play major roles in determining the spectrum of downstream signal-transducing molecules recruited by the IL-2R and other surface receptors expressed by T lymphocytes.

In addition to PI 3-K activation, IL-2R occupancy stimulates a PTK-dependent increase in the level of GTP-bound Ras proteins in T cells[11]. The Ras proteins are members of a growing family of GTPases which regulate a variety of cellular functions ranging from protein synthesis and transport to growth and differentiation. Ras proteins switch between "on" and "off" states by the replacement of GDP with GTP, and by the hydrolysis of the bound GTP to GDP, respectively. The rapid accumulation of GTP-bound Ras in IL-2-stimulated T cells is, therefore, thought to reflect the activation of Ras proteins in these cells. Ras proteins appear to function as molecular switches, relaying signals from receptor-activated PTKs in the plasma membrane to cytoplasmic serine-threonine kinases. Recent studies have outlined a

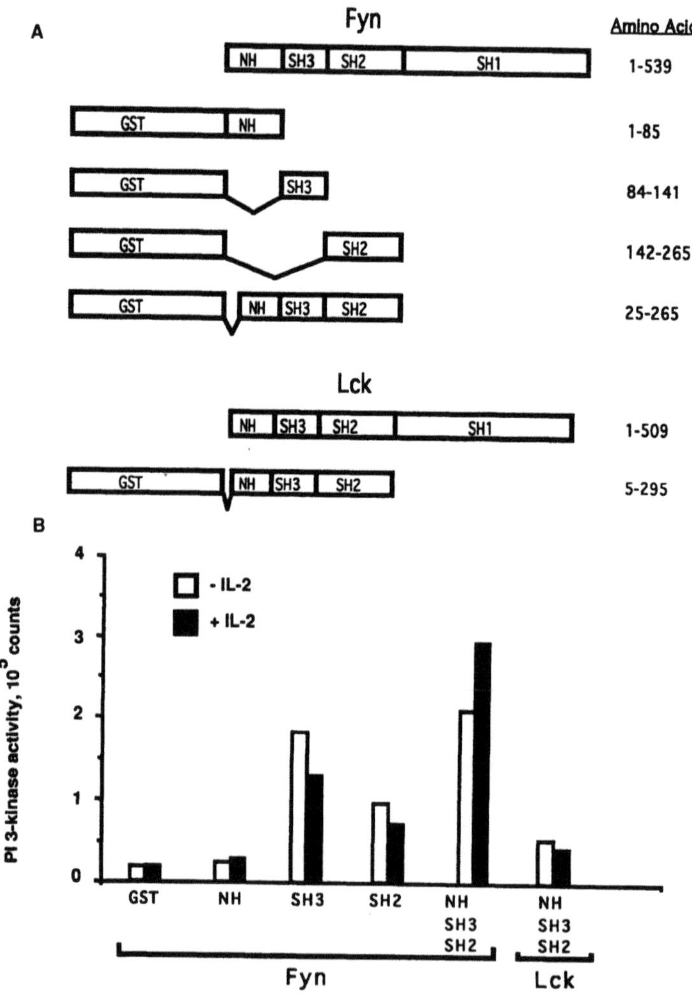


**Figure 3. Binding of PI 3-K activity to GST-Fyn fusion proteins.** (A) Schematic of GST-Fyn fusion proteins. (B) IL-2-deprived CTLL-2 cells were stimulated for 15 minutes with medium only or with IL-2. The cells were lysed, and anti-pY antibody-bound PI 3-K activities were assayed as described in reference 7.

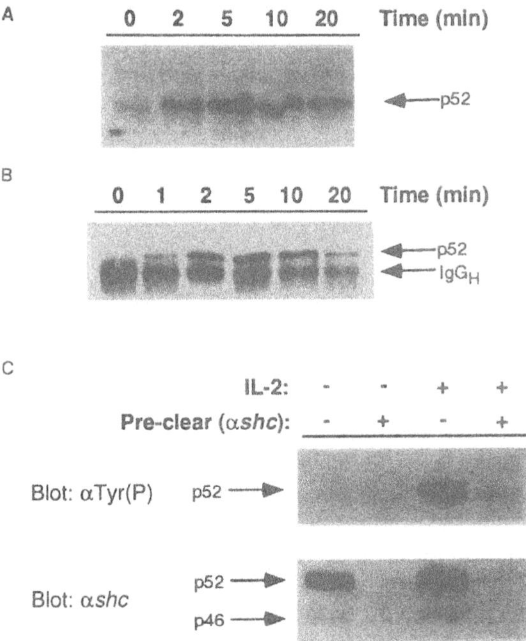


**Figure 4. IL-2-induced tyrosine phosphorylation of p52$^{shc}$.** CTLL-2 cells were cultured in basal medium for 4 h prior to stimulation with 200 units/ml IL-2 for the indicated times. (A) Anti-pY immunoblot of detergent-soluble cellular proteins. (B) Anti-pY immunoblot of anti-Shc immunoprecipitates. (C) Detergent-soluble lysates depleted of Shc proteins by immunoprecipitation with anti(α)-Shc antibodies show a concomitant depletion of the 52-kDa phosphoprotein. p46$^{shc}$, p52$^{shc}$, and IgG heavy chain (IgG$_H$) are indicated by *arrows*. Reproduced from reference 18 with the permission of the publisher.

trans-cytosolic signaling cascade from active, plasma membrane-bound Ras to the MAP/ERK kinases, and, ultimately, to the AP-1 transcription factor in the nucleus[12,13]. In spite of the intense interest in the Ras proteins as mitogenic signal transducers, surprisingly little information is available regarding either the mechanism of Ras activation by IL-2, or the function of the Ras signaling cascade in IL-2-dependent T-cell growth.

In earlier studies, we and others observed that IL-2 stimulation provoked a prominent increase in the tyrosine phosphorylation of a 50-55-kDa cytosolic protein in T lymphocytes[14,16]. The apparent molecular mass of this phosphoprotein was similar to that of the 52-kDa isoform of Shc, a known substrate for the epidermal growth factor receptor PTK[17]. Like the p85 subunit of PI 3-K, Shc is an SH2 domain-containing "adaptor" protein that functions to link pY-containing proteins, including activated PTKs, to downstream signaling enzymes (see below). To determine the relationship between the IL-2-inducible 50-55-kDa phosphoprotein and Shc, anti-Shc immunoprecipitates from unstimulated or IL-2-stimulated CTLL-2 cells were immunoblotted with anti-pY antibodies (Fig. 4B). Exposure to IL-2 resulted in a rapid increase in the tyrosine phosphorylation of Shc. The electrophoretic mobility of this Shc isoform was identical to that of the previously reported 50-55-kDa pY-containing protein (Fig. 4A). To confirm that the 50-55-kDa phosphoprotein was, in fact, Shc,

CTLL-2 cell extracts were immunodepleted with anti-Shc antibodies prior to SDS-PAGE and anti-pY antibody immunoblot analysis. The pre-cleared extracts displayed concomitant and specific reductions in both the 50-55-kDa protein and the Shc isoform, strongly suggesting that these proteins were identical. It is provocative that a similar 50-55-kDa protein has been shown to undergo tyrosine phosphorylation in various hematopoietic cell lines stimulated with IL-3, IL-5, GM-CSF, and erythropoietin (see reference 18 for discussion). Taken together with the present results, these observations provoke speculation that Shc phosphorylation is a widespread response to cytokine receptor stimulation.

Although the actual role of Shc phosphorylation in IL-2R signaling has not been defined, studies in other cell types suggest that Shc might participate in coupling the IL-2R to the Ras signaling pathway[19]. In the case of transmembrane receptors with intrinsic PTK activity, such as the EGF receptor, ligand binding triggers receptor dimerization and autophosphorylation. The autophosphorylation reaction creates binding sites for SH2 domain-containing cytoplasmic proteins, including the adaptor protein, Grb-2, which displays a single SH2 domain flanked by 2 SH3 domains[20]. The SH3 domains of Grb2 bind specifically to proline-rich sequences in the C-terminus of the Ras-guanine nucleotide-exchange factor, Sos1. The interaction of the autophosphorylated receptor with the Grb2 SH2 domain serves to deliver Sos1 to the plasma membrane, thereby approximating the exchange factor to Ras. In contrast to the EGF receptor, the IL-2R lacks an intrinsic PTK domain, and, therefore, apparently couples to the Ras pathway via non-receptor PTKs. The Shc adaptor protein may provide an alternative mechanism for linkage of the IL-2R and other cytokine receptors to Sos1, and, in turn, Ras. Recent studies have shown that fibroblasts transformed with v-*src* exhibit constitutive tyrosine phosphorylation of Shc, and association of this protein with the Grb2-Sos1 complex[20]. Similarly, IL-2R-mediated Lck or Fyn activation might lead to Shc phosphorylation and translocation of the Shc-Grb2-Sos1 complex to the T-cell plasma membrane. Supporting evidence for this model is provided by our recent observation that IL-2R occupancy triggers the association of the Grb2-Sos1 complex with tyrosine-phosphorylated Shc in CTLL-2 cells (Fig. 5).

The studies described above have yielded further insights into the mechanism whereby the proximal increase in PTK activity elicited by IL-2R occupancy couples the receptor to two downstream signal-transducers, PI 3-K and Ras. The identity of the

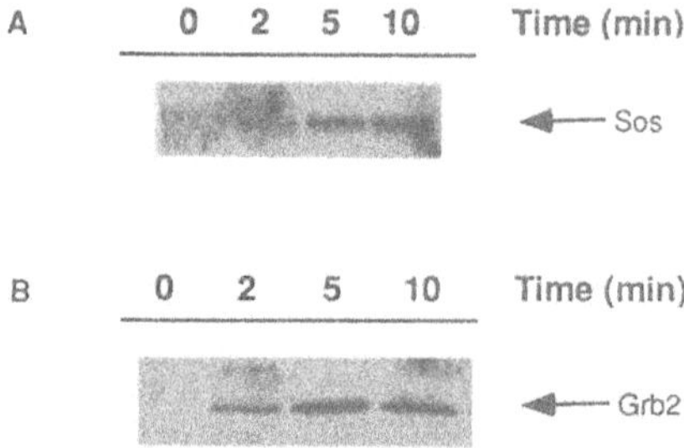


**Figure 5. IL-2-induced binding of Grb2-Sos complex to Shc.** Factor-deprived CTLL-2 cells were stimulated with IL-2 for the indicated times. Detergent-soluble proteins were immunoprecipitated with anti-She antibodies, the immunoprecipitates were subjected to SDS-PAGE, and the resolved proteins were immunoblotted with Sos *(upper panel)*- or Grb2 *(lower panel)*-specific antibodies.

PTK(s) responsible for linking the IL-2R to these signaling enzymes has not been definitively established. However, accumulating evidence suggests that the *src*-family members, Lck and Fyn, play partially or perhaps completely overlapping roles in the regulation of both PI 3-K and Ras activation. Whether other PTKs expressed in T cells, such as ZAP-70 or members of the JAK family, participate in this transduction pathway remains unclear. An even more salient question relates to the roles of PI 3-K and Ras in the regulation of IL-2-dependent T-cell mitogenesis. The signaling pathway triggered by the formation of D-3-phosphorylated phosphoinositides in growth factor-stimulated mammalian cells has thus far escaped definition. In contrast, compelling genetic data indicate that the Ras-to-MAP kinase cascade plays an obligatory role in the transmission of mitogenic signals from growth factor receptors expressed by fibroblastic cells. Nonetheless, the evidence that IL-2-dependent Ras activation is required for cell-cycle progression in T lymphocytes remains largely inferential. Clearly, much additional work will be required to understand the interface between the initial tyrosine phosphorylation signals provoked by IL-2 stimulation and the distal events leading to S-phase commitment in activated T cells.

## EFFECTS OF RAPAMYCIN ON CELL-CYCLE-RELATED EVENTS IN IL-2-STIMULATED T CELLS

Rapamycin (RAP) and the related macrolide ester, FK506, are potent immunosuppressive compounds that interfere with intracellular signaling pathways required for antigen-induced T-cell activation and growth (reviewed in 21). Whereas FK506 blocks a $Ca^{2+}$-dependent transcriptional event involved in IL-2 gene expression, RAP targets an undefined step needed for the entry of IL-2-stimulated T cells into S phase. The pharmacologic actions of RAP and FK506 are mediated through binding to a ubiquitously expressed family of cytosolic receptor proteins, termed FK506-binding proteins (FKBPs). The interaction of both drugs with FKBP results in a gain of function that leads to immunosuppression[22]. In the case of FK506, the FK506-FKBP complex binds specifically to the $Ca^{2+}$/calmodulin-regulated serine-threonine phosphatase, calcineurin. The resultant inhibition of calcineurin function interferes with the T-cell antigen receptor-dependent assembly of an active transcription complex at the IL-2 gene locus. The corresponding molecular target for the RAP-FKBP complex has not been defined; presumably, the function of this target molecule is required for the progression of IL-2 stimulated T cells from $G_1$- to S-phase.

Earlier studies indicated that RAP treatment blocked the entry of IL-2-stimulated T cells into S phase[23]. Initial studies were performed to localize more precisely the $G_1$-phase growth-arrest state induced by RAP. Factor-deprived CTLL-2 cells normally begin to enter S phase at 11-12 hours after restimulation of $G_1$-phase progression with IL-2. In contrast, RAP-treated cells begin to synthesize DNA as early as 2 hours after release from the drug-induced growth arrest, which indicated that RAP blocked IL-2-dependent cell-cycle progression in late $G_1$ phase, near the $G_1$-S boundary. Further evidence that the RAP affected a late event in the $G_1$-phase progression program was provided by the finding that drug treatment failed to inhibit the IL-2-induced expression of the immediate-early response genes, c-*myc* and c-*jun*, in CTLL-2 cells. Collectively, these results suggested that the critical mitogenic signaling event(s) targeted by RAP was located downstream of the early biochemical responses elicited by IL-2R occupancy. One logical approach to further understand the anti-proliferative mechanism of RAP was to examine the effect of this drug on the cell-cycle events related to the process of S-phase commitment in mammalian cells.

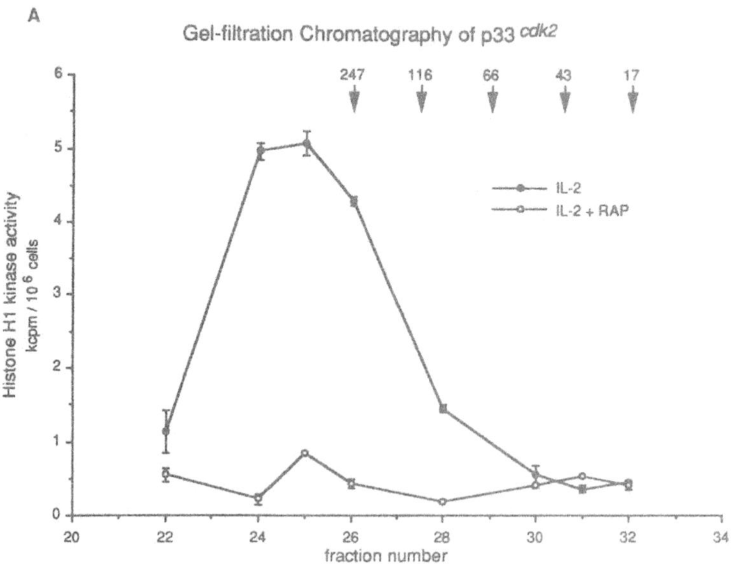


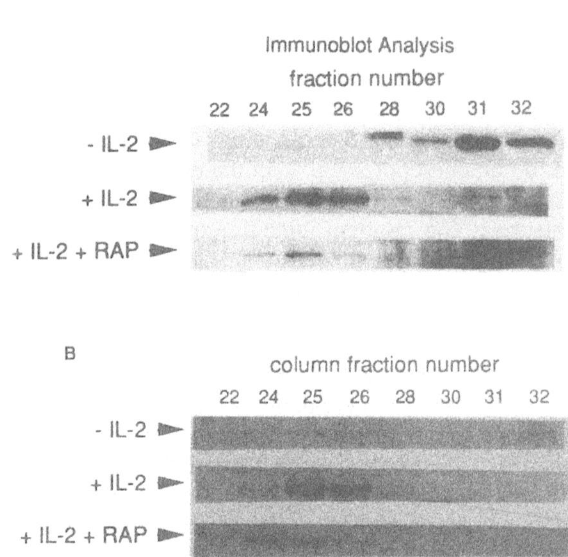


**Figure 6. Gel filtration chromatography of $p33^{cdk2}$ and cyclin E.** Factor-deprived CTLL-2 cells were cultured for 14 h in basal medium only (*-IL-2*), 50 units/ml IL-2 (*+IL-2*), or IL-2 + $10^{-9}$ M RAP (*+IL-2 + RAP*). Cultures were performed in the presence of 0.8 mM hydroxyurea (HU). The cells were subsequently lysed, and detergent extracts were subjected to gel filtration chromatography as described in the Fig. 1 legend. *Panel A*, the indicated chromatographic fractions were either assayed for $p33^{cdk2}$-specific histone H1 kinase activity (*upper panel*) or were immunoblotted with $p33^{cdk2}$-specific antibodies (*lower panel*). *Panel B*, the indicated chromatographic fractions were immunoblotted with cyclin E-specific antibodies. Reprinted from reference 26 with the permission of the publisher.

Accumulating evidence suggests that the passage of growth factor-stimulated cells through a late $G_1$-phase restriction point requires the formation of catalytically-active complexes containing cyclin E and $p33^{cdk2}$ [24]. Preliminary studies similar to those described above indicated that RAP arrested the growth of IL-2-stimulated CTLL-2 cells at or just prior to this restriction point. Subsequently, we examined the effects of RAP treatment on the activity of cyclin-associated $p33^{cdk2}$ in late $G_1$-phase CTLL-2 cells (Fig. 6). Factor-deprived CTLL-2 cells were restimulated to enter the cell cycle with IL-2 in the presence of hydroxyurea, which blocks cell-cycle progression in very early S phase. After 14 hours, the cells were extracted in detergent-containing buffer, and the cellular extracts were chromatographed over a Superose-12 gel filtration column. The column eluate was collected in 0.5 ml fractions, and the indicated fractions were immunoprecipitated with $p33^{cdk2}$-specific antibodies. IL-2-induced cell-cycle progression into late $G_1$-phase was accompanied by the appearance of high molecular weight complexes containing $p33^{cdk2}$-specific histone H1 kinase activity (Fig. 6A, *upper panel*). The chromatographic behavior of active $p33^{cdk2}$ likely reflects the IL-2-dependent association of the monomeric cdk with cyclins and other regulatory proteins. In contrast, extracts from cells stimulated to traverse $G_1$ in the presence of RAP contained markedly reduced levels of $p33^{cdk2}$ kinase activity. Immunoblot analysis of the same column fractions with anti-$p33^{cdk2}$ antibodies revealed that RAP treatment partially inhibited the entry of monomeric $p33^{cdk2}$ into the high molecular weight fractions which normally contain the active histone H1 kinase (Fig. 6A, *lower panel*).

The principal cyclin partners for $p33^{cdk2}$ during late $G_1$-phase are cyclins E and A[25]. In preliminary studies, we observed that IL-2 exposure only marginally increased the total amount of $p33^{cdk2}$ protein expressed in CTLL-2 cells, and that this response was not significantly altered by concomitant exposure of the cells to RAP (unpublished results). In contrast, expression of both cyclin E and cyclin A during $G_1$ phase was strictly dependent on IL-2 stimulation. Cyclin E protein and associated histone H1 kinase activity first appeared at 6-8 hours after re-addition of IL-2 to factor-deprived CTLL-2 cells, followed by the initial appearance of active cyclin A-containing complexes at 8-10 hours. The sequential expression of cyclin E and cyclin A-associated cdk activities is in keeping with the model that these cyclins regulate events associated with progression through late $G_1$- and S-phase, respectively[4].

Stimulation of factor-deprived CTLL-2 cells with IL-2 in the presence of RAP led to dramatic alterations in the expression of cyclin E- and cyclin A-associated $p33^{cdk2}$ kinase activities in late $G_1$ phase. As shown in Fig. 6A, *lower panel*, extracts from IL-2-plus RAP-treated cells contained a substantial pool of monomeric, catalytically-inactive $p33^{cdk2}$, relatively to those derived from cells treated with IL-2 only. A likely explanation for the partial inhibition of $p33^{cdk2}$ entry into high molecular weight complexes was provided by immunoblot analyses, which revealed that RAP treatment blocked the IL-2-dependent accumulation of cyclin A mRNA transcripts and protein in these cells[26]. In spite of the complete absence of cyclin A, gel-filtration chromatography indicated that a sub-population of $p33^{cdk2}$ molecules clearly persisted in high molecular complexes in RAP-treated cells (Fig. 6A, *lower panel*). Strikingly, the complexed form of $p33^{cdk2}$ was virtually devoid of histone H1 kinase activity. These results raised the possibility that cyclin E, the "upstream" partner for $p33^{cdk2}$ during $G_1$ phase, was expressed in RAP-treated CTLL-2 cells. The cyclin E immunoblot shown in Fig. 6B indicates that, in contrast to cyclin A, cyclin E protein expression was not blocked by RAP. Moreover, the IL-2-induced cyclin E was stoichiometrically incorporated into high molecular weight complexes, in the absence or presence of RAP. Direct evidence for the continued association of cyclin E with $p33^{cdk2}$ in RAP-treated cells was derived from co-precipitation experiments with anti-cyclin E antibodies (results not shown). Finally, histone H1 kinase assays performed with anti-cyclin E and -cyclin A antibody immunoprecipitates confirmed that RAP exposure

**Table 1.** Effects of RAP treatment on cyclin A- and cyclin E-associated histone H1 kinase activities

| Treatment | Cyclin A | Cyclin E |
|---|---|---|
| | *kcpm/10⁶ cells ± S.D.* | |
| -IL-2 | 0.74 ± 0.04 | 0.29 ± 0.04 |
| +IL-2 | 34.83 ± 0.63 | 7.03 ± 0.19 |
| +IL-2 + RAP | 2.89 ± 0.09 | 1.52 ± 0.02 |
| +IL-2 + RAP + FK | 16.52 ± 0.56 | 3.84 ± 0.06 |

Factor-deprived CTLL-2 cells were cultured for 14 h in basal medium containing 0.8 mM HU. Where indicated, the medium was supplemented with 50 units/ml IL-2 (+IL-2), $10^{-9}$ M RAP (+RAP), and 2 x $10^{-7}$ M FK506 (+FK). The cells were lysed in detergent-containing buffer, and cell extracts were immunoprecipitated with anti-cyclin A or anti-cyclin E antibodies. The immunoprecipitates were assayed in duplicate for associated histone H1 kinase activities using [$\gamma$-$^{32}$P]ATP as phosphate donor.

profoundly inhibited the IL-2-dependent increases in cdk activity associated with both cyclins (Table 1).

## CONCLUSIONS AND PERSPECTIVES

IL-2R occupancy triggers a proximal increase in protein tyrosine phosphorylation in T cells, due, at least in part, to the activation of the *src*-family kinases, Lck and Fyn. The available evidence suggests that these PTKs subserve partially or entirely overlapping functions during IL-2 signal transduction. Activation of *src*-family kinases apparently links the IL-2R to the downstream signal transducers, PI 3-K and Ras, via the phosphorylation of p85 and Shc, respectively. Ultimately, these signaling cascades converge on the nucleus, where they regulate the expression of growth-related genes, including the c-*fos* and c-*jun* protooncogenes. Although it is generally believed that Ras and PI 3-K participate in mitogenic signal transduction, an accumulating body of data suggest that activation of these enzymes is not sufficient to provoke a T-cell mitogenic response. At least one additional signaling pathway, perhaps linked to the "serine-rich" region of the IL-2R β-subunit[27], appears critical for IL-2-induced cell-cycle progression. Elucidating the biochemical nature of this Ras-independent transduction mechanism is clearly of paramount importance to our understanding of T-cell growth regulation.

The early biochemical responses provoked by IL-2 stimulation lead to the orderly expression of cyclins D, E, and A, and, in turn, to the activation of their associated protein kinases. These cell-cycle regulators carry out the functions needed to commit the $G_1$-phase T cell to enter S-phase and to execute the remainder of the mitogenic program. Rapamycin has proven a useful pharmacologic probe for studies of the distal events involved in the regulation of S-phase commitment in T lymphocytes. The present results suggest that interference with the activation of cyclin E-associated $p33^{cdk2}$ in mid/late $G_1$ phase might explain the T-cell growth-inhibitory action of this drug. Although the actual mechanism remains unclear, it is tempting to speculate that inappropriate expression of one of the recently described negative regulators cdk activity[28] is involved. It is striking that the anti-proliferative effect of RAP is relatively specific for bone marrow-derived cells. This argues against a direct action on a conserved component of the cell-cycle regulatory apparatus, such as cyclin E or $p33^{cdk2}$. Earlier studies indicated that activation of the 70 kd ribosomal protein S6 kinase

($p70^{S6K}$), which occurs within minutes of IL-2 stimulation, was blocked by RAP[29,30]. However, $p70^{S6K}$ activation is a virtually ubiquitous response of mammalian cells to mitogenic stimuli, and no clear-cut correlation exists between the inhibition of $p70^{S6K}$ and cell growth-arrest induced by RAP. $p70^{S6K}$ activity is also not inhibited directly by the RAP-FKBP complex[31], although the rapidity with which RAP inhibits this enzyme in intact cells suggests that a proximate upstream regulator of $p70^{S6K}$ may be the elusive target protein. The identification and functional analysis of the direct target of rapamycin in T lymphocytes is certain to yield novel insights into the mitogenic signal transduction through the IL-2R complex.

## ACKNOWLEDGMENTS

This research was supported by grants from the National Science Foundation and the National Institutes of Health. R. T. Abraham is a Leukemia Society of America Scholar and L. M. Karnitz is a Leukemia Society of America Special Fellow. The authors are grateful to Ms. Kathy Jensen for her assistance with the preparation of this manuscript.

## REFERENCES

1. D. Cosman, S.D. Lyman, R.L. Idzerda, M.P. Beckmann, L.S. Park, R.G. Goodwin, and C.J. March, A new cytokine receptor superfamily, *Trends Biochem. Sci.* 15:265 (1990).
2. D.A. Cantrell and K.A. Smith, The interleukin-2 T-cell system: A new cell growth model, *Science* 224:1312 (1984).
3. M.W. Kirschner, The biochemical nature of the cell cycle, *in:* "Important Advances in Oncology," V.T. DeVita, Jr., S. Hellman, and S.A. Rosenberg, eds., J.B. Lippincott Co., Philadelphia (1992).
4. C.J. Sherr, Mammalian $G_1$ cyclins, *Cell* 73:1059 (1993).
5. L.C. Cantley, K.R. Auger, C. Carpenter, B. Duckworth, A. Graziani, R. Kapeller, and S. Soltoff, Oncogenes and signal transduction, *Cell* 64:281 (1991).
6. C.P. Downes and A.N. Carter, Phosphoinositide 3-kinase: A new effector in signal transduction? *Cell. Signal.* 3:501 (1991).
7. J.A. Augustine, S.L. Sutor, and R.T. Abraham, Interleukin 2- and polyomavirus middle T antigen-induced modification of phosphatidylinositol 3-kinase activity in activated T lymphocytes, *Mol. Cell. Biol.* 11:4431 (1991).
8. L. Karnitz, S.L. Sutor, T. Torigoe, J.C.Reed, M.P. Bell, D.J. McKean, P.J. Leibson, and R.T. Abraham, Effects of p56[lck] deficiency on the growth and cytolytic effector function of an interleukin-2-dependent cytotoxic T-cell line, *Mol. Cell. Biol.* 12:4521 (1992).
9. H. MacArthur and G. Walter, Monoclonal antibodies specific for the carboxy-terminus of simian virus 40 large T antigen, *J. Virol.* 52:483 (1984).
10. R. Ren, B.J. Mayer, P. Cicchetti, and D. Baltimore, Identification of a ten-amino acid proline-rich SH3 binding site, *Science* 259:1157 (1993).
11. T. Satoh, M. Nakafuku, A. Miyajima, and Y. Kaziro, Involvement of *ras* p21 protein in signal-transduction pathways from interleukin 2, interleukin 3, and colony-stimulating factor, but not from interleukin 4, *Proc. Natl. Acad. Sci. USA.* 88:3314 (1991).
12. J. Posada and J.A. Cooper, Molecular signal integration. Interplay between serine, threonine, and tyrosine phosphorylation, *Mol. Biol. Cell.* 3:583 (1992).
13. C.M. Crews and R.L. Erikson, Extracellular signals and reversible protein phosphorylation: What to mek of it all, *Cell* 74:215 (1993).
14. W.L. Farrar and D.K. Ferris, Two-dimensional analysis of interleukin 2-regulated tyrosine kinase activation mediated by the p70-75 $\beta$ subunit of the interleukin 2 receptor, *J. Biol. Chem.* 264:12562 (1989).
15. J.A. Augustine, J.W. Schlager and R.T. Abraham, Differential effects of interleukin-2 and interleukin-4 on protein tyrosine phosphorylation in factor-dependent murine T cells, *Biochim. Biophys. Acta* 1052:313 (1990).

16. E.M. Saltzman, S.M. Luhowskyj, and J.E. Casnellie, The 75,000-dalton interleukin-2 receptor transmits a signal for the activation of a tyrosine protein kinase, *J. Biol. Chem.* 264:19979 (1989).
17. G. Pelicci, L. Lanfrancone, F. Grignani, J. McGlade, F. Cavallo, G. Forni, I. Nicoletti, F. Grignani, T. Pawson, and P.G. Pelicci, A novel transforming protein (SHC) with an SH2 domain is implicated in mitogenic signal transduction, *Cell* 70:93 (1992).
18. L.A. Burns, L.M. Karnitz, S.L. Sutor, and R.T. Abraham, Interleukin-2-induced tyrosine phosphorylation of $p52^{shc}$ in T lymphocytes, *J. Biol. Chem.* 268:17659 (1993).
19. S.E. Egan, B.W. Giddings, M.W. Brooks, L. Buday, A.M. Sizeland, and R.A. Weinberg, Association of Sos Ras exchange protein with Grb2 is implicated in tyrosine kinase signal transduction and transformation, *Nature* 363:45 (1993).
20. J. Schlessinger, How receptor tyrosine kinases activate Ras, *Trends Biochem. Sci.* 18:273 (1993).
21. N.H. Sigal and F.J. Dumont, Cyclosporin A, FK-506, and rapamycin: Pharmacologic probes of lymphocytes signal transduction, *in:* "Annual Review of Immunology," W.E. Paul, ed., Annual Reviews, Inc., Palo Alto (1992).
22. S.L. Schreiber and G.R. Crabtree, The mechanism of action of cyclosporin A and FK506, *Immunol. Today* 13:136 (1992).
23. F.J. Dumont, M.R. Melino, M.J. Staruch, S.L. Koprak, P.A. Fischer, and N.H. Sigal, The immunosuppressive macrolides FK-506 and rapamycin act as reciprocal antagonists in murine T cells, *J. Immunol.* 144:1418 (1990).
24. A. Koff, M. Ohtsuki, K. Polyak, J.M. Roberts, and J. Massagué, Negative regulation of G1 in mammalian cells: Inhibition of cyclin E-dependent kinase by TGF-β, *Science* 260:536 (1993).
25. A. Koff, A. Giordano, D. Desai, K. Yamashita, J.W. Harper, S. Elledge, T. Nishimoto, D.O. Morgan, B.R. Franza, and J.M. Roberts, Formation and activation of a cyclin E-cdk2 complex during the $G_1$ phase of the human cell cycle, *Science* 257:1689 (1992).
26. W.G. Morice, G. Wiederrecht, G.J. Brunn, J.J. Siekierka, and R.T. Abraham, Rapamycin inhibition of interleukin-2-dependent $p33^{cdk2}$ and $p34^{cdc2}$ kinase activation in T lymphocytes, *J. Biol. Chem.* 268:22737 (1993).
27. T. Kono, Y. Minami, and T. Taniguchi, The interleukin-2 receptor complex and signal transduction: role of the β-chain, *Sem. Immunol.* 5:299 (1993).
28. T. Hunter, Braking the cycle, *Cell* 75:839 (1993).
29. J. Chung, C.J. Kuo, G.R. Crabtree, and J. Blenis, Rapamycin-FKBP specifically blocks growth-dependent activation of and signaling by the 70 kd S6 protein kinases, *Cell* 69:1227 (1992).
30. V. Calvo, C.M. Crews, T.A. Vik, and B.E. Bierer, Interleukin 2 stimulation of p70 S6 kinase activity is inhibited by the immunosuppressant rapamycin, *Proc. Natl. Acad. Sci. USA* 89:7571 (1992).
31. C.J. Kuo, J. Chung, D.F. Fiorentino, W.M. Flanagan, J. Blenis, and G.R. Crabtree, Rapamycin selectively inhibits interleukin-2 activation of p70 S6 kinase, *Nature* 358:70 (1992).

# THE IL-4 RECEPTOR - SIGNALING MECHANISMS

Achsah Keegan, Keats Nelms and William E. Paul

Laboratory of Immunology, National Institute of Allergy and Infectious Diseases,
Bethesda, MD 20892

## INTRODUCTION

Interleukin-4 (IL-4) is a multifunctional cytokine that is a member of the hematopoietin family[1]. Its structure consists of a bundle of four left handed α helices in which the first and second and the third and fourth helices are connected by long over hand loops[2,3,4,5]. IL-4 is produced by CD4+ T cells[1] and by basophils and mast cells[6] in response to receptor-mediated stimulation. It mediates its functions by binding to high affinity receptors expressed on a wide range of hematopoeitic and non-hematopoietic cell types[7]. The molecule defined as the IL-4 receptor (IL-4R) possesses an extracellular domain that is a member of the hematopoeitin-receptor family[8]. Its cytosolic domain is ~500 amino acids in length and lacks both endogenous tyrosine kinase and nucleotide-binding motifs[9]. The IL-4R does possess two acid rich regions and sequences homologous to the Box 1 region defined for the gp130 component of the IL-6 receptor[10].

### The IL-4 Receptor and the IL-2 Receptor γ Chain Participate in IL-4 Binding and Are Required For Signal Transduction

Based on the paradigm developed from the study of signalling through the growth hormone receptor[11], it has been postulated that each of the hematopoietins would stimulate target cells through dimerization of receptor subunits[12]. It has recently been shown that IL-4 interacts not only with the IL-4 receptor (IL-4R) but forms a complex with that receptor and with the γ chain of the IL-2 receptor (now designated $\gamma_{common}$ or $\gamma_c$)[13,14]. L cells, which express the IL-4R but lack $\gamma_c$, fail to transduce an IL-4-induced signal; stable L cell tansfectants that expresses $\gamma_c$ do respond to IL-4 with signal transduction[13].

### 4PS is Tyrosine Phosphorylated in Hematopoietic Cells in Response to IL-4 and Insulin

In order to study signal transduction through the IL-4R/$\gamma_c$ complex, we have utilized the responses of a set of IL-3-dependent hematopoietic cell lines. Upon exposure

*Mechanisms of Lymphocyte Activation and Immune Regulation V*
Edited by S. Gupta *et al.*, Plenum Press, New York, 1994

to IL-4, cells of these lines take up $^3$H-thymidine, although they generally fail to grow long term. IL-4 causes rapid tyrosine phosphorylation of a predominant substrate of ~170 kDa in size[15,16,17]. This substrate, which has been designated the IL-4 phosphorylation substrate (4PS)[17], is distinct from the substrates phosphorylated in these cells in response to IL-3. In addition, IL-4 also causes tyrosine phosphorylation of the IL-4R. However, cells such as the myeloid lines FDC-P1 and FDC-P2 also show DNA synthesis in response to insulin and insulin-like growth factor-1 (IGF-1). Insulin and IGF-1 also cause predominant phosphorylation of a 170 kDa substrate in FDC-P1 and FDC-P2 cells, but fail to stimulate phosphorylation of the IL-4R. The tyrosine phosphorylated peptides in 4PS and the 170 kDa substrate phosphorylated in response to insulin had indistinguishable mobilities, suggesting that they were the same[18].

## 4PS Appears To Be Related to IRS-1

Tyrosine phosphorylation in response to insulin and IGF-1 has been extensively studied in fibroblasts. The principal substrate is a 180 kDa molecule designated insulin receptor substrate-1 (IRS-1)[19]. IRS-1 has been molecularly cloned[20]; it has more than 20 potential tyrosine phosphorylation sites of which have 9 YMXM or YXXM motifs, consistent with their being targets for the SH2-domain of phosphatidyl inositol-3-kinase (PI-3 kinase). The mobilities of the peptides that are tyrosine phosphorylated in response to insulin in FDC-P2 cells and CHO cells that have been transfected with a cDNA for IRS-1 are not identical suggesting that 4PS, not IRS-1, is the major substrate for insulin in hematopoeitic cells.[18] Some anti-IRS-1 antibodies show partial cross-reactivity with tyrosine phosphorylated 4PS while others do not. This suggests that 4PS and IRS-1 are related molecules. However, until 4PS has been molecularly cloned, the degree of relatedness of these molecules cannot be determined.

Although IRS-1 does not appear to be a major substrate for IL-4 or insulin in FDC-P2 cells or other hematopoeitic cells, if expressed in such cells as a result of transfection, it will be phosphorylated[18]. L cells, which express IRS-1, will phosphorylate this molecule in response to IL-4, but only if they have been transfected with a cDNA for $\gamma_c$[13]. Thus, 4PS and IRS-1 appear to function in hematopoietic and non-hematopoeitic cells, respectively, as the major substrate tyrosine phosphorylated in response to IL-4, insulin or IGF-1.

## 4PS Become Associated With The Regulatory Subunit of PI-3 Kinase

Upon phosphorylation of 4PS and IRS-1 in response to either IL-4 or insulin, these molecules become associated with the regulatory subunit of PI-3 kinase[17]. Small amounts of PI-3 kinase can be co-precipitated with the IL-4R, possibly reflecting the binding of some tyrosine phosphorylated 4PS to the IL-4R.

## IRS-1/4PS Is Essential For 32D Cells To Grow In Response to IL-4

Not only is 4PS/IRS-1 strikingly phosphorylated in response to IL-4, IGF-1 or insulin, these molecules appear to play a role in cell growth in response to such stimuli. This has been shown by examining the growth characteristics of 32D cells. 32D cells are members of an immature hematopoeitic cell line that can be propagated in IL-3. These cells show little or no DNA synthesis in response to IL-4, IGF-1 or insulin, although they do express some receptors for these three factors. 32D cell lines that have been stably transfected with a cDNA for IRS-1 demonstrate both IRS-1 phosphorylation and cell growth in response to IL-4 or insulin[21,22]. This indicates that expression of substrate (and presumably its phosphorylation) is critical to cell growth in response to IL-4 or insulin in 32D cells.

## Truncation Mutants Of The IL-4R Indicate That The Region Between aa437 And 557 Are Important For Signalling Phosphorylation of IRS-1

Wild type and mutant human IL-4R were transfected into 32D cells that over-expressed IRS-1 (32D-IRS-1 cells). Since human IL-4 does not interact with the mouse IL-4R, this allowed us to study the structural requirement of the cytosolic domain of the IL-4R for IRS-1/4PS phosphorylation in response to IL-4. Mutants truncated at amino acids 657 and 557 transmitted signals for both IRS-1 phosphorylation and DNA synthesis in response to IL-4. By contrast a mutant terminated at aa437 showed little or no growth or IRS-1 phosphorylation in response to IL-4[22]. The aa437 truncation mutant (d437) contains no tyrosines and lacks the more membrane distal of the two acidic domains of the IL-4R. d557, which appears to respond normally to IL-4, contains only one of the five cytosolic domain tyrosines and both of the acidic domains.

## The IL-4R Contains A Motif (I4R) Also Found In The Insulin/IGF-1 Receptors That Plays A Major Role In IL-4-Induced Phosphorylation And Cell Growth

Examination of the sequence surround the single tyrosine in the interval between aa437 and aa557 reveals that it has considerable homology to a sequence surrounding a critical tyrosine in the insulin and IGF-1 receptors. This motif, $XPLX_4NPXYXSXSDXX$, has been designated the insulin receptor/ IL-4 receptor (I4R) motif[22]. It had been previously shown that mutating the tyrosine within this motif in the insulin receptor to a phenylalanine inhibits the phosphorylation of IRS-1 in response to insulin although it does not block autophosphorylation[23]. In order to substantiate the importance of the I4R motif within the IL-4R, a Y-> F mutation at position 497 (Y497F) of the human IL-4R was tested by transfection of mutant cDNA into 32D/IRS-1 cells[22]. Stable cell lines expressing 1,000 to 6,000 mutant receptors were obtained. These lines failed to detectably phosphorylate IRS-1 in response to human IL-4 although they did show phosphorylation in response to mouse IL-4. Most, but not all, of the Y497F-expressing cell lines also failed to synthesize DNA in response to human IL-4.

## The Intact IL-4R And Fusion Protein Containing aa 424-561 Associate With A Tyrosine Kinase And Mediate The Phosphorylation Of IRS-1

Immunoprecipitates of the IL-4R from FDC-P2 cells or glutathione-S-transferase fusion proteins containing aa424 to 561 of the IL-4R incubated with FDC-P2 cell extracts were capable of tyrosine phosphorylating a set of substrates in the presence of ATP[22]. This implies that a tyrosine kinase is associated with the IL-4R and that this kinase (or one such kinase) is associated with the 424-561 region. Indeed, one of these substrates is IRS-1, suggesting that both a kinase and IRS-1 can interact with the region between aa424 and aa561. As noted above, this region contains the I4R motif.

## Proposed Mechanism Of Signal Transduction Through The IL-4R/$\gamma_c$ Complex

Based on the experiments described here, we propose a speculative model for signaling though the IL-4R/$\gamma_c$ complex. When IL-4 binds to cells expressing both IL-4R and $\gamma_c$, it causes heterodimerization of these receptor chains. $\gamma_c$ is critical to the response so that $\gamma_c$ or molecules associated with it must somehow regulate the signaling competence of the IL-4R. In order for the receptor to transduce a significant growth signal, a receptor region (aa437-557) containing the I4R motif must be present. We propose that a tyrosine kinase is associated with this region; possibly its first substrate upon activation is Y497 of the receptor, although has not been directly shown. Tyrosine phosphorylation of Y497 may make it a target for 4PS/IRS-1, which docks to this region either directly or through an adapter molecule. In this site, 4PS/IRS-1 becomes phosphorylated; 4PS/IRS-1 phosphorylated on a critical residue(s) diffuses away from the receptor, presumably making way for more 4PS/IRS-1 molecules and thus allowing a striking amplification of the signal. Phosphorylated 4PS/IRS-1 is capable of binding a set of SH2 domain-containing proteins, including the regulatory subunit of PI-3 kinase. Through these interactions, further down stream signalling events occur.

Many issues remain to be clarified. Among these are whether tyrosine phosphorylation of Y497 is essential for function; what is the role of $\gamma_c$; what is the IL-4R tyrosine kinase; do IL-4 and insulin phosphorylate 4PS/IRS-1 identically or are there differences in pattern of tyrosine phosphorylation that are associated with differences in the biology of these cells. The study of this signaling pathway may give important insights into signal transduction in general and may offer specific suggestions about the design of potential inhibitors of IL-4.

## Acknowledgment

We wish to thank our colleagues, Drs. Ling-Mei Wang and Jacalyn Pierce, for their major contributions to our joint work.

## REFERENCES

1. W.E. Paul, Interleukin-4: a prototypic immunoregulatory lymphokine, *Blood.* 77:1859 (1991).
2. R. Powers, D.S. Garrett, C.J. March, E.A. Frieden, A.M. Gronenborn, and G.M. Clore, The high-resolution, three-dimensional solution structure of human interleukin-4 determined by multidimensional heteronuclear magnetic resonance spectroscopy, *Biochemistry.* 32:6744 (1993).
3. M.R. Walter, W.J. Cook, B.G. Zhao, R.J. Cameron, S.E. Ealick, *et al.*, Crystal structure of recombinant human interleukin-4, *J Biol Chem.* 267:20371 (1992).
4. A. Wlodaver, A. Pavlovsky, and A. Gustchina, Crystal structure of human recombinant interleukin-4 at 2.25 A resolution, *Febs Lett.* 309:59 (1992).
5. L. J. Smith, C. Redfield, J. Boyd, G.M. Lawrence, R.G. Edwards, R.A. Smith, and C.M. Dobson, Human interleukin 4. The solution structure of a four-helix bundle protein, *J Mol Biol.* 224:899 (1992).
6. W.E. Paul, R.A. Seder, and M. Plaut, Lymphokine and cytokine production by Fc epsilon RI+ cells, *Adv Immunol.* 53:1 (1993).
7. J. Ohara and W.E. Paul, Receptors for B-cell stimulatory factor-1 expressed on cells of haematopoietic lineage, *Nature.* 325:537 (1987).
8. J.F. Bazan, Haemopoietic receptors and helical cytokines, *Immunol Today.* 11:350 (1990).
9. B. Mosley, M.P. Beckmann, C.J. March, R.L. Idzerda, S.D. Gimpel, *et al.*, The murine interleukin-4 receptor: molecular cloning and characterization of secreted and membrane bound forms, *Cell.* 59:335 (1989).
10. M. Murakami, M. Narazaki, M. Hibi, H. Yawata, K. Yasukawa, M. Hamaguchi, T. Taga, and T. Kishimoto, Critical cytoplasmic region of the interleukin 6 signal transducer gp130 is conserved in the cytokine receptor family, *Proc Natl Acad Sci U S A.* 88:11349 (1991).
11. B.C. Cunningham, M. Ultsch, V.A. De, M.G. Mulkerrin, K.R. Clauser, and J.A. Wells, Dimerization of the extracellular domain of the human growth hormone receptor by a single hormone molecule, *Science.* 254:821 (1991).
12. J.L. Boulay and W.E. Paul, The interleukin-4-related lymphokines and their binding to hematopoietin receptors, *J Biol Chem.* 267:20525 (1992).

13. S.M. Russell, A.D. Keegan, N. Harada, Y. Nakamura, M. Noguchi, *et al.*, Interleukin-2 receptor gamma chain: a functional component of the interleukin-4 receptor, *Science*. 262:1880 (1993).
14. M. Kondo, T. Takeshita, N. Ishii, M. Nakamura, S. Watanabe, K. Arai, and K. Sugamura, Sharing of the interleukin-2 (IL-2) receptor gamma chain between receptors for IL-2 and IL-4, *Science*. 262:1874 (1993).
15. A.O. Morla, J. Schreurs, A. Miyajima, and J.Y. Wang, Hematopoietic growth factors activate the tyrosine phosphorylation of distinct sets of proteins in interleukin-3-dependent murine cell lines, *Mol Cell Biol.* 8:2214 (1988).
16. R.J. Isfort and J.N. Ihle, Multiple hematopoietic growth factors signal through tyrosine phosphorylation, *Growth Factors*. 2:213 (1990).
17. L.M. Wang, A.D. Keegan, W.E. Paul, M.A. Heidaran, J.S. Gutkind, and J.H. Pierce, IL-4 activates a distinct signal transduction cascade from IL-3 in factor-dependent myeloid cells, *EMBO J.* 11:4899 (1992).
18. L.M. Wang, A.D. Keegan, W. Li, G.E. Lienhard, S. Pacini, *et al.*, Common elements in interleukin 4 and insulin signaling pathways in factor-dependent hematopoietic cells, *Proc Natl Acad Sci U S A*. 90:4032 (1993).
19. M.F. White and C.R. Kahn, The insulin signaling system, *J Biol Chem* 269:1 (1994).
20. X.J. Sun, P. Rothenberg, C.R. Kahn, J.M. Backer, E. Araki, P.A. Wilden, D.A. Cahill, B.J. Goldstein, and M.F. White, Structure of the insulin receptor substrate IRS-1 defines a unique signal transduction protein, *Nature*. 352:73 (1991).
21. L.M. Wang, M.J. Myers, X.J. Sun, S.A. Aaronson, M. White, and J.H. Pierce, IRS-1: essential for insulin- and IL-4-stimulated mitogenesis in hematopoietic cells, *Science*. 261:1591 (1993).
22. A.D. Keegan, K. Nelms, M.F. White, L.-M. Wang, J.H. Pierce, and W.E. Paul, A region of the interleukin-4 receptor containing a motif found in the insulin receptor is important for IL-4 mediated IRS-1 phosphorylation and proliferation, *Cell*. 76:in press (1994).
23. M.F. White, J.N. Livingston, J.M. Backer, V. Lauris, T.J. Dull, A. Ullrich, and C.R. Kahn, Mutation of the insulin receptor at tyrosine 960 inhibits signal transmission but does not affect its tyrosine kinase activity, *Cell* 54:641 (1988).

# FUNCTION OF THE COMMON β SUBUNIT OF THE GM-CSF/IL-3/IL-5 RECEPTORS

Alice Mui[1], Akihiko Muto[2], Kazuhiro Sakamaki[3], Noriko Sato[2], Taisei Kinoshita[1], Sumiko Watanabe[2], Takashi Yokota[2], Kenichi Arai[2] and Atsushi Miyajima[1]

[1] DNAX Research Institute of Molecular and Cellular Biology, 901 California Avenue, Palo Alto, CA 94304
[2] Institute of Medical Science, The University of Tokyo, Shirokanedai Minatoku, Tokyo 108, Japan
[3] Pharmaceutical Basic Research Laboratories, JT Inc., Fukuura, Kanazawa-ku, Yokohama 236, Japan

## INTRODUCTION

Interleukin 3 (IL-3) and granulocyte-macrophage colony stimulating factor (GM-CSF) stimulate various lineage-committed cells as well as early multipotential progenitors while interleukin 5 (IL-5) stimulates eosinophils and basophils[1]. These three cytokines induce similar intracellular signals and exhibit similar functions in their common target cells. Although primary amino acid sequences of IL-3, IL-5 and GM-CSF show no obvious homology, they consist of four α-helices and their gross tertiary structures are similar. Interestingly, binding of a human cytokine to its high affinity receptor is inhibited by another cytokine on their common target cells, e.g. high affinity IL-3 binding to its receptor is inhibited by GM-CSF and vice versa[2].

These common features of the three cytokines are now well explained by the receptor structure. The high affinity receptors for these cytokines consist of α and β subunits[3] (Fig. 1). The α subunits are specific to each cytokine and bind their specific ligand with low affinity. By contrast, there is only one β subunit (common β, or $\beta_c$) in the human receptors. The $\beta_c$ subunit does not bind any cytokine by itself, but is required for formation of a high affinity receptor for either IL-3, GM-CSF or IL-5 and provides a molecular explanation for cross-competition in receptor binding, i.e. competition between different α subunits for the shared $\beta_c$ subunit. As described in detail below, the $\beta_c$ subunit is not only required for high affinity binding of a cytokine, but is an essential component of signal transduction. Thus, the shared β subunit is responsible for common biological function of IL-3, GM-CSF, and IL-5 while expression of the α subunits is responsible for cytokine specificity[3].

Both α and β subunits are members of the class I cytokine receptor family and have one and two units, respectively, of the common motif of the class I receptors in their extracellular domains[3]. The cytoplasmic domains of the α subunits consist of about 50 amino acid residues and that of the $\beta_c$ subunit is 432 amino acid residues. Although the three cytokines rapidly induce protein tyrosine phosphorylation, none of these subunits have any intrinsic enzymatic activity such as kinases. Hence, it is presumed that non-receptor tyrosine kinases are involved in signal transduction by these cytokine receptors.

## REQUIREMENT OF THE CYTOPLASMIC DOMAINS

To define the function of the intracellular domains of the receptors, cytoplasmic deletions of the human GM-CSF receptor α and $\beta_c$ subunits were generated and expressed in mouse IL-3 dependent proB cell line BaF3[4]. Truncation of the cytoplasmic domain of either α or β subunit does not affect the binding affinity of the receptor. However, complete truncation of the cytoplasmic domains of either α or β subunit abrogated their ability to induce proliferation. As $\beta_c$ is common among the three cytokine receptors and has a large cytoplasmic domain of 432 amino acid residues, we generated a series of cytoplasmic deletions of the human $\beta_c$ and expressed each with the full length human GM-CSF receptor α subunit in BAF3 cells. Truncations at 763 and 626 ($\beta_{763}$, $\beta_{626}$) showed almost the same proliferative response to GM-CSF as the normal $\beta_c$ subunit. The truncation at 517 ($\beta_{517}$), which possesses only about 60 cytoplasmic amino acid residues, still retained the ability to induce a weak proliferation signal in medium containing 10% fetal calf serum (FCS) and the truncation at 455 ($\beta_{455}$) that lacks the entire cytoplasmic domain did not respond to GM-CSF at all[4] (Fig. 2).

**Table 1.** Function of $\beta_c$ deletion mutants

| | $\beta_c$ mutants | | | | | |
|---|---|---|---|---|---|---|
| Function | $\beta_{455}$ | $\beta_{517}$ | $\beta_{626}$ | $\beta_{763}$ | $\beta_{825}$ | $\beta_c$ |
| Growth | - | +* | + | + | + | + |
| Tyrosine phosphorylation of $\beta_c$ | - | - | - | + | + | + |
| Tyrosine phosphorylation of Shc | - | - | - | + | + | + |
| Ras activation | - | - | - | + | + | + |
| Raf phosphorylation | - | - | - | + | + | + |
| MAP kinase activation | - | - | - | + | + | + |
| p70/p90 S6 kinases activation | - | - | - | + | + | + |
| Induction of c-fos/c-jun | - | - | - | + | + | + |
| Induction of c-myc | - | + | + | + | + | + |
| Induction of Pim-1 | - | + | + | + | + | + |

* Proliferation is significantly weak compared with the full length $\beta_c$

To find the role of the cytoplasmic domain of $\beta_c$, various intracellular signals induced by these deletion mutants were examined. First, we examined tyrosine phosphorylation induced by GM-CSF in BaF3 transfectants. The major tyrosine phosphorylated proteins induced by either mIL-3 or hGM-CSF in BaF3/GMR$\alpha\beta_c$ were 120, 95, and 60 kD proteins. Immunoprecipitation with anti-$\beta_c$ antibody revealed that the 120 kD phosphoprotein is $\beta_c$ itself. The tyrosine phosphorylated $\beta_c$ became smaller by deletions such as $\beta_{825}$ and $\beta_{763}$

and undetectable by $\beta_{626}$, suggesting that the tyrosine phosphorylation site is between 763 and 626[4]. In fact, the tyrosine residue at 750 is surrounded by acidic amino acids and is most likely to be the phosphorylation site. Tyrosine phosphorylation of the other major bands at 95 and 60 kD were significantly reduced in cells with $\beta_{626}$ (BaF3/GMR$\alpha\beta_{626}$) and no significant tyrosine phosphorylation was induced by hGM-CSF in BaF3 cells expressing $\beta_{517}$ (BaF3/GMR$\alpha\beta_{517}$). Although western blot analysis of cell lysates with anti-phosphotyrosine showed no significant phosphorylation, proliferation of BaF3/GMR$\alpha\beta_{517}$ induced by hGM-CSF was sensitive to a tyrosine kinase inhibitor herbimycin A[4]. This result suggests that tyrosine phosphorylation is still induced by $\beta_{517}$ and required for proliferation.

## DIFFERENTIAL REQUIREMENT OF THE β CYTOPLASMIC DOMAINS FOR DIFFERENT SIGNALS

Various growth signals are mediated by a GTP binding protein Ras and many cytokines including IL-2, IL-3, GM-CSF, IL-5 and erythropoietin (Epo) are known to activate Ras[5, 6]. Activity of Ras is regulated by the bound guanine nucleotide: The GDP-bound form is inactive and the GTP-bound Ras is the active form that transmits signals to the downstream signaling molecules such as Raf. Extracellular growth signals either stimulate exchange of GDP with GTP via an exchange factor such as SOS or inhibits the Ras GTPase activity, resulting in accumulation of the GTP-bound form. GM-CSF induced activation of Ras in BaF3/GMR$\alpha\beta_{763}$ as well as BaF3/$\alpha\beta_c$[7]. However, BaF3 cell with deletion mutants, $\beta_{626}$, $\beta_{517}$, or $\beta_{455}$ did not activate Ras, indicating that the region between 626 and 763 is required for activation of Ras. Growth factor receptors such as EGF and PDGF receptors activate Ras through a nucleotide exchange factor SOS via SH2 containing adapter molecules, Grb-2 or Shc[8, 9]. Shc is an SH2 containing protein and is tyrosine phosphorylated by EGF. We therefore examined the possibility that Shc is involved in Ras activation by the GM-CSF receptor. GM-CSF stimulation resulted in tyrosine phosphorylation of Shc by $\beta_{763}$ but not by $\beta_{626}$, indicating that the region between 763 and 626 is required for tyrosine phosphorylation of Shc[7]. Therefore it is very likely that Shc mediates the signal from the receptor to the nucleotide exchange factor. However, it is unknown at present which tyrosine kinase phosphorylates Shc upon GM-CSF stimulation. Involvement of Grb-2 is also suggested by the fact that GM-CSF or IL-3 stimulates association of Grb-2 with tyrosine phosphorylated proteins (unpublished). While it still remains to be proven that SOS is involved in cytokine-mediated Ras activation, involvement of Shc and Grb2 suggests this possibility. In addition, as a hematopoietic specific protein Vav is known to stimulate the exchange of the Ras-bound GDP with GTP[10] and is tyrosine phosphorylated by GM-CSF, IL-3 and Epo[11], Vav may also involve in activation of Ras by cytokines. Thus, there may be two alternative pathways to Ras activation from the cytokine receptors in hematopoietic cells.

Activation of Ras leads to activation of serine/threonine kinases Raf-1 and MAP kinase. Recently activated Ras was found to directly interact with Raf[12] and activated Raf in turn activates MAP kinase through MAP kinase kinase[13]. Therefore we examined the activation of Raf and MAP kinase using the deletion mutants. As expected, the $\beta_{763}$ mutant activated Raf and MAP kinase in response to GM-CSF. However, $\beta_{626}$ mutant did not activate Raf and MAP kinase, indicating that the region between 763 and 626 is also responsible for activation of Raf and MAP kinase[7]. These results strongly suggest that the cascade Ras --> Raf --> MAP kinase known for growth factor signaling is also activated by cytokines.

Two distinct serine/threonine kinases, pp90$^{rsk}$ and p70-S6, are known to phosphorylate the 40S ribosomal protein S6 and are implicated in cell proliferation[14]. We examined phosphorylation of pp90$^{rsk}$ and p70-S6 kinases and found that the region between 763 and 626 is required for phosphorylation of both kinases[7]. Because MAP kinase directly phosphorylates pp90$^{rsk}$, but not p70-S6 in vitro, activation of p70-S6 kinase may be mediated by a pathway distinct from MAP kinase. Phosphatidylinositol 3 kinase (PI3K) is associated with growth factor receptors through the SH2 domain and activation of PI3K by cytokines such as IL-4 has been demonstrated[15]. Although the association of PI3K to the GM-CSF and IL-3 receptors was insignificant, we found that PI3K activity in anti-phosphotyrosine immunoprecipitates was significantly increased after stimulation with IL-3 or GM-CSF. GM-CSF-induced activation of PI3K was dramatically reduced by $\beta_{626}$. However, unlike Ras activation, $\beta_{626}$ was still capable of activating PI3K at a low level[7]. Pim-1 is a serine/threonine kinase which is relatively specific to hematopoietic cells. Kinase activity of this protein is difficult to measure, but Pim-1 expression is highly inducible by cytokines such as IL-2, IL-3 and GM-CSF. Interestingly, induction of Pim-1 by GM-CSF was observed even in $\beta_{517}$, but not in $\beta_{455}$ transfectants[7]. While the role of this kinase is still unknown, since Pim-1 and Myc are known to cooperatively induce lymphomas, this kinase may be involved in IL-3-mediated proliferation[16].

Signals are eventually transmitted to the nucleus and regulate gene expression. IL-3 and GM-CSF stimulate induction of nuclear protooncogenes, c-myc, c-fos and c-jun, which are implicated in proliferation and differentiation. The $\beta_{763}$ mutant as well as the full length $\beta_c$ were able to induce c-myc, c-fos and c-jun in response to GM-CSF. Although $\beta_{626}$ and $\beta_{517}$ induced c-myc in response to GM-CSF, they did not induce c-fos and c-jun and $\beta_{455}$ induced none of these[7]. Thus, the membrane proximal region is responsible for c-myc induction and is indispensable for proliferation. By contrast, the distal region between 626 and 763 is required for induction of c-fos and c-jun, which is downstream of MAP kinase. Since BaF3 cells expressing $\beta_{517}$ proliferated in response to GM-CSF, the signaling pathways activated by the distal region such as the Ras pathway is dispensable for proliferation. However, it should be noted that the proliferation assays of BaF3 cells were carried out in the presence of FCS and FCS might stimulate the Ras pathway. This possibility was supported by our recent results that FCS induced MAP kinase activation as well as c-fos expression in BaF3 cells and that GM-CSF did not support long term proliferation of BaF3/$\alpha\beta_{517}$ in the absence of FCS (Sakamaki et al., in preparation). Thus, signals induced by FCS appear to alleviate the requirement of the pathways activated by the distal region of the $\beta$ subunit in BaF3 cells.

## COMMON VS SPECIFIC SIGNALING PATHWAYS

There are significant overlap between signaling pathways activated by growth factors and cytokines. As described above, both induce activation of Ras and the downstream pathways that lead to induction of c-fos. In fact, the EGF receptor ectopically expressed in hematopoietic cells activated Ras, induced c-fos, and supported short term survival in response to EGF[5, 17]. On the other hand, the high affinity GM-CSF receptor ectopically expressed in NIH3T3 cells induced tyrosine phosphorylation, transcription of c-myc, c-fos and c-jun, and stimulated proliferation in response to GM-CSF[18]. Thus, it is clear that the signaling pathways present in nonhematopoietic cells can be coupled to the GM-CSF receptor when the receptor is expressed. However, it does not necessary imply that the signaling pathways in hamatopoietic cells and nonhematopoietic cells are identical. As proliferation is fundamental to every cell, a similar signaling pathway may lead to proliferation in response to various stimulation. By contrast, cytokines exhibit a number of unique functions and these functions may be mediated by hematopoietic specific signaling molecules.

One possible mechanism that provides cytokine specific function may be direct activation of cytoplasmic transcription factors by a tyrosine kinase that associates with a cytokine receptor. In the case of interferon (IFN) signaling, it is known that the Tyk-2 tyrosine kinase activated by the IFN receptor phosphorylates the latent cytoplasmic transcription factor ISGF3$\alpha$. The phosphorylated ISGF3$\alpha$ translocates to the nucleus and activates genes by binding to the IFN response elements[19]. Tyk-2 together with JAK-1 and JAK-2 form a subfamily of tyrosine kinases and JAK-2 was recently found to be activated by Epo, growth hormone (GH), IL-3, GM-CSF, G-CSF and IFN$\gamma$[20-22].

Using the series of deletion mutants of the $\beta_c$ subunit, we tested if JAK2 kinase is tyrosine phosphorylated by GM-CSF. GM-CSF induced JAK2 phosphorylation through $\beta_{763}$ and $\beta_{626}$, although the level of phosphorylation by $\beta_{626}$ was lower than that by $\beta_{763}$. Interestingly, even $\beta_{517}$ was able to stimulate phosphorylation of JAK2 at a very low level (N. Sato unpublished). Thus, the requirement of the $\beta_c$ cytoplasmic domain for JAK2 phosphorylation was distinct from that for the Ras pathways, suggesting that JAK2 and Ras are on different pathways. Based on the model of IFN-mediated signal transduction[19], it is reasonable to assume that JAK2 tyrosine-phosphorylates latent cytoplasmic transcription factors and stimulates their translocation to the nucleus. Furthermore, as the receptor-associated kinase such as JAK is directly involved in activation of transcription, this system may provide a way to deliver cytokine specific signals. In fact, we found that tyrosine phosphorylated protein of about 90 kD appeared in the nucleus upon GM-CSF stimulation. This protein was distinct from ISGF3$\alpha$ and the appearance of this protein in the nucleus was correlated with the level of JAK2 phosphorylation, i.e. inducion of the 90 kD protein phosphorylation by $\beta_{517}$ was much weaker than that by the $\beta$ deletions with a longer cytoplasmic tail (A. Mui, unpublished). Currently the nature of this 90 kD protein is unknown and identification of the transcription factors directly activated by JAKs and their target genes is obviously important for understanding the role of the JAK-mediated signaling pathway.

## MECHANISM OF RECEPTOR ACTIVATION

Activation of receptors with intrinsic tyrosine kinase is initiated by ligand-induced dimerization and transphosphorylation of receptor tyrosine kinases[23]. Activation of cytokine receptors without an intrinsic kinase also requires association of multiple receptor subunits. The IL-6-related cytokines such as IL-6, LIF, OSM, and CNTF induce either homodimerization of gp130 or heterodimerization of gp130 and LIF binding component[24, 25]. Interestingly the JAK family members associate with these molecules before ligand binding and are activated upon ligand-induced receptor dimerization[26]. Thus, the activation mechanism of the receptors with an intrinsic kinase may be applicable to cytokine receptors without an intrinsic kinase.

As association of the $\beta$ subunit with JAK2 has been demonstrated[22], two possible mechanisms for activation of the GM-CSF receptor may be considered. One is that each $\alpha$ and $\beta$ subunit associates with a JAK kinase and GM-CSF-induced heterodimerization of $\alpha$ and $\beta$ leads to activation of JAKs. A second possibility is that JAK associates only with $\beta$ and GM-CSF binding induces oligomerization of $\beta$ through an $\alpha\beta$ heterodimer. To test these possibilities, chimeric receptors between $\alpha$ and $\beta$ subunits were constructed by exchanging the cytoplasmic domains. The $\alpha/\beta$ chimeric subunit has the $\alpha$ extracellular domain and the $\beta$ intracellular domain, and the $\beta/\alpha$ chimera has the $\beta$ extracellular and the $\alpha$ intracellular domain. GM-CSF induced proliferation through the chimeric receptor consisting of $\alpha/\beta$ and $\beta/\alpha$. Interestingly, GM-CSF also stimulated growth through the chimeric receptor with $\alpha/\beta$ and the normal $\beta$, whereas the chimeric receptor with the normal $\alpha$ and $\beta/\alpha$ failed to mediate signals. These results suggest that ligand-induced association of the intracellular domain of $\beta$ is important for growth promoting signals and the $\alpha$ cytoplasmic domain is not required for the activation by the chimeric receptors (A. Muto et al., submitted). Thus, the activation mechanism of the GM-CSF receptor appears to be similar to that of the IL-6 receptor. However, in contrast to the IL-6 receptor which does not require the cytoplasmic domain of the IL-6 receptor $\alpha$ subunit for signal transduction, the $\alpha$ cytoplasmic domain of the GM-

CSF receptor is necessary for induction of growth signals by the normal receptor composed of α and β. Although the exact mechanism is unknown, the α cytoplasmic domain may help β to form a dimer or an oligomer in the normal GM-CSF receptor. In addition, the α cytoplasmic domains of the GM-CSF, IL-3 and IL-5 receptors may play a role in induction of a cytokine-specific signal.

## REFERENCES


1. K. Arai, F. Lee, A. Miyajima, S. Miyatake, N. Arai , and T. Yokota, Cytokines: coordinators of immune and inflammatory responses. *Annu. Rev. Biochem.* 59: 783 (1990).
2. A.F. Lopez, M.V. Vadas, J.M. Woodcock, S.E. Milton, A. Lewis, M.J. Elliott, D. Gillis, R. Ireland, E. Olwell, and L.S. Park, Interleukin-5, interleukin-3, and granulocyte-macrophage colony-stimulating factor cross-compete for binding to cell surface receptors on human eosinophils. *J. Biol. Chem.* . 266: 24741 (1991).
3. A. Miyajima, A. L-F. Mui, T. Ogorochi, and K. Sakamaki, Receptors for granulocyte-macrophage colony-stimulating factor, interleukin 3 and interleukin 5. *Blood* . 82: 1960 (1993).
4. K. Sakamaki, I. Miyajima, T. Kitamura, and A. Miyajima, Critical cytoplasmic domains of the common beta subunit of the human GM-CSF, IL-3 and IL-5 receptors for growth signal transduction and tyrosine phosphorylation. *EMBO J* . 11: 3541 (1992).
5. T. Satoh, M. Nakafuku, A. Miyajima, and Y. Kaziro, Involvement of ras p21 protein in signal-transduction pathways from interleukin 2, interleukin 3, and granulocyte/macrophage colony-stimulating factor, but not from interleukin 4. *Proc Natl Acad Sci U S A* . 88: 3314 (1991).
6. V. Duronio, M.J. Welham, S. Abraham, P. Dryden, and J.W. Schrader, $p21^{ras}$ activation via hemopoietin receptors and c-kit requires tyrosine kinase activity but not tyrosine phosphorylation of $p21^{ras}$ GTPase-activating protein. *Proc. Natl. Acad. Sci. USA.* . 89: 1587 (1992).
7. N. Sato, K. Sakamaki, N. Terada, K. Arai, and A. Miyajima, Signal transduction by the high affinity GM-CSF receptor: two distinct cytoplasmic regions of the common β subunit responsible for differentiation. *EMBO J.* . 12: 4181 (1993).
8. E. J. Lowenstein, R.J. Daly, A.G. Batzer, W. Li, B. Margolis, R. Lammers, A. Ullrich, E.Y. Skolnik, S.D. Bar, and J. Schlessinger, The SH2 and SH3 domain-containing protein GRB2 links receptor tyrosine kinases to ras signaling. *Cell* . 70: 431 (1992).
9. G. Pelicci, L. Lanfrancone, F. Grignani, J. McGlade, F. Cavallo, G. Forni, I. Nicoletti, F. Grignani, T. Pawson, and P.G. Pelicci, A novel transforming protein (SHC) with an SH2 domain is implicated in mitogenic signal transduction. *Cell* . 70: 93 (1992).
10. E. Gulbins, K. Coggeshall, G. Baier, S. Katzav, P. Burn, and A. Altman, Tyrosine kinase-stimulated guanine nucleotide exchange activity of vav in T cell activation. *Science* . 260: 822 (1993).
11. A. Mui, R. Cutler, M. Alai, X. Bustelo, M. Barbacid, and G Krystal, Steel factor and interleukin-3 stimulate the tyrosine phosphorylation of p95vav in hemopoietic cell lines. *Exp Hematol* . 20: 752a (1992).
12. L. Van Aelst, M. Barr, S. Marcus, A. Polverino, and M. Wigler, Complex formation between RAS and RAF and other protein kinases. *Proc. Natl. Acad. Sci. USA* . 90: 6213 (1993).

13. E. Nishida and Y. Gotoh, The MAP kinase cascade is essential for diverse signal transduction pathways. *Trends Biochem Sci* . 18: 128 (1993).
14. J. Blenis, J. Chung, E. Erikson, D.A. Alcorta, and R.L. Erikson, Distinct mechanisms for the activation of the RSK kinases/MAP2 kinase/pp90rsk and pp70-S6 kinase signaling systems are indicated by inhibition of protein synthesis. *Cell Growth Diff.* . 2: 279 (1991).
15. L.M. Wang, A.D. Keegan, W.E. Paul, M.A. Heidaran, J.S. Gutkind, and J.H. Pierce, IL-4 activates a distinct signal transduction cascade from IL-3 in factor-dependent myeloid cells. *EMBO J* . 11: 4899 (1992).
16. J. Domen, N.M.T. van der Lugt, P. Laird, C.J.M. Saris, A.R. Clarke, M.L. Hooper, and A. Berns, Impaired interleukin-3 response in *Pim*-1-deficient bone marrow-derived mast cells. *Blood* . 82: 1445 (1993).
17. H.-M. Wang, M. Collins, K. Arai, and A. Miyajima, EGF induces differentiation of an IL-3-dependent cell line expressing the EGF receptor. *EMBO J* . 8: 3677 (1989).
18. S. Watanabe, A. L. Mui, A. Muto, J.X. Chen, K. Hayashida, T. Yokota, A. Miyajima, and K. Arai, Reconstituted human granulocyte-macrophage colony-stimulating factor receptor transduces growth-promoting signals in mouse NIH3T3 cells: Comparison with signaling in BA/F3 pro-B cells. *Mol Cell Biol* . 13: 1440 (1993).
19. X-Y. Fu, A transcription factor with SH2 and SH3 domains is directly activated by an interferon α-induced cytoplasmic protein tyrosine kinase(s). *Cell* . 70: 323 (1992).
20. L.S. Artgetsinger, G.S. Cambell, X. Yang, B.A. Witthuhn, O. Silvennoinen, J.N. Ihle, and C. Carter-Su, Identification of JAK2 as a growth hormone receptor-associated tyrosine kinase. *Cell* . 74: 237 (1993).
21. B.A. Witthuhn, F.W. Quelle, O. Silvennoinen, T. Yi, B. Tang, O. Miura, and J.N. Ihle, JAK2 associates with the erythropoietin receptor and is tyrosine phosphorylated and activated following stimulation with erythropoietin. *Cell* . 74: 227 (1993).
22. O. Silvennoinen, B.A. Witthuhn, F.W. Quelle, J.L. Cleveland, T. Yi, and J.N. Ihle, Structure of the murine JAK2 protein tyrosine kinase and its role in IL-3 signal transduction. *Proc. Natl. Acad. Sci. USA.* . 90: 8429 (1993).
23. J. Schlessinger and A. Ullrich: Growth factor signaling by receptor tyrosine kinases. *Neuron* . 9: 383 (1992).
24. S. Davis, T.H. Aldrich, N. Stahl, L. Pan, T. Taga, T. Kishimoto, I.N. Y., and G.D. Yancopoulos, LIFRβ and gp130 as heterodimerizing signal transducers of the tripartite CNTF receptor. *Science* . 260: 1805 (1993).
25. M. Murakami, M. Hibi, N. Nakagawa, T. Nakagawa, K. Yasukawa, K. Yamanishi, T. Taga, and T. Kishimoto, IL-6-induced homodimerization of gp130 and associated activation of a tyrosine kinase. *Science* . 260: 1808 (1993).
26. N. Stahl, T.G. Boulton, T. Farruggella, N.Y. Ip, S. Davis, B.A. Witthuhn, F.W. Quelle, O. Silvennoinen, G. Barbieri, S. Pellegrini, J.N. Ihle, and G.D. Yancopoulos, Association and activation of Jak-Tyk kinases by CNTF-LIF-OSM-IL-6 β receptor components. *Science* . 263: 92 (1994).

# SHARING OF A COMMON $\gamma$ CHAIN, $\gamma_c$, BY THE IL-2, IL-4, AND IL-7 RECEPTORS: IMPLICATIONS FOR X-LINKED SEVERE COMBINED IMMUNODEFICIENCY (XSCID)

Warren J. Leonard, Masayuki Noguchi, and Sarah M. Russell

Section on Pulmonary and Molecular Immunology
National Heart, Lung, and Blood Institute
National Institutes of Health
Bethesda, MD 20892

## SUMMARY

X-linked severe combined immunodeficiency (XSCID) is a disease characterized by profoundly diminished cellular and humoral immunity[1,2]. XSCID is by far the most common form of SCID, accounting for at least half of all cases. It is characterized by the presence of few or no T cells; B cells are present at relatively normal levels but are nonfunctional. In this manuscript, we will first summarize the studies that led to the discovery that the molecular basis for XSCID is mutation of the IL-2 receptor $\gamma$ chain (IL-2R$\gamma$)[3]. Because defects in humans with XSCID are greater than those found in IL-2 deficient mice or humans, we speculated that the IL-2R$\gamma$ was likely to be a component of more than one cytokine receptor[3]. This led to the discovery that IL-2R$\gamma$ is in fact a common $\gamma$ chain, $\gamma_c$, which is also a component of both the IL-4[4,5] and IL-7[6] receptors. Given the importance of IL-2, IL-4, and IL-7 in B-cell and T-cell function, this finding helps to explain the basis for the immunological defects found in XSCID patients. Moreover, it provides important insights into understanding the molecular basis as to how these cytokines can exhibit both overlapping and competing activities.

## INTRODUCTION

The interaction of IL-2 and IL-2 receptors critically regulates the magnitude and duration of the T cell immune response following antigen activation[7]. It is now clear that there are three classes of IL-2 receptors. Resting lymphocytes express intermediate IL-2 receptors, while activated lymphocytes express high and low affinity receptors[7]. Low affinity receptors contain the IL-2 receptor α chain, intermediate affinity receptors contain the β and γ chains, and high affinity receptors contain all three chains[7-9]. The intermediate and high affinity receptors are functional, indicating that the β and γ chains are both required for mediating an IL-2 signal. The α chain was characterized[10-12] and cloned[13-15] in the early 1980s, while the β chain was characterized[16-18] and cloned[8] in the late 1980s. The γ chain was the most recently identified subunit[9,19-22]. Both β and γ are members of a cytokine receptor superfamily, characterized principally by a conserved WSXWS motif and four conserved cysteine residues[23], whereas α is not similar to any other cytokine receptor.

## MUTATION OF THE IL-2 RECEPTOR γ CHAIN RESUILTS IN XSCID IN HUMANS

An investigation of the chromosomal loci of the components of the IL-2 receptor system revealed that whereas the α and β chains are both encoded by autosomal genes on chromsomes 10[24] and 22[25,26], respectively, the γ chain gene is located on the X chromosome[3]. The finding of this important immunological receptor protein on chromosome X immediately suggested the possibility that this gene was responsible for an X-linked immunodeficiency. The use of fluorescent in situ hybridization methodology allowed refinement of the mapping to Xq13[3]. This location was striking in that it was in the same region as the locus for XSCID, as had been determined by linkage analysis within XSCID pedigrees[27,28]. The IL-2Rγ mapping was further refined using a genetic linkage analysis made possible by the discovery of single stranded conformation polymorphisms within the IL-2Rγ gene. This analysis revealed tight linkage to X-chromosomal loci previously reported to be tightly linked to the XSCID locus[3]. Thus, IL-2Rγ was a strong candidate to be the gene which when mutated results in XSCID.

To determine whether IL-2Rγ was in fact the gene responsible for XSCID, it was necessary to directly sequence DNA derived from XSCID patients. The first three patients whose DNA was sequenced each had IL-2Rγ genes with different nonsense mutations, resulting in a different premature stop codon in each patient. These data proved that IL-2Rγ mutations result in XSCID in humans. Additional XSCID patients have now been sequenced; not surprisingly, a wider range of mutations has now been found, including for example, single amino acid changes and splice junction mutations[29]. The discovery that IL-2Rγ is the defective gene in XSCID immediately opens up the possibility of prenatal and postnatal diagnosis, carrier female identificaton, and eventually gene therapy for XSCID.

In addition to the clinical ramifications, one of the most striking features of the discovery was that whereas XSCID patients do not have T-cells, IL-2 deficient mice[30] and humans[31] have normal numbers of T cells. Thus, IL-2Rγ deficiency was more severe than IL-2 deficiency[3]. This suggested that IL-2Rγ might be part of more than one cytokine receptor system so that when it was mutated, it would result in the simultaneous inactivation of several cytokine signaling pathways[3]. This hypothesis has now been confirmed.

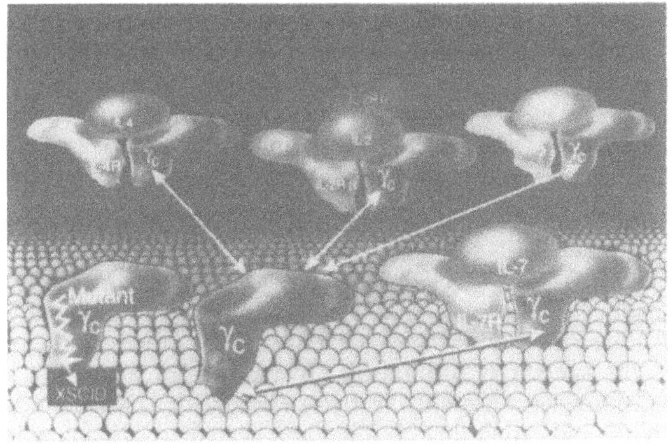


**Figure 1**. A common $\gamma$ chain, $\gamma_c$, is a component of at least three cytokine receptors. When $\gamma_c$ is mutated, this can result in XSCID. This figure is a black and white version of a Figure published in Science vol. 262, p. 1818, 1993 (reproduced with permission of Science).

## IL-2R$\gamma$ IS A COMMON $\gamma$ CHAIN, $\gamma_c$, THAT IS ALSO A COMPONENT OF THE IL-4 AND IL-7 RECEPTORS

There are now definitive data indicating that the $\gamma$ chain is also a functional component of both IL-4 and IL-7 receptors. Our laboratory has therefore proposed that IL-2R$\gamma$ should be denoted as the common $\gamma$ chain, $\gamma_c$, given its role in multiple cytokine systems[6] (see Figure 1). Using chemical cross-linking methodology, our laboratory directly demonstrated that $\gamma_c$ is physically associated with the human IL-7 receptor, augments IL-7 binding affinity, and is required for efficient internalization of IL-7[6]. For IL-4, we and our collaborators demonstrated that $\gamma_c$ is physically associated with the human IL-4 receptor, augments IL-4 binding affinity, and is required for IL-4 induced phosphorylation of insulin receptor substrate-1 (IRS-1)[4], a substrate that plays a vital role in IL-4 mitogenesis in hematopoietic cells[32]. Additionally, Kondo et al.[5] demonstrated that an anti-murine $\gamma_c$ antibody could block IL-4 induced proliferation of murine CTLL-2 cells.

## IMPLICATIONS OF A COMMON $\gamma$ CHAIN

The knowledge that IL-2, IL-4, and IL-7 each use $\gamma_c$ helps to clarify the nature of the XSCID defect. First, the roles of IL-7 as a growth factor for pre-B cells[33,34], IL-4 as the major B-cell growth factor[35,36], and IL-2 as a factor that can act on B cells[37] helps to explain the nonfunctional B cells and non-random X chromosome inactivation pattern seen in mature B cells of carrier females (i.e., only the B cells expressing the normal $\gamma_c$ chain have matured in terminally differentiated B cells)[38]. Second, the roles of IL-7 as a thymocyte[39,40] and T-cell growth factor[41,42], and of IL-2[7] and IL-4[33,34] as T-cell growth factors (with some activities on thymocytes as well) helps to explain the profound abnormality of T cell development. IL-7 might be the most significant factor in this regard[3] since treatment of mice with anti-IL-7 antibodies results in dramatically diminished T cell maturation[43].

There are at least two other sets of cytokines which share common receptor subunits. First, the IL-6, IL-11, leukemia inhibitory factor, oncostatin M, and ciliary neurotrophic factor receptors all share gp130 as a signaling molecule[44-46]. Second, the IL-3, IL-5, and GM-CSF receptors all share a common $\beta$ chain, denoted $\beta_c$[47,48]. In each of these cases, the shared molecule has a large cytoplasmic domain and is viewed as the major signaling molecule. In the case of IL-3, IL-5, and GM-CSF, the signal transduced by each cytokine is quite similar. This suggests that $\beta_c$ determines the specificity of the signal and that the other chain of each receptor determines the specificity of binding. In the case of IL-2, IL-4, and IL-7, the situation is somewhat different. In these cases, each cytokine induces a rather different signal, indicating that both binding specificity and signaling specificity are significantly controlled by IL-2R$\beta$, IL-4R, or IL-7R, the chains with extensive cytoplasmic domains[4,6]. Nevertheless, $\gamma_c$ is essential for signaling in each system. $\gamma_c$ presumably allows coupling to critical cellular signaling systems; truncation of its cytoplasmic domain is sufficient to suppress IL-2 signaling[3,49]. Given that some of the actions of these cytokines, particularly IL-2 and IL-4, can work in opposite directions, it is reasonable to assume that $\gamma_c$ might be differentially recruited into one cytokine receptor system and simultaneously sequestered from another[4]. In other words, IL-4 might inhibit IL-2 effects not only by transducing a different signal but by actively blocking the IL-2 signal by sequestering $\gamma_c$. In addition to the sharing of $\gamma_c$ by IL-2, IL-4, and IL-7, it seems likely that IL-13 will share $\gamma_c$ since there are data indicating that IL-4 and IL-13 share a receptor component[50]. In addition, IL-9 has been noted to be structurally related to IL-7[51], making it reasonable to also hypothesize that this cytokine might use $\gamma_c$[4]. Thus, the sharing of $\gamma_c$ by multiple cytokine receptors not only helps to explain the range of immunological defects in XSCID, but also helps to elucidate the basis for a number of features of overlapping and opposing signals that have been attributed to various cytokines.

## REFERENCES


1. M.D. Cooper, M.D. and J.L. Butler, Primary immunodeficiency diseases, *in*: "Fundamental Immunology," W.E. Paul, ed., Raven Press, New York, p. 1034 (1989).
2. M.E. Conley, Molecular approaches to analysis of X-linked immunodeficiencies, *Annu. Rev. Immunol.* 10:215 (1992).
3. M. Noguchi, H. Yi, H.M. Rosenblatt, A.H. Filipovich, S. Adelstein, W.S. Modi, O.W. McBride, and W.J. Leonard, Interleukin-2 receptor $\gamma$ chain mutation results in X-linked severe combined immunodeficiency in humans, *Cell* 73:147 (1993).
4. S.M. Russell, A.D. Keegan, N. Harada, Y. Nakamura, M. Noguchi, P. Leland, M.C. Friedmann, A. Miyajima, R. Puri, W.E. Paul, and W.J. Leonard, Interleukin-2 receptor $\gamma$ chain: a functional component of the interleukin-4 receptor, *Science*, 262:1880 (1993).

5. M. Kondo, T. Takeshita, N. Ishii, M. Nakamura, S. Watanabe, K-i. Arai and K. Sugamura, Sharing of the interleukin-2 (IL-2) receptor γ chain between receptors for IL-2 and IL-4, *Science* 262:1874 (1993).
6. M. Noguchi, Y. Nakamura, S.M. Russell, S.F. Ziegler, M., Tsang, X. Cao, and W.J. Leonard, Interleukin-2 receptor γ chain: a functional component of the interleukin-7 receptor, *Science*, 262:1877 (1993).
7. W.J. Leonard, M. Noguchi, S.M. Russell, and O.W. McBride, The molecular basis of X-linked severe combined immunodeficiency: the role of the interleukin-2 receptor γ chain as a common γ chain, $\gamma_c$, *Immunological Reviews*, in press (1994).
8. M. Hatakeyama, M. Tsudo, S. Minamoto, T. Kono, T. Doi, T. Miyata, M. Miyasaka, and T. Taniguchi, Interleukin-2 receptor β chain gene; generation of three receptor forms by cloned human α and β cDNAs. *Science* 244, 551 (1989).
9. T. Takeshita, H. Asao, K. Ohtani, N. Ishii, S. Kumaki, N. Tanaka, H. Munakata, M. Nakamura, and K. Sugamura, Cloning of the γ chain of the human IL-2 receptor, *Science* 257:379 (1992).
10. W.J. Leonard, J.M. Depper, T. Uchiyama, K.A. Smith, T.A. Waldmann, and W.C. Greene, A monoclonal antibody that appears to recognize the receptor for human T-cell growth factor; partial characterization of the receptor, *Nature* 300:267 (1982).
11. W.J. Leonard, J.M. Depper, R.J. Robb, T.A. Waldmann, and W.C. Greene, Characterization of the human receptor for T-cell growth factor, *Proc. Natl. Acad. Sci. USA* 80:6957 (1983).
12. W.J. Leonard, J.M. Depper, M. Kronke, R.J. Robb, T.A. Waldmann, and W.C. Greene, The human receptor for T-cell growth factor, *J. Biol. Chem.* 260:1872 (1985).
13. W.J. Leonard, J.M. Depper, G.R. Crabtree, S. Rudikoff, J. Pumphrey, R.J. Robb, M. Kronke, P.B. Svetlik, N.J. Peffer, T.A. Waldmann, and W.C. Greene, Molecular cloning and expression of cDNAs for the human interleukin-2 receptor, *Nature* 311:625 (1984).
14. T. Nikaido, A. Shimizu, N. Ishida, H. Sabe, K. Teshigawara, M. Maeda, T. Uchiyama, J. Yodoi, and T. Honjo, Molecular cloning of cDNA encoding human interleukin-2 receptor, *Nature* 311:631 (1984).
15. D. Cosman, D.P. Ceretti, A. Larsen, L. Park, C. March, S. Dower, S. Gillis, and D. Urdal, Cloning, sequence and expression of human interleukin 2 receptor, *Nature* 321:768 (1984).
16. M. Sharon, R.D. Klausner, B.R. Cullen, R. Chizzonite, and W.J. Leonard, Novel interleukin-2 receptor subunit detected by cross-linking under high affinity conditions, *Science* 234:859 (1986).
17. M. Tsudo, R.W. Kozak, C.K. Goldman, and T.A. Waldmann, Demonstration of a non-Tac petide that binds interleukin-2: a potential participant in a multichain interleukin-2 receptor complex, *Proc. Natl. Acad. Sci. USA* 83:9694 (1986).
18. K. Teshigawara, H.M. Wang, K. Kata, and K.A. Smith, Interleukin-2 high affinity receptor expression requires two distinct binding proteins, *J. Exp. Med.* 165:223 (1987).
19. T. Takeshita, H. Asao, J. Suzuki, and K. Sugamura, An associated molecule, p64, with high-affinity interleukin 2 receptor, *Int. Immunol.* 2:477 (1990).

20. T. Takeshita, K. Ohtani, H. Asao, S. Kumaki, M. Nakamura, and K. Sugamura, An associated molecule, p64, with IL-2 receptor β chain: its possible involvement in the formation of the functional intermediate-affinity IL-2 receptor complex, *J. Immunol.* 148:2154 (1992).
21. S.D. Voss, P.M. Sondel, and R.J. Robb, Characterization of the interleukin 2 receptors (IL-2R) expressed on human natural killer cells activated in vivo by IL-2: association of the p64 IL2R γ chain with the IL-2R β chain in functional intermediate-affinity IL2R, *J. Exp. Med.* 176:531 (1992)
22. N. Arima, M. Kamio, K. Imada, T. Hori, T. Hattori, M. Tsudo, M. Okuma, and T. Uchiyama, Pseudo-high affinity interleukin 2 (IL-2) receptor lacks the third component that is essential for functional IL-2 binding and signaling, *J. Exp. Med.* 176:1265 (1992).
23. J.F. Bazan, Structural design and molecular evolution of a cytokine receptor superfamily, *Proc. Natl. Acad. Sci. USA* 87:6934 (1990).
24. W.J. Leonard, T.A. Donlon, R.V. Lebo, and W.C. Greene, The gene encoding the human interleukin-2 receptor is located on chromosome 10, *Science* 228:1547 (1985)
25. J.R. Gnarra, H. Otani, H., M.G. Wang, O.W. McBride, M. Sharon, and W.J. Leonard, Human interleukin 2 receptor β chain gene: chromosomal localization and identification of 5' regulatory sequences, *Proc. Natl. Acad. Sci. USA* 87:3440 (1990).
26. H. Shibuya, M. Yoneyama, Y. Nakamura, H. Harada, M. Hatakeyama, S. Minamoto, T. Kono, T. Doi, R. White, and T. Taniguchi, The human interleukin-2 receptor β chain gene: genomic organization, promoter analysis and chromosomal assignment, *Nucl. Acids. Res.* 18:3697 (1990).
27. G.D. de Saint Basile, B. Arveiler, I. Oberle, S. Malcom, R.J. Levinsky, Y.L. Lau, M. Hofker, M. Debre, C. Griscelli, and J.L. Mandel, Close linkage of the locus for X chromosome-linked severe combined immunodeficiency to polymorphic DNA markers in Xq11-Xq13, *Proc. Natl. Acad. Sci. USA* 84:7576 (1987).
28. J.M. Puck, R.L. Nussbaum, D.L. Smead, and M.E. Conley, X-linked severe combined immunodeficiency; localization within the region Xq13.1-q21.1 by linkage and deletion analysis, Am. J. Hum. Genet. 44:724 (1989).
29. J.M. Puck, S.M. Deschenes, J.C. Porter, A.S. Dutra, C.J. Brown, H.F. Willard, and P.S. Henthorn, The interleukin-2 receptor γ chain maps to Xq13.1 and is mutated in X-linked severe combined immunodeficiency, SCIDX1, *Human Molec. Genet.* 2:1099 (1993).
30. H. Schorle, T. Holtschke, T. Hunig, A. Schimpl, and I. Horak, Development and function of T cells in mice rendered interleukin-2 deficient by gene targeting, *Nature* 352:621 (1991).
31. K. Weinberg and R. Parkman, Severe combined immunodeficiency due to a specific defect in the production of interleukin-2, *N. Eng. J. Med.* 322:1718 (1990) .
32. L.-M. Wang, M.G. Myers, Jr., S.-J. Sun, S.A. Aaronson, M. White, and J.H. Pierce, IRS-1: essential for insulin-and IL-4- stimulated mitogenesis in hematopoietic cells, *Science* 261:1591 (1993).
33. A.E. Namen, A.E. Schmierer, C.J. March, R.W. Overell, L.S. Park, D.L. Urdal, and D.Y. Mochizuki, B cell precursor growth-promoting activity: purification and characterization of a growth factor active on lymphocyte precursors, *J. Exp. Med.* 167:988 (1988).

34. A.E. Namen, S. Lupton, K. Hjerrild, J. Wignall, D.Y. Mochizuki, A. Schmierer, B. Mosley, C.J. March, D. Urdal, S. Gillis, D. Cosman, and R.G. Goodwin, Stimulation of B-cell progenitors by cloned murine interleukin-7, *Nature* 333:571 (1988).
35. J.-L. Boulay and W.E. Paul, The interleukin-4-related lymphokines and their binding to hematopoietin receptors, *J. Biol. Chem.* 267:20525 (1992).
36. W.E. Paul, Interleukin-4: A prototypic immunoregulatory lymphokine. Blood 77:1859 (1991).
37. M.C. Mingari, F. Gerosa, G. Carra, R.S. Accolla, A. Moretta, R.H. Zubler, T.A. Waldmann, and L. Moretta, Human interleukin-2 promotes proliferation of activated B cells via surface receptors similar to those of activated T cells, *Nature* 312:641 (1984).
38. M.E. Conley, A. Lavoie, C. Briggs, P. Brown, C. Guerra, and J.M. Puck, Nonrandom X chromosome inactivation in B cells carriers of X chromosome-linked severe combined immunodeficiency, *Proc. Natl. Acad. Sci. USA* 85:3090 (1988).
39. J.D. Watson, P.J. Morrissey, A.E. Namen, P.J. Conlon, and M.B. Widmer, Effect of IL-7 on the growth of fetal thymocytes in culture, *J. Immunol.* 143:1215 (1989).
40. F.M. Uckun, L. Tuel-Ahlgren, V. Obuz, R. Smith, I. Dibirdik, M. Hanson, M.-C. Langli, and J.A. Ledbetter, Interleukin 7 receptor engagement stimulates tyrosine phosphorylation, inositol phospholipid turnover, proliferation, and selective differentiation to the CD4 lineage by human fetal thymocytes, *Proc. Natl. Acad. Sci. USA* 88:6323 (1991).
41. G.D. Chazen, G.M.B. Pereira, G. LeGros, S. Gillis, and E.M. Shevach, Interleukin 7 is a T-cell growth factor, *Proc. Natl. Acad. Sci. USA* 86:5923 (1989).
42. M. Londei, A. Verhoef, C. Hawrylowicz, J. Groves, P. De Berardinis, and M. Feldmann, Interleukin 7 is a growth factor for mature human T cells, *Eur. J. Immunol.* 20:425 (1990).
43. K.H. Grabstein, T.J. Waldschmidt, F.D. Finkelman, B.W. Hess, A.R. Alpert, N.E. Boiani, A.E. Namen, and P.J. Morrissey, Inhibition of murine B and T lymphopoiesis in vivo by an anti-interleukin 7 monoclonal antibody, *J. Exp. Med.* 178:257 (1993).
44. D.P. Gearing, M.R. Comeau, D.J. Friend, S.D. Gimpel, C.J. Thut, J. McGourty, J., K.K. Brasher, J.A. King, S. Gillis, B. Mosley, S.F. Ziegler, and D. Cosman, The IL-6 signal transducer, gp130; an oncostatin M receptor and affinity converter for the LIF receptor, *Science* 255:1434 (1992).
45. T. Taga, M. Narazaki, K. Yasukawa, T. Saito, D. Miki, M. Hamaguchi, S. Davis, M. Shoyab, G.D. Yancopoulos, and T. Kishimoto, Functional inhibition of hematopoietic and neurotrophic cytokines by blocking the interleukin 6 signal transducer gp130, *Proc. Natl. Acad. Sci. USA* 89:10998 (1992).
46. T. Yin, T. Taga, M.L. Tsang, K. Yasukawa, T. Kishimoto, and Y.C. Yang, Involvement of IL-6 signal transducer gp130 in IL-11-mediated signal transduction, *J Immunol* 151:2555 (1993).
47. T. Kitamura, N. Sato, K. Arai, and A. Miyajima, Expression cloning of the human IL-3 receptor cDNA reveals a shared β subunit for the human IL-3 and GM-CSF receptors, *Cell* 66:1165 (1991).
48. J. Tavernier, R. Devos, S. Cornelis, T. Tuypens, J. van der Heyden, W. Fiers, and G. Plaetinck, A human high affinity interleukin-5 receptor (IL-5R) is

composed of an IL-5-specific α chain and a β chain shared with the receptor for GM-CSF, *Cell* 66:1174 (1991).

49. H. Asao, T. Takeshita, N. Ishii, S. Kumaki, M. Nakamura, and K. Sugamura, Reconstitution of functional interleukin 2 receptor complexes on fibroblastoid cells: involvement of the cytoplasmic domain of the γ chain in two distinct signaling pathways. *Proc. Natl. Acad. Sci. USA* 90:4127 (1993).
50. S.M. Zurawski, F. Vega, B. Huyghe, and G. Zurawski, Receptors for interleukin-13 and interleukin-4 are complex and share a novel component that functions in signal transduction, *EMBO J.* 12:2663 (1993).
51. J.-L. Boulay and W.E. Paul, Hematopoietin sub-family classicification based on size, gene organization and sequence homology, *Current Biol.* 3:573 (1993).

# X-LINKED AGAMMAGLOBULINEMIA AND BRUTON'S TYROSINE KINASE

Satoshi Tsukada and Owen N. Witte

Howard Hughes Medical Institute
Department of Microbiology and Molecular Genetics
University of California, Los Angeles
Los Angeles, CA 90024-1662

The genetic defect associated with human X-linked agammaglobulinemia (XLA) and murine X-linked immunodeficiency (XID) was recently identified as the deficiency of function of a new cytoplasmic tyrosine kinase called Bruton's tyrosine kinase (Btk)[1,2,3,4]. The phenotypes associated with these immunodeficiencies indicate that Btk plays a crucial role in B lymphocyte development.

## XLA AND XID

XLA is the prototypic inherited humoral immunodeficiency described in 1952 by Bruton[5]. Affected males develop recurrent bacterial infections early in life. Circulated immunoglobulin levels are profoundly decreased and there is a severe deficit in B cells and their plasma cell progeny. B cells in female obligate carriers of XLA exhibit non-random X chromosome inactivation implying that XLA results from an autonomous B-lineage defect[6]. The XLA defect may interfere at several levels in B-lineage development. The first bottleneck is the failure of pre-B cells to thrive and undergo the normal dramatic clonal expansion responsible for the production of immature B-lineage cells[7]. Subsequently the "leaky" B cells (small number of peripheral blood B cells) in XLA patients exhibit an immature phenotype[8]. Although the gene involved in the pathogenesis of XLA remained unidentified for 40 years since the disease was first described, the XLA locus was recently mapped to chromosome Xq22 by linkage analysis[9,10,11,12,13,14].

The XID defect results in a failure of B cells to become phenotypically and functionally diverse[15]. B cells from XID mice do not respond to thymus-independent type II antigens[15], and have abnormal responses to a variety of activation signals including immunoglobulin cross-linking, IL-5[16], IL-10[17], and anti-CD38 (M. Howard personal communication). This indicates that XID B cells lack essential signals for B cell activation and maturation. In contrast to XLA, XID mice produce relatively large

*Mechanisms of Lymphocyte Activation and Immune Regulation V*
Edited by S. Gupta *et al.*, Plenum Press, New York, 1994 

**Table 1.** Comparison of XLA and XID

| | XLA (human X-linked agammaglobulinemia) | XID (murine X-linked immunodeficiency CBA/N mouse) |
|---|---|---|
| Immunodeficiency | intrinsic to B lineage cells (non-random X chromosome inactivation in B cells) | |
| Disease phenotype | severe | mild |
| B cell numbers | less than 1% of normal | 30-50% of normal |
| B cell phenotype | immature | immature<br>reduced or absent response to thymus-independent type II antigen, IL-5, IL-10, anti-IgM, anti-CD38<br>response normally to thymus dependent antigens |
| Locus on X chromosome | Xq22 | syntenic region with human Xq22 |

numbers of peripheral B cells and immunoglobulin isotypes (TABLE 1 shows a comparison of XLA and XID). Despite their differences, XLA and XID have several common features. First, both XLA and XID B cells have immature phenotypes[8,15]. Second, both disorders appear to result from an intrinsic B-lineage defect, as evidenced by non-random X chromosome inactivation limited to B cells[6,18]. Finally, the XID locus maps to a region of the X chromosome which shares homology with the human XLA locus at Xq22[19,20]. These observations suggested that the pathogenesis of both deficiencies could have a common origin.

## BRUTON'S TYROSINE KINASE (Btk)

To study the mechanism that regulates B cell growth and differentiation, our laboratory has developed an in vitro culture system that can selectively enrich B-cell progenitors[21,22]. With the intention of isolating tyrosine kinases involved in early B cell development, a cDNA library prepared from these progenitors was screened with the tyrosine kinase domain sequence of human ltk gene using reduced stringency. A gene was identified with sequence homology to the catalytic domain of known tyrosine kinases. The gene encoding this novel tyrosine kinase was formerly designated BPK (B cell progenitor kinase and hereafter is referred to as Btk)[1]. Btk is expressed in all stages of B-lineage development and in myeloid cell types. Like the Src subfamily

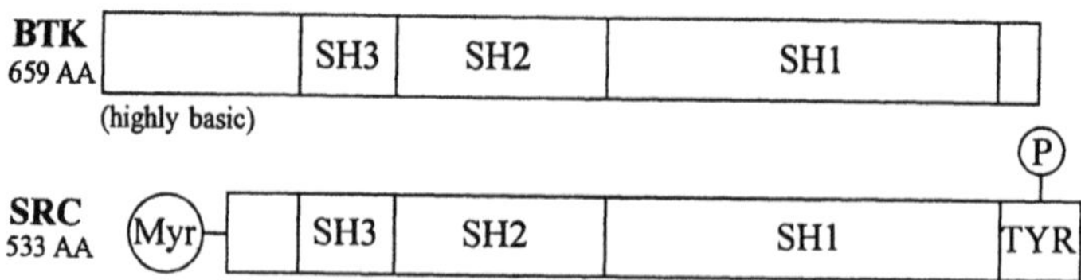


**Figure 1.** Comparison of Btk and c-Src.

kinases Btk contains classical SH1, SH2 and SH3 domains (Fig. 1 shows a comparison of Btk to c-Src). However, unlike the Src subfamily kinases an amino terminal myristylation signal essential for posttranslational modification is absent in Btk. The majority of Btk protein is located in the cytoplasm. The short carboxyl terminal segment following the catalytic domain of Btk lacks the residue equivalent to tyrosine 527 of c-Src (a regulatory phosphorylation site). In addition, Btk contains an unusually long, basic amino terminal unique region domain upstream of the SH3 domain. Btk has strong amino acid homology with several other cytoplasmic tyrosine kinases; Dsrc28c[23], TecII[24], and Itk[25] (also known as Tsk)[26]. In addition to a high percentage identity in sequence, these kinases share several structural features described above, which suggest that these kinases form a novel subfamily of cytoplasmic tyrosine kinases. One of the striking features of this new subfamily (we refer to as the Btk subfamily) is that a significant amino acid identity is present among their amino terminal unique regions (in the case of Src subfamily kinases homology is not found among their unique regions). Recently it was shown that the amino terminal unique regions of the Btk subfamily contain a PH domain (pleckstrin homology domain)[27]. This newly recognized homology domain is present in a broad array of signaling proteins including several serine-threonine kinases, GTPases, GTP activating proteins and nucleotide exchange factors (GRF)[28,29]. While the function of the PH domain is still poorly understood, the presence of PH domains in Btk subfamily kinases may suggest that these kinases are involved in common protein interactions within divergent signaling pathways.

## INVOLVEMENT OF Btk IN XLA AND XID

In a collaborative effort with other laboratories, we mapped the human Btk gene to Xq22 (XLA locus known by linkage analysis) by fluorescence in situ hybridization and X chromosome somatic cell hybrid analysis. The Btk locus was further analyzed by screening a panel of YAC clones forming a partial contig map of the Xq22 region, which showed that the Btk locus was within 100 kb of the DXS178 probe defining the polymorphism most closely linked to XLA. This mapping data, along with the restricted expression of Btk protein to the B and myeloid-lineages, made Btk a strong candidate for the XLA gene. Subsequently, using antisera against the amino terminal unique region of Btk, we demonstrated reduced or absent kinase activity of Btk in EB virus-transformed B cell lines from several XLA patients[1]. At the same time, Vetrie et al. reported the identical gene (formerly atk) as the candidate for the XLA gene using a positional cloning strategy[2]. They detected intragenic deletions and restriction site alterations in the genomic DNA coding Btk gene in 8/33 XLA patients. Furthermore, they reported two point mutations in the Btk catalytic domains of XLA patients, which was predicted to abolish its kinase activity. These two reports, which used contrasting strategies, strongly indicated that Btk corresponds to the gene involved in XLA. This is the first case that a cytoplasmic tyrosine kinase, which is an important regulator of cell growth and differentiation, is involved in the pathogenesis of human hereditary disease. Since the initial characterization of Btk as the XLA gene, several mutations of this gene in XLA patients have been identified, including an amino acid substitution in the SH2 domain which resulted in an unstable Btk protein[30], and a deletion of one exon encoding a portion of the SH3 domain (H. Ochs, personal communication).

The murine Btk gene was mapped to the XID locus of the murine X chromosome by interspecific backcross analysis, making Btk a strong candidate for the gene which is defective in XID. In contrast to XLA, Btk mRNA expression and protein kinase activity in vitro were unaltered in XID B cells[3]. The Btk coding sequence of XID was sequenced to identify a mutation that might alter its ability as a signal transducer. A missense mutation predicted to alter a single residue (Arg28 to Cys) in the amino terminal unique region was identified[3,4]. As described above, the unique regions of Btk subfamily members exhibit a high degree of identity. This Arg 28 is located in the first highly conserved block and is present in all the Btk subfamily members. Interestingly, this Arg28 falls into the second conserved region of the PH domain and this Arg residue is one of the most conserved residues in this motif (in 45 molecules with a PH domain, this Arg is conserved in 30 molecules)[27]. The Arg28 to Cys substitution in the highly conserved region of Btk unique segment may perturb recognition of an as yet unidentified Btk-associated protein(s), resulting in the alteration of B cell signaling in XID mice.

There may be two alternative explanations for the difference in disease phenotypes exhibited by typical XLA (severe phenotype) and XID (mild phenotype). One explanation is that the amino acid substitution at the Btk unique region in XID mice may only partially abolish the Btk activity and some Btk signal transduction pathways may remain partially active. An alternative explanation is that the distinct phenotypes in XLA and XID may represent a species difference between human and murine B cell development. Generation of Btk knock-out mice, or alternatively identification of unique region mutations in XLA patients will provide opportunities to test these possibilities.

The phenotypes manifested in these two immunodeficiencies suggest that Btk has a crucial role in B cell development. Identification of molecules interacting with Btk should help to define its possible roles in signal transduction pathways. Mutation analysis of Btk in XLA patients will provide a unique opportunity to study the function of cytoplasmic tyrosine kinase and identify its important residues. Btk gene therapy may be useful in the management of XLA patients.

## REFERENCES

1. Tsukada, S., Saffran, D. C., Rawlings, D. J., Parolini, O., Allen, R. C., Klisak, I., Sparkes, R. S., Kubagawa, H., Mohandas, T., Quan, S., Belmont, J. W., Cooper, M. D., Conley, M. E., and Witte, O. N., Deficient expression of a B cell cytoplasmic tyrosine kinase in human X-linked agammaglobulinemia, *Cell* 72. 279-290 (1993).

2. Vetrie, D., Vorechovsky, I., Sideras, P., Holland, J., Davies, A., Flinter, F., Hammarström, L., Kinnon, C., Levinsky, R., Bobrow, M., Smith, C. I. E., and Bentley, D. R., The gene involved in X-linked agammaglobulinaemia is a member of the *src* family of protein-tyrosine kinases, *Nature* 361. 226-233 (1993).

3. Rawlings, D. J., Saffran, D. C., Tsukada, S., Largaespada, D. A., Grimaldi, J. C., Cohen, L., Mohr, R. N., Bazan, J. F., Howard, M., Copeland, N. G., Jenkins, N. A., and Witte, O. N., Mutation of the amino-terminal unique region of bruton's tyrosine kinase in murine X-linked immunodeficiency, *Science* 358. 358-361 (1993).

4. Thomas, J. D., Sideras, P., Smith, C. I. E., Vorechovsky, I., Chapman, V., and Paul, W. E., Colocalization of X-linked agammaglobulinemia and X-linked immunodeficiency genes, *Science* 261. 355-358 (1993).

5. Bruton, O. C., Agammaglobulinemia, *Pediatrics* 9. 722-727 (1952).

6. Singer, J. W., and Fialkow, P. J., Expression of the gene defect in X-linked agammaglobulinemia, *N. Engl. J. Med.* 315. 564-567 (1986).

7. Campana, D., Farrant, J., Inamdar, N., Webster, A. D. B., and Janossy, G., Phenotypic features and proliferative activity of B cell progenitors in X-linked agammaglobulinemia, *J. Immunol.* 145. 1675-1680 (1990).

8. Conley, M. E., B cells in patients with X-linked agammaglobulinemia, *J. Immunol.* 134. 3070-3074 (1985).

9. Kwan, S.-P., Kunkel, L., Bruns, G., Wedgewood, R. J., Latt, S., and Rosen, F. S., Mapping of the X-linked agammaglobulinemia locus by use of restriction fragment-length polymorphism, *J. Clin. Invest.* 77. 649-652 (1986).

10. Mensink, E. J. B. M., Thompson, A., Schot, J. D. L., van de Greef, W. M. M., Sandkuyl, L. A., and Schuurman, R. K. B., Mapping of a gene for X-linked agammaglobulinemia and evidence for genetic heterogeneity, *Hum. Genet.* 73. 327-332 (1986).

11. Malcolm, S., de Saint Basile, G., Arveiler, B., Lau, Y. L., Szabo, P., Fischer, A., Griscelli, C., Debre, M., Mandel, J. L., Callard, R. E., Robertson, M. E., Goodship, J. A., Pembrey, M. E., and Levinsky, R. J., Close linkage of random DNA fragments from Xq 21.3-22 to X-linked agammaglobulinemia (XLA), *Hum. Genet.* 77. 172-174 (1987).

12. Guioli, S., Arveiler, B., Bardoni, B., Notarangelo, L. D., Panina, P., Duse, M., Ugazio, A., de Saint Basile, G., Mandel, J. L., and Camerino, G., Close linkage of probe p212 (DXS178) to X-linked agammaglobulinemia, *Hum. Genet.* 84. 19-21 (1989).

13. Kwan, Sau-P., Terwilliger, J., Parmley, R., Raghu, G., Sandkuyl, L. A., Ott, J., Ochs, H., Wedgwood, R., and Rosen, F., Identification of closely linked DNA marker, DXS178, to further refine the X-linked agammaglobulinemia locus, *Genomics* 6. 238-242 (1990).

14. Parolini, O., Hejtmancik, J. F., Allen, R. C., Belmont, J. W., Lassiter, G. L., Henry, M. J., Barker, D. F., and Conley, M. E., Linkage analysis and physical mapping near the gene for X-linked agammaglobulinemia at Xq22, *Genomics* 15. 342-349 (1993).

15. Scher, I., The CBA/N mouse strain: an experimental model illustrating the influence of the X-chromosome on immunity, *Adv. Immunol.* 33. 1-71 (1982).

16. Hitoshi, Y., Sonoda, E., Kikuchi, Y., Yonebara, S., Nakauchi, H., and Takatsu, K., Interleukin 5 receptor positive B cells, but not eosinophils, are functionally and numerically influenced in the mice carrying the X-linked immune defect, *Int. Immunol.* 5. 1183-1190 (1993).

17. Go, N. F., Castle, B. E., Barrett, R., Kastelein, R., Dang, W., Mosmann, T. R., Moore, K. W., and Howard, M., Interleukin 10, a novel B cell stimulatory factor: Unresponsiveness of X chromosome-linked immunodeficiency B cells, *J. Exp. Med.* 172. 1625-1631 (1990).

18. Nahm, M. H., Paslay, J. W., and Davie, J. M., Unbalanced X chromosome mosaicism in B cells of mice with X-linked immunodeficiency, *J. Exp. Med.* 158. 920-931 (1983).

19. Hillyard, A. L., Doolittle, D. P., Davisson, M. T., and Roderick, T. H., Locus map of the mouse, *Mouse Genome* 90. 8-21 (1992).

20. Copeland, N. G., and Jenkins, N. A., Development and applications of a molecular genetic linkage map of the mouse genome, *TIG* 7. 113 (1991).

21. Scherle, P. A., Dorshkind, K., and Witte, O. N., Clonal lymphoid progenitor cell lines expressing the BCR/ABL oncogene retain full differentiative function, *Proc. Natl. Acad. Sci. USA* 87. 1908-1912 (1990).

22. Saffran, D. C., Faust, E. A., and Witte, O. N., Establishment of a reproducible culture technique for the selective growth of B-cell progenitors, *Curr. Top. Microbiol. Immunol.* 182. 34-44 (1992).

23. Gregory, R. J., Kammermeyer, K. L., Vincent, W. S., III, and Wadsworth, S. G., Primary sequence and developmental expression of a novel *Drosophila melanogaster src* gene, *Mol. Cell. Biol.* 7. 2119-2127 (1987).

•24. Mano, H., Mano, K., Tang, B., Koehler, M., Yi, T., Gilbert, D. J., Jenkins, N. A., Copeland, N. G., and Ihle, J. N., Expression of novel form of *Tec* kinase in hematopoietic cells and mapping of the gene to chromosome 5 near *Kit*, *Oncogene* 8. 417-424 (1993).

25. Siliciano, J. D., Morrow, T. A., and Desiderio, S., V, *itk*, a T-cell-specific tyrosine kinase gene inducible by interleukin 2, *Proc. Natl. Acad. Sci. USA* 89. 11194-11198 (1992).

26. Heyeck, S. D., and Berg, L. J., Developmental regulation of a murine T-cell-specific tyrosine kinase gene, *Tsk*, *Proc. Natl. Acad. Sci. USA* 90. 669-673 (1993).

27. Musacchio, A., Gibson, T., Rice, P., Thompson, J., and Saraste, M., The PH domain is a common piece in the structural patchwork of signalling proteins, *TIBS* 18. 343-348 (1993).

28. Mayer, B. J., Ren, R., Clark, K. L., and Baltimore, D., A putative modular domain present in diverse signaling protein, *Cell* 73. 629-630 (1993).

29. Haslam, R. J., Kolde, H. B., and Hemmings, B. A., Pleckstrin domain homology, *Nature* 363. 309-310 (1993).

30. Saffran, D. C., Parolini, O., Fitch-Hilgenberg, M. E., Rawlings, D. J., Afar, D. E. H., Witte, O. N., and Conley, M. E., A point mutation in the SH2 domain of Bruton's tyrosine kinase resulting in protein instability and atypical X-linked agammaglobulinemia, . in press (1993).

# THE ROLE OF CD40 LIGAND IN HUMAN DISEASE

Melanie K. Spriggs

Molecular Biology
Immunex Research and Development Corporation
Seattle, WA 98101

CD40 is a 50kDA surface glycoprotein expressed predominantly on B cells, monocytes, dendritic cells, thymic epithelium and certain carcinomas.[1-3] It is a member of the tumor necrosis factor receptor (TNFR) superfamily,[4,5] a group of related type I transmembrane molecules which, in addition to CD40, includes both forms of TNFR, the low affinity nerve growth factor (NGF) receptor, CD27, CD30, OX40, 4-1BB, and Fas.[6-9] Members of this family are characterized by the presence of multiple cysteine-rich repeats consisting of approximately 40 amino acids in the extracellular amino terminal domain.[5] The average sequence homology between family members in the extracellular domain is around 25%.

The development of mAb specific for CD40 led to the demonstration that ligation of CD40 could mediate a wide range of activities on human B cells. These include the initiation of homotypic adhesion,[10,11] induction of short-term proliferation in the presence of costimuli,[11,12] and long-term proliferation when CD40 mAb are immobilized on human FcγRII expressed by transfected murine L cells.[13] In addition, CD40 mAb costimulate secretion of IgE in the presence of IL-4[14-16] and other Ig isotypes with IL-2 or IL-10.[17-19] Furthermore, CD40 mAb have been shown to rescue germinal center B cells from undergoing spontaneous apoptosis in vitro.[20]

The chromosomal location of the *CD40l* locus in the mouse was determined by interspecific backcross analysis. Southern blotting was used to identify restriction fragment length polymorphisms.[21,22] The *CD40l* locus mapped to the X chromosome approximately 1.5 cM distal of the *hprt* gene. This suggested that the human CD40L gene might also be linked to the HPRT locus and would map to Xq26. This was confirmed using in situ

*Mechanisms of Lymphocyte Activation and Immune Regulation V*
Edited by S. Gupta *et al.*, Plenum Press, New York, 1994

hybridization in our laboratory,[22] that of Aruffo and colleagues,[23] and by Graf et al.[24]

The biological activity of recombinant CD40L protein in *in vitro* assays, together with its provocative chromosomal location, suggested that defects in the CD40L gene might be responsible for a known X-linked immunodeficiency. At the time, there were at least five described X-linked immunodeficiencies for which the underlying defect was unknown. These included severe combined immunodeficiency (XSCID), agammaglobulinemia, Wiskott-Aldrich syndrome, X-linked lymphoproliferative syndrome (XLP), and X-linked hyper-IgM (HIGM). Of these, XLP, X-linked HIGM, Wiskott-Aldrich, and XSCID were syndromes which seemed likely to involve a defect in T cell function. X-linked HIGM was reported to map to the region of Xq24-Xq27,[25,26] suggesting this syndrome might be a strong candidate for a disorder associated with defects in the CD40L gene. PCR analysis of cDNA derived from patient PBL mRNA quickly showed that the X-linked HIGM patients, but not other X-linked immunodeficiency patients, exhibited point mutations in the extracellular domain of the CD40L.

X-linked HIGM syndrome is a rare disorder characterized by normal to elevated levels of IgM and decreased or absent serum levels of IgG, IgA and IgE.[27,28] Diagnosis is frequently made following recurrent bacterial infections in the first few years of life.[27] Intravenous administration of gamma globulin and aggressive use of antibiotics are used, but there is a general consensus that the functional immune deficiency is relatively severe. Interestingly, affected males exhibit an unusually high incidence of pneumocystis carinii infection, suggesting defective helper T cells. In addition, many patients exhibit a severe recurrent neutropenia.[28,29]

sIgM and $CD19^+$ or $CD20^+$ cells are present in HIGM patients, but B cells bearing mature isotypes are absent.[30] *In vitro* T cell function tests are generally normal, and T cell subpopulations are also typical. In the peripheral lymphoid tissue germinal centers are diminished or absent.

The following paragraphs provide a summary of the findings of several laboratories which reported the association of defects in the CD40L gene and X-linked HIGM.

To date, nucleotide sequence results of the CD40L mRNA derived from thirteen patients have been reported.[22,23,31,32] Eight of these patients exhibit point mutations within the extracellular domain, three contain deletions, one contains a point mutation within the transmembrane domain, and one patient does not appear to have any changes in the coding region, suggesting a defect at the translational regulation level. Many of these changes are relatively conservative and might easily be considered polymorphisms. Subsequent experiments, described briefly below and in more detail in the cited publications, have proven that the mutations are in fact deleterious and result in CD40L proteins which are blocked during intracellular transport or are expressed but do not bind to CD40. This suggests that the secondary and/or tertiary structure of the ligand is critical to its biological activity and very sensitive to mutation.

Flow cytometric analysis using a soluble CD40.Fc fusion protein on activated peripheral blood T cells (PB T) indicated that twelve of the thirteen patients studied showed no CD40-Fc binding.[22,23,31,32] One patient did show weak, but reproducible, binding.[22] In contrast, normal donor activated PBL bound CD40-Fc well, indicating that any of the mutations described to date are sufficient to interfere with biological activity.

The literature describing HIGM syndrome and its possible causes prior to the discovery of the CD40L defect was inconclusive in resolving which lymphoid cell type carried the defect responsible for HIGM.[33,34] To determine whether patient B cells could respond normally to wild type CD40L, proliferative assays on four of the HIGM patients were performed. Cultures containing T depleted peripheral blood mononuclear cells (PBMC) and CV-1/EBNA cells transfected with CD40L or with a control plasmid were analyzed for $^3$H-TdR uptake. All four patients' B cell cultures showed levels of proliferation comparable to those of normal donor controls.[22] In addition, Aruffo et al. noted similar levels of patient and control proliferation in culture systems containing PBMC, CD40 mAb and IL-4.[23]

The diminished IgG, IgA, and IgE phenotype of X-linked HIGM patients suggested that their B cells should respond to wild type CD40L and cytokines by producing downstream immunoglobulin isotypes. *In vitro* culture of normal PBMC in the presence of IL-4 results in detectable IgE secretion after ten days.[35] Cultures of HIGM PBMC with IL-4 or with IL-4 and CV-1/EBNA cells transfected with vector alone result in no detectable IgE secretion. However, co-culture of patient PBMC with CV-1/EBNA cells expressing the CD40L does result in IgE secretion. Similar results are obtained from culture systems using CD40 mAb and IL-4.[14-16,22] Korthauer et al. have demonstrated also that the HIGM patient B cells cultured in the presence of CD40L plus IL-10 can be induced to secrete IgG and IgA.[32]

The combination of *in vitro* data with the known phenotype of X-linked HIGM patients clearly indicate that although CD40L functions *in vitro* as a potent mitogen for B cells, other mechanisms can and will function *in vivo* to allow proliferation and IgM secretion from B cells.

Additional PB T activation antigens were examined by flow cytometry to insure that HIGM patient T cells could respond to polyclonal activation. In one instance, the p55 chain of the IL-2R was tested,[22] in another, CD69.[23] In both cases, comparable surface expression was seen on HIGM activated PB T as was evident on normal donor tissue. These data support, but do not prove, the idea that the defect in the T cell population of HIGM patients is restricted to CD40L expression. Recent unpublished findings[36,37] indicate that the CD40L can act as a co-stimulus for the proliferation of T cells. Thus, additional detailed studies are necessary before the full extent of defective CD40L expression is understood.

Recently, analysis of Hodgkin's disease (HD) cells and cultured Hodgkin's Reed-Sternberg (H-RS) cells revealed that significant levels of CD40 were constitutively expressed on their cell surface. HD is defined as a group of malignant lymphomas which express a heterogeneous pattern of cytokines and cytokine receptors. Recombinant CD40L was found to induce IL-8 secretion and to enhance IL-6, TNF$\alpha$ and TNF$\beta$ production from H-RS cells. In addition, CD40L enhanced the surface expression of ICAM-1 and B7-

1 and downregulated the expression of HD-associated antigen CD30.[38] Thus, the CD40-CD40L may be important in regulating the cytokine network in HD.

## REFERENCES


1. E.A. Clark and P.J. Lane, Regulation of human B-cell activation and adhesion, *Annu. Rev. Immunol.* 9:97 (1991).
2. A.H. Galy and H. Spits, CD40 is functionally expressed on human thymic epithelial cells, *J. Immunol.* 149:775 (1992).
3. M.R. Alderson, R.J. Armitage, T.W. Tough, L. Strockbine, W.C. Fanslow, and M.K. Spriggs, CD40 expression by human monocytes: Regulation by cytokines and activation of monocytes by the ligand for CD40., *J. Exp. Med.* 178:669 (1993).
4. C.A. Smith, T. Davis, D. Anderson, L. Solam, M.P. Beckmann, R. Jerzy, S.K. Dower, D. Cosman, and R.G. Goodwin, A receptor for tumor necrosis factor defines an unusual family of cellular and viral proteins., *Science* 248:1019 (1990).
5. S. Mallett and A.N. Barclay, A new superfamily of cell surface proteins related to the nerve growth factor receptor., *Immunol. Today* 12:220 (1991).
6. D. Camerini, G. Walz, W.A.M. Loenen, J. Borst, and B. Seed, The T cell activation antigen CD27 is a member of the NGF/TNF receptor gene family., *J. Immunol.* 147:3165 (1991).
7. H. Dürkop, U. Latza, M. Hummel, F. Eitelbach, B. Seed, and H. Stein, Molecular cloning and expression of a new member of the nerve growth factor receptor family that is characteristic for Hodgkin's disease., *Cell* 68:421 (1992).
8. N. Itoh, S. Yonehara, A. Ishii, M. Yonehara, S.-I. Mizushima, M. Sameshima, A. Hase, Y. Seto, and S. Nagata, The polypeptide encoded by the cDNA for human cell surface antigen Fas can mediate apoptosis., *Cell* 66:233 (1991).
9. M. Baens, M. Chaffanet, J.J. Cassiman, H. van den Berghe, and P. Marynen, Construction and evaluation of a hncDNA library of human 12p transcribed sequences derived from a somatic cell hybrid, *Genomics* 16:214 (1993).
10. T.B. Barrett, G. Shu, and E.A. Clark, CD40 signaling activates CD11a/CD18 (LFA-1)-mediated adhesion in B cells., *J. Immunol.* 146:1722 (1991).
11. J. Gordon, M.J. Millsum, G.R. Guy, and J.A. Ledbetter, Resting B lymphocytes can be triggered directly through the $CD_W40$ (Bp50) antigen., *J. Immunol.* 140:1425 (1988).
12. E.A. Clark and J.A. Ledbetter, Activation of human B cells mediated through two distinct cell surface differentiation antigens, Bp35 and Bp50., *Proc. Natl. Acad. Sci. USA* 83:4494 (1986).
13. J. Banchereau, P. de Paoli, A. Valle, E. Garcia, and F. Rousset, Long term human B cell lines dependent on interleukin-4 and antibody to CD40., *Science* 251:70 (1991).

14. H.H. Jabara, S.M. Fu, R.S. Geha, and D. Vercelli, CD40 and IgE: Synergism between anti-CD40 monoclonal antibody and interleukin-6 in the induction of IgE synthesis by highly purified human B cells., *J. Exp. Med.* 172:1861 (1990).
15. K. Zhang, E.A. Clark, and A. Saxon, CD40 stimulation provides and IFN-$\gamma$-independent and IL-4-dependent differentiation directly to human B cells for IgE production., *J. Immunol.* 146:1836 (1991).
16. H. Gascan, J.-F. Gauchat, G. Aversa, P. van Vlasselaer, and J.E. de Vries, Anti-CD40 monoclonal antibodies or $CD4^+$ T cell clones and IL-4 induce IgG4 and IgE switching in purified human B cells via different signalling pathways., *J. Immunol.* 147:8 (1991).
17. F. Rousset, E. Garcia, and J. Banchereau, Cytokine-induced proliferation and immunoglobulin production of human B lymphocytes triggered through their CD40 antigen, *J. Exp. Med.* 173:705 (1991).
18. F. Rousset, E. Garcia, T. Defrance, C. Peronne, N. Vezzio, D.H. Hsu, R. Kastelein, K.W. Moore, and J. Banchereau, Interleukin 10 is a potent growth and differentiation factor for activated human B lymphocytes, *Proc. Natl. Acad. Sci. USA* 89:1890 (1992).
19. T. Defrance, B. Vanbervliet, F. Briere, I. Durand, F. Rousset, and J. Banchereau, Interleukin 10 and transforming growth factor beta cooperate to induce anti-CD40-activated naive human B cells to secrete immunoglobulin A, *J. Exp. Med.* 175:671 (1992).
20. Y.J. Liu, D.E. Joshua, G.T. Williams, C.A. Smith, J. Gordon, and I.C.M. MacLennon, Mechanism of antigen-driven selection in germinal centres., *Nature* 342:929 (1989).
21. N.G. Copeland and N.A. Jenkins, Development and applications of a molecular genetic linkage map of the mouse genome, *Trends Genet.* 7:113 (1991).
22. R.C. Allen, R.J. Armitage, M.E. Conley, H. Rosenblatt, N.A. Jenkins, N.G. Copeland, M.A. Bedell, S. Edelhoff, C.M. Disteche, D.K. Simoneaux, W.C. Fanslow, J. Belmont, and M.K. Spriggs, CD40 ligand gene defects responsible for X-linked hyper-IgM syndrome., *Science* 259:990 (1993).
23. A. Aruffo, M. Farrington, D. Hollenbaugh, X. Li, A. Milatovich, S. Nonoyama, J. Bajorath, L.S. Grosmaire, R. Stenkamp, M. Neubauer, R.L. Roberts, R.J. Noelle, J.A. Ledbetter, U. Francke, and H.D. Ochs, The CD40 ligand, gp39, is defective in activated T cells from patients with X-linked hyper-IgM syndrome., *Cell* 72:291 (1993).
24. D. Graf, U. Korthauer, H.W. Mages, G. Senger, and R.A. Kroczek, Cloning of TRAP, a ligand for CD40 on human T cells, *Eur. J. Immunol.* 22:3191 (1992).
25. E.J. Mensink, A. Thompson, L.A. Sandkuyl, M.E. Kraakman, J.D. Schot, T. Espanol, and R.K. Schuurman, X-linked immunodeficiency with hyperimmunoglobulinemia M appears to be linked to the DXS42 restriction fragment length polymorphism locus, *Hum. Genet.* 76:96 (1987).

26. M. Padayachee, C. Feighery, A. Finn, C. McKeown, R.J. Levinsky, C. Kinnon, and S. Malcolm, Mapping of the X-linked form of hyper-IgM syndrome (HIGM1) to Xq26 by close linkage to HPRT, *Genomics* 14:551 (1992).
27. F.S. Rosen, S.V. Kevy, E. Merler, C.A. Janeway Jr., and D. Gitlin, Recurrent bacterial infections and dysgammaglobulinemia: Deficiency of 7S gamma-globulins in the presence of elevated 19S gamma-globulins., *Pediatrics* 28:182 (1961).
28. L.D. Notarangelo, M. Duse, and A.G. Ugazio, Immunodeficiency with hyper-IgM (HIM), *Immunodefic. Rev.* 3:101 (1992).
29. M.E. Conley, Molecular approaches to analysis of X-linked immunodeficiencies., *Annu. Rev. Immunol.* 10:215 (1992).
30. L. Mayer, S.P. Kwan, C. Thompson, H.S. Ko, N. Chiorazzi, T. Waldmann, and F. Rosen, Evidence for a defect in "switch" T cells in patients with immunodeficiency and hyperimmunoglobulinemia M, *N. Engl. J. Med.* 314:409 (1986).
31. J.P. DiSanto, J.Y. Bonnefoy, J.F. Gauchat, A. Fischer, and G. de Saint Basile, CD40 ligand mutations in X-linked immunodeficiency with hyper-IgM, *Nature* 361:541 (1993).
32. U. Korthauer, D. Graf, H.W. Mages, F. Briere, M. Padayachee, S. Malcolm, A.G. Ugazio, L.D. Notarangelo, R.J. Levinsky, and R.A. Kroczek, Defective expression of T-cell CD40 ligand causes X-linked immunodeficiency with hyper IgM, *Nature* 361:539 (1993).
33. D. Levitt, P. Haber, K. Rich, and M.D. Cooper, Hyper IgM immunodeficiency. A primary dysfunction of B lymphocyte isotype switching, *J. Clin. Invest.* 72:1650 (1983).
34. R.W. Hendriks, M.E. Kraakman, I.W. Craig, T. Espanol, and R.K. Schuurman, Evidence that in X-linked immunodeficiency with hyperimmunoglobulinemia M the intrinsic immunoglobulin heavy chain class switch mechanism is intact, *Eur. J. Immunol.* 20:2603 (1990).
35. W.C. Fanslow, D. Anderson, K.H. Grabstein, E.A. Clark, D. Cosman, and R.J. Armitage, Soluble forms of CD40 inhibit biological responses of human B cells., *J. Immunol.* 149:655 (1992).
36. R.J. Armitage, T.W. Tough, B.M. Macduff, W.C. Fanslow, M.K. Spriggs, F. Ramsdell, and M.R. Alderson, CD40 ligand is a T-cell growth factor, *Eur. J. Immunol.* 23:2326 (1993).
37. W.C. Fanslow, K.N. Clifford, M. Seaman, M.R. Alderson, M.K. Spriggs, R.J. Armitage, and F. Ramsdell, Recombinant CD40-ligand exerts potent biological effects on T cells, *J. Immunol., in press* (1994).
38. H.-J. Gruss, D. Hirschstein, B. Wright, D. Ulrich, M.A. Caligiuri, L. Strockbine, R.J. Armitage, and S.K. Dower, Expression and function of CD40 on Hodgkin and Reed-Sternberg cells and the possible relevance for Hodgkin's disease, *Blood, in press* (1994).

# INVOLVEMENT OF THE PROTEIN TYROSINE PHOSPHATASE PTP1C IN CELLULAR PHYSIOLOGY, AUTOIMMUNITY AND ONCOGENESIS

John McCulloch[1] and Katherine A.Siminovitch[1,2]

[1]Samuel Lunenfeld Research Institute
[2]Departments of Medicine and Immunology
University of Toronto
Mount Sinai Hospital
600 University Avenue
Toronto, Ontario M5G 1X5

## INTRODUCTION

Control of a wide range of cellular functions is governed by the phosphorylation of cellular proteins on tyrosine residues. Activities regulated in such a manner include proliferation, differentiation and many specific cellular functions. The phosphorylation status of intracellular proteins is determined by the activity of two mutually antagonistic classes of enzyme - protein tyrosine kinases (PTKs) and protein tyrosine phosphatases (PTPs). This article will concentrate on current knowledge concerning the haemopoietic cell phosphatase PTP1C and will describe our recent findings concerning the *in vivo* and *in vitro* roles of this protein from PTP1C-mutant mice.

## ACTIVITIES OF PTPs IN HAEMOPOIETIC CELLS

In recent years at least 20 mammalian PTPs have been isolated and structurally characterised. Of these proteins, many appear to have a ubiquitous distribution pattern, but some, such as CD45, PTP1C and HePTPase, are restricted primarily to cells of haematopoietic origin.[1,2] It is not certain whether these PTPs all play major roles in aspects of haemopoiesis, but their restricted expression may indicate that this is the case. Although the physiological roles of PTPs are not as well characterised as those of PTKs, the effects of PTP activity on cell function are becoming clearer. The results of studies using inhibitors of PTP activity, for example, indicate that PTP blockade can lead to unrestricted cell growth in the absence of growth factors, a

*Mechanisms of Lymphocyte Activation and Immune Regulation V*
Edited by S. Gupta *et al.*, Plenum Press, New York, 1994

finding which suggests that PTPs can negatively regulate cell growth.[3] Similarly, there is strong evidence suggesting that PTPs can suppress the growth of transformed cells.[4] Haemopoietic PTPs may also positively influence cell growth. The receptor PTP CD45, for example, has been shown to be critical for antigen receptor-mediated signaling, and hence proliferation, in both B- and T-cells.[5,6] In a recent study, transgenic mice bearing a truncated CD45 molecule were found to have a block in T-cell maturation leading to a paucity of mature, functional T-cells in the periphery.[7] Together, these data suggest that PTPs play both positive and negative modulatory roles in relation to influencing cell functions.

## THE *MOTHEATEN* MOUSE

The *motheaten* (*me*) mutation, and its less severe allelic form, *viable motheaten* (*me*$^{v}$), arose spontaneously in mice of the C57BL/6J background and were first identified at the Jackson Laboratory.[8,9] A variety of haemopoietic cell abnormalities are found in animals homozygous for the *me* or *me*$^{v}$ mutations. These defects result in a patchy alopecia (which causes a "motheaten" appearance), arthritis, glomerulonephritis, limb necrosis and death at 3 weeks (*me/me*) or 9 weeks (*me*$^{v}$/*me*$^{v}$) of age consequent to haemorrhagic pneumonitis.

Lymphopoiesis is severely impaired in *motheaten* mice as shown by the data in Figure 1. B-cell precursors (surface B220+) are greatly reduced in number, although the proportion of CD5+ B-cells is increased.[10] In wild type mice these latter cells constitute a minor, more primitive B-cell lineage and are restricted primarily to the peritoneal cavity. CD5+ B-cells have been implicated in the enhanced production of lamda Ig light chain, IgM and IgG3 isotypes, and autoantibodies in *me* and *me*$^{v}$ mice.[11,12] However, the relevance of these cells to the pathology observed in the mutant mice appears doubtful since mice doubly homozygous for *me* and *xid* (a mutation which blocks the development of CD5+ B-cells) manifest essentially the same morbidity as mice homozygous for *me* alone.[13] In contrast to these B-cell defects, T-cell precursors appear to be numerically normal in young *motheaten* mice (Figure 1), despite the fact that the thymus involutes prematurely.[14-16] Lymphocyte function has not been thoroughly investigated as yet, but it seems clear that the B-cell population possesses a more severe functional impairment than the T-cell population.[17,18]

*Me* and *me*$^{v}$ mice also show abnormalities of the natural killer cell and erythroid cell precursor populations,[19-21] but the most obvious haematological abnormality in these animals is a huge proliferation of monocytic and granulocytic cells (see Mac-1 positive cells, Figure 1). These cells appear to be responsible for most of the observed pathology. For instance, homozygotes suffer from splenomegaly largely as the result of expansion of these populations, and the characteristic patchy alopecia appears driven by large focal infiltrates of these cells. Furthermore, the premature death of the mutant homozygotes is caused by a fulminant haemorrhagic pneumonitis triggered by massive infiltration of monocytic and granulocytic cells. Experiments using cultured leucocytes from *me* or *me*$^{v}$ mice are often characterised by an overgrowth of macrophage-like cells. These unusual cells are able to proliferate at an enhanced rate and achieve this independently of the myeloid growth factor CSF-1.[22,23] This escape from normal growth constraints suggests a breakdown in the cell signaling processes governing proliferation.

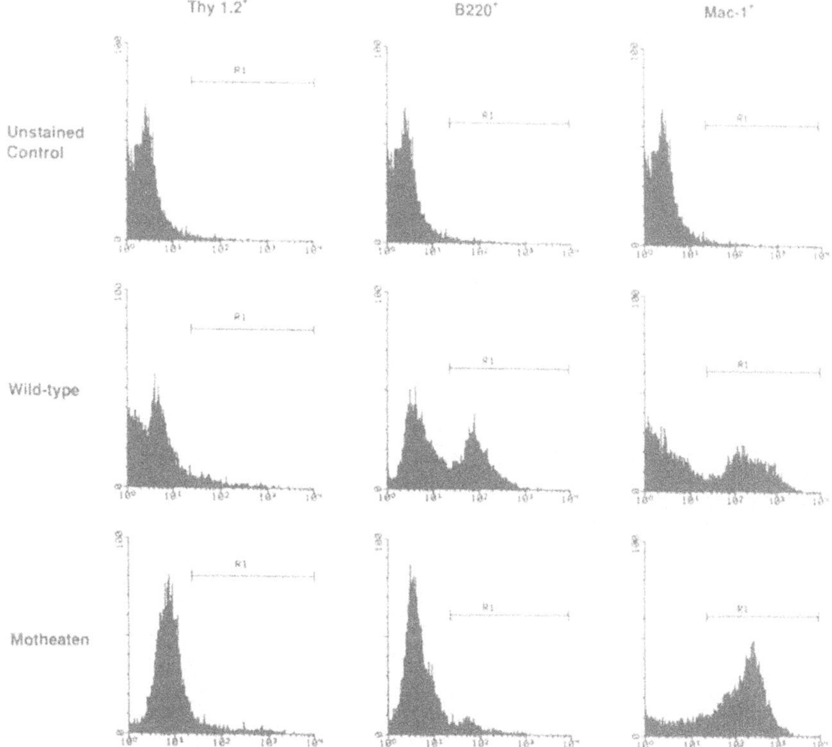


**Figure 1.** Comparative flow cytometry analysis of wild type and *motheaten* bone marrow cells. The histograms depict log specific fluorescence (X axis) versus cell number (Y axis). This analysis shows that in *motheaten* mice T-cell (Thy 1.2+) numbers are normal, B-cell precursor (B220+) numbers are much reduced, and monocyte/macrophage (Mac-1+) numbers are enhanced compared to the wild type.

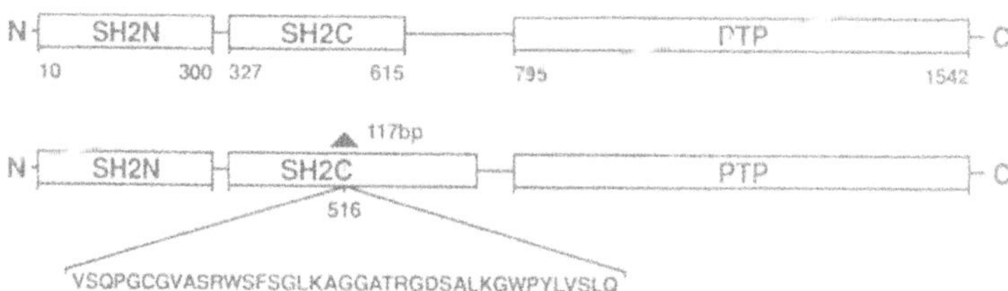


**Figure 2.** Structure of wild type PTP1C transcripts. The three domain structure of PTP1C, consisting of two N-terminal SH2 domains and one C-terminal PTP domain, is shown. Numbers refer to nucleotide position. Two forms of wild type (+/+) PTP1C exist, one being 117 bp longer (black triangle) alternative splicing in the distal SH2 domain.

## Background

The above findings are interesting from the standpoint of autoimmunity and immunodeficiency, but what is the relevance to tyrosine phosphorylation? Recently, a haemopoietic cell tyrosine phosphatase named PTP1C (also known as HCP, SHP and SHPTP1) was independently identified by four groups.[24-27] This ≈ 67 kDa molecule contains two N-terminal SH2 domains and a C-terminal catalytic PTP domain. Variation in molecular weight appears to be due to alternative splicing of a 117 basepair (bp) insertion in the second SH2 domain at position 516 (Figure 2).[28,29] While the presence of SH2 domains strongly implicates PTP1C in the formation of protein-protein complexes essential in cell signaling processes,[30,31] the physiological substrate(s) for PTP1C is unknown. The results of *in vitro* studies, however, have shown that PTP1C can interact with and dephosphorylate the EGF, PDGF, Kit and IL-3 receptors.[25,28,32,33] Investigation of PTP1C expression in wild type mice has shown expression to be strongest in cells of haemopoietic origin (Figure 3) indicating a role for PTP1C in the growth and/or differentiation of such cells. PTP1C is also expressed in liver and lung tissue, possibly reflecting expression by epithelial cells and/or Kupffer cells and bronchus-associated lymphoid tissue. The level of this expression is, however, significantly less than that observed in lymphoid organs (Figure 3).

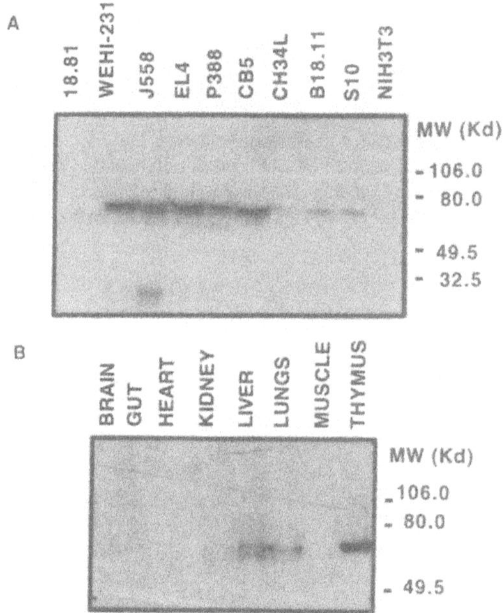


**Figure 3.** Expression of PTP1C. A. Lysates of various cell lines were separated by SDS-PAGE, blotted onto nitrocellulose and probed with a PTP1C-specific antiserum. PTP1C expression was strongest in haemopoietic cell lines, weaker or absent in cells derived from other lineages (18.81, pre-B lymphoma; WEHI-231, B cell lymphoma; J558, B cell myeloma; EL4, thymoma; P388, lymphoid neoplasm; CB5, erythroid line ; CH34L, CD5+ B-cell hybridoma; Bl8.11, B cell hybridoma; S10, epithelial cell line; NIH 3T3, fibroblast line). B. Tissue lysates were prepared from +/+ mice and processed as above. PTP1C expression was extremely strongest in thymus, weaker in liver and lungs, and absent in other tissues. This pattern is consistent with PTP1C being predominantly expressed by haemopoietic cells.

A possible role for PTP1C in the development of the *motheaten* phenotype was first suggested by the finding that the human PTP1C gene locates to 12p12-13,[25] a region which is syntenic to the region on chromosome 6 which contains the *me* gene.[6,8] Since PTP1C had been shown to be strongly expressed by haemopoietic cells and the *motheaten* phenotype was known to be transferable by bone marrow cells to lethally-irradiated syngeneic mice, it seemed reasonable to investigate mutations in the PTP1C gene as the cause of the *motheaten* phenotype.[25-27] Our analysis of the PTP1C gene in *me* and *me*$^v$ mice has indeed confirmed this hypothesis.[34] The process whereby we elucidated the nature of the PTP1C genetic lesions in *motheaten* mice is discussed below.

**The *me* and *me*$^v$ Mutations**

To assess PTP1C expression in *me* mice, expression of the SH2 domain region and the PTP domain segment was examined in wild type and mutant bone marrow cell RNA transcripts using the reverse transcription-polymerase chain reaction (RT-PCR) and appropriate primers.[34] While no differences were detected between the wild type and *me* PTP1C catalytic domain sequences, analysis of this region in *me*$^v$ mice revealed the presence of two aberrant PTP1C transcripts, one having a 69 bp in-frame insertion and the other a 15 bp in-frame deletion (Figure 4, Figure 5). A subsequent analysis of the SH2 domain region revealed sequence abnormalities in the *me* PTP1C transcripts as well, the majority showing a 101 bp deletion in the N-terminal SH2 domain. Two forms of the 101 bp-deleted transcript were found corresponding to alternative splicing of the 117 bp sequence as also detected in the wild type carboxy SH2 domain transcript (Figure 4, Figure 5). In addition, a minor species was detected which contained a single nucleotide deletion (C at position 228) within the N-terminal SH2 domain (Figure 4). By contrast, no abnormalities were detected in the *me*$^v$ SH2 domain region. Together these observations strongly suggested that PTP1C expression was impaired in mutant mice and that this abnormality likely reflected mutations in the PTP1C gene.

To investigate this possibility, we conducted sequencing analysis of the regions of normal, *me* and *me*$^v$ PTP1C genes corresponding to the sites of transcript abnormalities. Analysis of the *me* PTP1C gene showed absence of a cytidine residue at position 228 in the N-terminal SH2 domain (Figure 5). This deletion changes the normal CTGGTCGAGT sequence to CTGGTGAGT, thus creating a more favourable context for use of a GT dinucleotide as a 5′ splice site. Use of this splice site combined with a 3′ acceptor splice site 100 bp downstream results in the 101 bp deletion. As this transcript truncation results in frameshifts which activate stop signals 81 bp or 69 bp downstream, no functional PTP1C protein is produced in *me/me* mice.

A PTP1C gene defect was also detected in *me*$^v$ mice, but in this case, the mutation, a single basepair substitution (T to A), mapped to position 1074 within the catalytic domain region (Figure 5). As shown in the figure, this transversion disrupts the ususal 5′ donor splice site GT dinucleotide resulting in activation of a cryptic splice site either 15 bp upstream or 69 bp downstream. The use of these alternative splice sites results in the production of mutant PTP1C protein with a 5 amino acid deletion or 23 amino acid insertion in the PTP domain, respectively.

The effects of *me* and *me*$^v$ mutations on PTP1C protein expression are shown in Figure 6. As is consistent with the genetic data predicting two PTP1C isoforms, two PTP1C species of 67 and 70 kDa were detected in wild type haemopoietic tissues. While these bands were also detected in *me*$^v$ tissues, the presence of an additional ≈ 34 kDa species in the latter tissues suggests that the *me*$^v$ PTP1C proteins are not normal and are subject to degradation. No PTP1C protein was detected in *me* tissues.

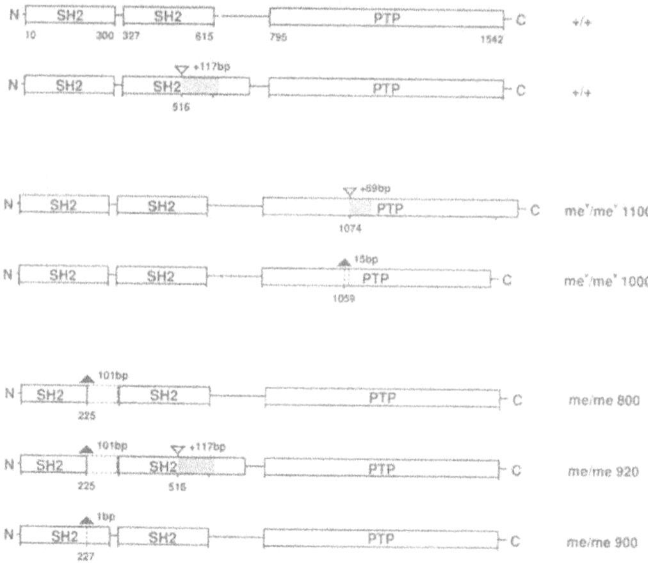


**Figure 4.** Summary of the structural features of wild type, *mc heaten* and *viable motheaten* PTP1C transcripts.

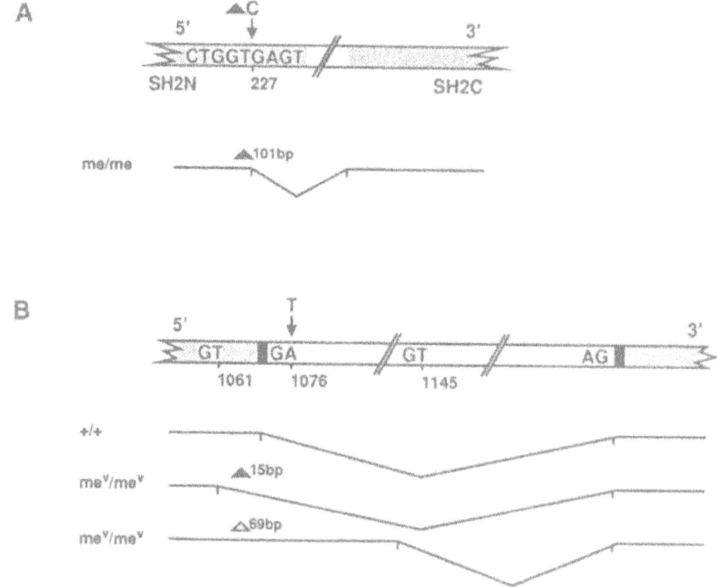


**Figure 5.** Mutant PTP1C genes and their transcripts. PTP1C genes are shown double-lined, transcripts as single lines. Deletions are represented by black triangles, insertions by white triangles. A. The *motheaten* mutation consists of a single nucleotide genomic deletion (C at position 227) which creates an aberrant splice site leading to a 101 bp deletion. B. The *viable motheaten* mutation consists of a genomic T to A substitution. This also causes faulty splicing leading to either a 15 bp deletion or 69 bp insertion in the PTP domain.

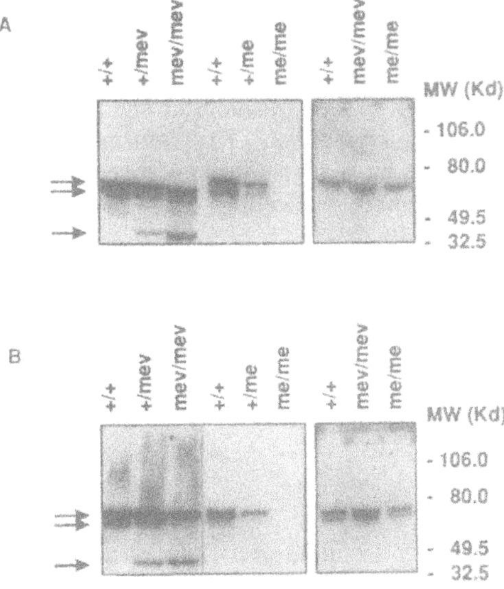


**Figure 6.** Expression of PTP1C (left) and Syp (right) in wild type, *motheaten* and *viable motheaten* spleen (A) and thymus (B) revealed by western blotting of tissue lysates. The main features to note are as follows: +/+ PTP1C is found as 67 and 70 kDa isomers; the $me^{v}$ PTP1C includes two species of similar weights to +/+ PTP1C, but also an additional 34 kDa species; *me* mice fail to produce levels of PTP1C detectable by western analysis; all mice examined have comparable expression of Syp.

By contrast, expression of the structually similar PTP, Syp, was normal in all mice studied, indicating that the *me* and $me^{v}$ phenotypes are not due to a general loss of PTP activity but reflect the selective loss of PTP1C.

These data, subsequently confirmed by others,[31] together with the recent co-localisation of the *me* gene and the PTP1C gene by linkage analysis,[25,29] indicate that mutations within the PTP1C gene are responsible for the *motheaten* and *viable motheaten* phenotypes.

## A ROLE FOR PTP1C IN ONCOGENESIS?

As noted earlier, previous data from several groups have revealed that substances which block PTP activity promote uncontrolled cell growth and that PTPs can inhibit the growth of transformed cells. It would therefore be predicted that loss of PTP1C activity might predispose to malignant transformation. Similarly, the overgrowth of myeloid/granulocytic cells in *me* and $me^{v}$ mice reveals the association of PTP1C deficiency with a disruption of the normal constraints for cell growth. While the early mortality of the *me* homozygous mice might preclude detection of tumours, we have noted that mice heterozygous for *me* or $me^{v}$, which appear normal at birth, have a very high incidence of neoplasia in later life. The *me*/+ heterozygotes succumb to adenocarcinomas, whereas $me^{v}$/+ mice develop lymphomas. These assessments were made on the basis of independent pathology scoring and immunofluorescence analysis (lymphomas only). The role of PTP1C in the genesis of these tumours is now being explored, but it would seem possible that these mice have a mutation or deletion in their normal PTP1C allele leaving the cells with either no

PTP1C or mutant PTP1C protein. If true, this finding would strongly indicate that PTP1C functions as a tumour suppressor gene.

In summary, the detection of PTP1C mutation in *motheaten* mice should advance our understanding of the physiological and pathological roles of PTP1C in haemopoiesis and growth regulation. This information, in turn, should provide important insights into the molecular processes governing both the normal activities and transformation of haemopoietic cells.

**Acknowledgements**

This work was supported by grants from the Medical Research Council and the National Cancer Institute of Canada (NCIC). K.A.Siminovitch is an Ontario Ministry of Health Career Scientist and an Arthritis Society of Canada Research Scientist.

## REFERENCES

1. Q.Yang and N.K.Tonks. Isolation of a clone encoding a human protein-tyrosine phosphatase with homology to the cytoskeletal-associated proteins band 4.1, ezrin, and talin. *Proc.Natl.Acad.Sci.USA* 88:5949 (1991)
2. B.Zanke, H.Suzuki, K.Kishihara, L.Mizzen, M.Minden, T.Pawson and T.W.Mak. Cloning and expression of an inducible lymphoid-specific, protein tyrosine phosphatase (HePTPase). *Eur.J.Immunol.* 22:235 (1992)
3. J.K.Karlund. Transformation of cells by an inhibitor of phosphatase acting on phosphotyrosine in proteins. *Cell* 41:707 (1985)
4. S.Brown-Shimer, K.A.Johnson, D.E.Hill and A.M.Bruskin. Effect of protein tyrosine phosphatase 1B expression on transformation by the human *neu* oncogene. *Cancer Res.* 52:478 (1992)
5. J.T.Pingel and M.L.Thomas. Evidence that the leucocyte-common antigen is required for antigen-induced T-cell proliferation. *Cell* 58:1055 (1989)
6. M.F.Gruber, J.M.Bjorndahl, S.Nakamura, S.M.Fu. Anti-CD45 inhibition of human B cell proliferation depends on the nature of activation signals and the state of B cell activation. A study with anti-IgM and anti-CDw40 antibodies. *J.Immunol.* 142:4144 (1989)
7. K.Kishihara, J.Penninger, V.A.Wallace, T.M.Kundig, K.Kawai, A.Wakeman, E.Timms, K.Pfeffer *et al.* Normal B cell development but impaired T cell maturation in CD45-exon 6 protein tyrosine phosphatase deficient mice. *Cell* 74:143 (1993)
8. M.C.Green and L.D.Shultz. *Motheaten*, an immunodeficient mutant of the mouse. *J.Hered.* 66:250 (1975)
9. L.D.Shultz, D.R.Coman, C.L.Bailey, W.G.Beamer and C.L.Sidman. *Viable motheaten*, a new allele at the *motheaten* locus. I. Pathology. *Am.J.Pathol.* 116:179 (1984)
10. C.L.Sidman, L.D.Shultz, R.R.Hardy, K.Hayakawa and L.A.Herzenberg. Production of immunoglobulin isotypes by Ly-1+ B cells in *viable motheaten* and normal mice. *Science* 232:1423 (1986)
11. L.A.Herzenberg, A.M.Stall, P.A.Lalor, C.Sidman, W.A.Moore, D.R.Parks and L.A.Herzenberg. The Ly-1 B cells lineage. *Immunol.Rev.* 93:81 (1986)
12. A.B.Kantor. The development and repertoire of B-1 cells (CD5 B cells). *Immunol.Today* 12:389 (1991)

13. C.L.Scribner, C.T.Hansen, D.M.Klinman and A.D.Steinberg. The interaction of the *xid* and *me* genes. *J.Immunol.* 138:3611 (1987)
14. I.Goldschneider, K.L.Komschlies and D.L.Greiner. Studies of thymocytopoiesis in rats and mice. I. Kinetics of appearance of thymocytes using a direct intrathymic adoptive transfer assay for thymocyte precursors. *J.Exp.Med.* 163:1 (1986)
15. D.L.Greiner, I.Goldschneider, K.L.Komschlies, E.S.Medlock, F.J.Bollum and L.D.Shultz.Defective lymphopoiesis in bone marrow of *motheaten* (*me/me*) and *viable motheaten* ($me^{v}/me^{v}$) mutant mice. I. Analysis of development of prothymocytes, early B lineage cells, and terminal deoxynucleotidyl transferase-positive cells. *J.Exp.Med.* 164:1129
16. S.M.Hayes, L.D.Shultz and D.R.Greiner. Thymic involution in *viable motheaten* ($me^{v}$) mice is associated with a loss of thymic precursor activity. *Dev.Immunol.* 2:191 (1992)
17. C.L.Sidman, L.D.Shultz and E.R.Unanue. The mouse mutant "motheaten." II. Functional studies of the immune system. *J.Immunol.* 121:2399 (1987)
18. W.F.Davidson, H.C.Morse, S.O.Sharrow and T.M.Chused. Phenotypic and functional effects of the *motheaten* gene on murine B and T lymphocytes. *J.Immunol.* 122:884 (1979)
19. E.A.Clark, L.D.Shultz and S.B.Pollack. Mutations in mice that influence natural killer (NK) cell activity. *Immunogenetics* 12:601 (1981)
20. G.C.Koo, C.L.Manyak, J.Dasch, L.Ellingsworth and L.D.Shultz. Suppressive effects of monocytic cells and transforming growth factor-ß on natural killer cell differentiation in autoimmune *viable motheaten* mice. *J.Immunol.* 147:1194 (1991)
21. G.van Zant and L.D.Shultz. Hematologic abnormalities of the immunodeficient mouse mutant, *viable motheaten* ($me^{v}$). *Exp.Hematol.* 17:81 (1989)
22. K.L.McCoy, E.Chi, D.Engel and J.Clagett. Accelerated rate of mononuclear phagocyte production *in vitro* by splenocytes from autoimmune *motheaten* mice. *Am.J.Path.* 112:18 (1983)
23. K.L.McCoy, K.Neilson and J.Clagett. Spontaneous production of colony stimulating activity by splenic Mac-1 antigen positive cells from autoimmune *motheaten* mice. *J.Immunol.* 132:272 (1984)
24. S-H.Shen, L.Bastien, B.I.Posner and P.Chretien. A protein tyrosine phosphatase with sequence similarity to the SH2 domain of the protein-tyrosine kinases. *Nature* 352:736 (1991)
25. T.Yi, J.L.Cleveland and J.N.Ihle. Protein tyrosine phosphatase containing SH2 domains: characterisation, preferential expression in hematopoietic cells, and localization to human chromosome 12p12-13. *Mol.Cell.Biol.* 12:836 (1992)
26. J.Plutzky, B.G.Necl, and R.D.Rosenberg. Isolation of a novel src homology 2 - SH2 containing tyrosine phosphatase. *Proc.Natl.Acad.Sci.USA* 89:1123 (1992)
27. R.J.Matthews, D.B.Browne, E.Flores and M.L.Thomas. Characterization of hematopoietic intracellular protein tyrosine phosphatases: description of a phosphatase containing an SH2 domain and another enriched in proline-, glutamic acid-, serine-, and threonine-rich sequences. *Mol.Cell.Biol.* 12 2396 (1992)
28. M.Kozlowski, I.Mlinaric-Rascan, G-S.Feng, R.Shen, T.Pawson and K.A.Siminovitch. Expression and catalytic activity of the tyrosine phosphatase PTP1C is severly impaired in *motheaten* and *viable motheaten* mice. *J.Exp.Med.* 178:2157 (1993)

29. L.D.Shultz, P.A.Schweitzer, T.V.Rajan, T.Yi, J.N.Ihle, R.J.Matthews, M.L.Thomas and D.R.Beier. Mutations at the murine *motheaten* locus are within the hematopoietic cell protein-tyrosine phosphatase (Hcph) gene. *Cell* 73:1445 (1993)
30. C.A.Koch, D.Anderson, M.F.Moran, C.Ellis and T.Pawson.SH2 and SH3 domains: elements that control interactions of cytoplasmic signaling proteins. *Science* 252:668 (1991)
31. T.Pawson and G.D.Gish. SH2 and SH3 domains: from structure to function. *Cell* 71:359 (1992)
32. W.Vogel, R.Lammers, J.Huang and A.Ullrich. Activation of phosphotyrosine phosphatase by tyrosine phosphorylation. *Science* 259:1611 (1993)
33. T.Yi and J.N.Ihle. Association of hematopoietic cell phosphatase with c-kit after stimulation with c-kit ligand. *Mol.Cell.Biol.* 13:3350 (1993)
34. H.W.Tsui, K.A.Siminovitch, L.de Souza and F.W.L.Tsui. *Motheaten* and *viable motheaten* mice have mutations in the haematopoietic cell phosphatase gene. *Nature Genetics* 4:124 (1993)

## CONTRIBUTORS

ROBERT T. ABRAHAM--Department of Immunology and Pharmacology, Mayo Clinic and Foundation, Rochester, MN 55905-0001

JOSEPH BOLEN--Department of Molecular Biology, Bristol-Myers Squibb Pharmaceutical Research Institute, Princeton, NJ 08543

DOREEN A. CANTRELL--Lymphocyte Activation Laboratory, Imperial Cancer Research Fund, London WC2 3PX, United Kingdom

EDWARD A. CLARK--Department of Microbiology, University of Washington, Seattle, WA 98195-0001

GERALD R. CRABTREE--Department of Pathology, Stanford University, Stanford, CA 94305

ANTHONY L. DeFRANCO--Department of Microbiology and Immunology, University of California, San Francisco, CA 94143-0552

PHYLLIS I. GARDNER--Department of Medicine, Stanford University, Stanford, CA 94305

ROBERT GEAHLEN--Department of Medicinal Chemistry and Pharmacognosy, Purdue University, Lafayette, IN 47907

SUDHIR GUPTA--Department of Medicine, University of California, Irvine, CA 92717-4069

MICHAEL J. LENARDO, M.D.--Laboratory of Immunology, NIAID, National Institutes of Health, Bethesda, MD 20812

WARREN J. LEONARD--NHLBI, National Institutes of Health, Bethesda, MD 20892-0001

DAN R. LITTMAN--Department of Microbiology and Immunology, University of California, San Francisco, CA 94143-0414

BERNARD MALISSEN--Centre d'Immunologie, INSERM-CNRS de Marseille-Luminy, Marseille, France

JAMEY D. MARTH--Biomedical Research Centre, University of British Columbia, Vancouver, B.C., Canada V6T 1Z3

HENRY METZGER--NIAMS, National Institutes of Health, Bethesda, MD 20892-0001

ATSUSHI MIYAJIMA--DNAX Research Institute, Palo Alto, CA 94304-1104

WILLIAM E. PAUL--NIAID, National Institutes of Health, Bethesda, MD 20892-0001

ROGER M. PERLMUTTER--Department of Immunology, University of Washington, Seattle, WA 98915

JEFFREY V. RAVETCH--Laboratory of Biochemical Genetics, Memorial Sloan-Kettering Cancer Center, New York, NY 10021-6094

MICHAEL RETH--Max-Planck Institut für Immunologie, D-7800 Freiburg, Germany

BRIAN SEED--Department of Molecular Biology, Massachusetts General Hospital, Boston, MA 02114

BARTHOLOMEW M. SEFTON--Molecular Biology and Virology Laboratory, The Salk Institute, San Diego, CA 92186

KATHERINE A. SIMINOVITCH--Departments of Medicine and Immunology, University of Toronto, Toronto, Ontario, Canada M5G 1X5

MELANIE K. SPRIGGS--Department of Molecular Biology, Immunex Research & Development Corporation, Seattle, WA 98101

ARTHUR WEISS--Division of Rheumatology/Immunology, University of California, San Francisco, CA 94143-0724

OWEN WITTE--Department of Microbiology, University of California, Los Angeles, CA 90024-1662

# INDEX

www.ingramcontent.com/pod-product-compliance
Ingram Content Group UK Ltd.
Pitfield, Milton Keynes, MK11 3LW, UK
UKHW020112200726
13856UKWH00002B/502